D1499014

BIOLOGICAL MEMBRANES

BIOLOGICAL MEMBRANES

Edited by

DENNIS CHAPMAN

Department of Chemistry
The University
Sheffield, England

and

DONALD F. H. WALLACH

Tufts University School of Medicine
Department of Therapeutic Radiology
Boston, Massachusetts
U.S.A.

VOLUME 2

1973

ACADEMIC PRESS 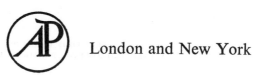 London and New York

ACADEMIC PRESS INC. (LONDON) LTD.
24/28 Oval Road
London NW1

United States Edition published by
ACADEMIC PRESS INC.
111 Fifth Avenue
New York, New York 10003

Library of Congress Catalog Card Number: 67-19850
ISBN: 0-12-168542-X

PRINTED IN GREAT BRITAIN BY
ADLARD AND SON LTD., BARTHOLOMEW PRESS, DORKING

Contributors

J. AVRUCH, *Department of Medicine, Washington University, School of Medicine, St. Louis, Missouri 63110, U.S.A.*

D. CHAPMAN, *Department of Chemistry, The University, Sheffield S37 HF, England.*

E. FERBER, *Max Planck-Institut für Immunbiologie, Freiburg/Brg., Germany.*

H.-D. KLENK, *Institut für Virologie, Justus Liebig-Universitat, Giessen, Germany.*

S. POHL, *Department of Medicine, Washington University, School of Medicine, St. Louis, Missouri 63110, U.S.A.*

G. G. SHIPLEY, *Biophysics Division, Department of Medicine, Boston University School of Medicine, Boston, Massachusetts 02118, U.S.A.*

D. F. H. WALLACH, *Tufts University School of Medicine, Department of Therapeutic Radiology, 136 Harrison Avenue, Boston, Massachusetts 02111, U.S.A.*

Introduction

The first volume, *Biological Membranes 'Physical Fact and Function'*, edited by one of us (D.C.), was produced in 1968 with the stated aim of providing a timely review on biological membranes, directed particularly at the physical aspects, and with the expressed hope that it would help to stimulate further advances in our knowledge of membranes.

The intervening years have seen a remarkable growth of interest in these important cell structures, an interest shared by biologists, biochemists, physiologists, botanists, physicists, chemists and physicians. Certain aspects of membrane systems which were still uncertain in 1968 and discussed at that ime are now much more clear. Thus it is now generally accepted that many membranes contain regions of bilayer of lipid, although the relationship of protein to lipid remains uncertain in most cases and is still being actively studied. The importance of the fluid nature of membranes has been increasingly appreciated and has been shown to be relevant to many biological situations, although recent work suggests that in some membranes both gel and fluid regions can exist at the cell's growth temperature (see this Volume p. 136). Early suggestions that the lipids diffuse along a bilayer system, mentioned in the first Volume, have led to quantitative measurement of diffusion rates.

There are, however, still very many aspects of membrane systems which require study and discussion. This new volume, *Biological Membranes 2*, is intended to review some of these aspects and to bridge the gaps between the physical and biomedical. The present book is arranged so that the first two chapters cover studies using physical techniques such as X-ray techniques, thermal techniques and various spectroscopic methods. The advantages arising from recent advances in NMR instrumentation using higher magnetic fields and the use of C^{13} nuclei are pointed out. The middle chapters, more biological in nature, cover discussion of virus membranes, hormone action and the dynamics of phospholipids in plasma membranes. The final chapter addresses the role of the plasma membrane in disease processes and is intended to stimulate further interest in those aspects of membranes which are relevant to health problems, e.g. immunology, demyelination, transport defects and metal toxicity effects.

An understanding of biological membranes is now seen to be of prime importance for the detailed interpretation of many biological and disease situations. As a result of the intense interest and activity in membranes many further advances can be expected in this field. We hope that this volume is relevant, topical and will stimulate further activity in this fascinating field.

We wish to thank Mrs. June Campbell for her assistance with the preparation of the Subject Index and for her secretarial help.

D. CHAPMAN
January 1973   D. F. H. WALLACH

ACKNOWLEDGEMENTS

We acknowledge permission to reproduce material in this book from the following publishers: Elsevier Publishing Co.; North Holland Publishing Co.; Plenum Press; Macmillan Journals Ltd.; Rockerfeller Institute Press; National Academy of Science, Washington; Verlag der Zeitschrift für Naturforschung.

Contents

Chapter 1

Recent X-ray Diffraction Studies of Biological Membranes and Membrane Components

G. G. SHIPLEY

Biophysics Division, Department of Medicine, Boston University School of Medicine, Boston, Massachusetts, U.S.A.

I. Introduction

The discovery by M. von Laue in 1912 of the diffraction of X-rays by crystals followed by the structural determination of simple inorganic salts by W. H. Bragg and W. L. Bragg provided the cornerstone for advances made on structural aspects in the fields of physics, chemistry, geology and minerology. However, perhaps the most spectacular achievements have resulted from the application of X-ray diffraction methods to biology, where the pioneering work and enthusiasm of W. L. Bragg, J. D. Bernal and W. Astbury has culminated in the detailed structural determinations of complex biomacromolecules, e.g. proteins, enzymes and nucleic acids.

1

In view of these outstanding successes of X-ray diffraction and their contribution to the field of molecular biology it is perhaps surprising that relatively little attention has been paid to the structure of the membranes surrounding cells and cell organelles. It is probably true to say that electron microscopy, providing for the first time a visual picture of cells and cellular components and a seemingly invariant morphology for the membranes associated with cells and cell organelles, this morphology being consistent with the existing model for membrane structure, acted as a deterrent to investigation by diffraction techniques. A more cogent reason is probably related to the complex composition of a given membrane, the variability in composition between different membranes and, most important, our inability to obtain crystals of membranes.

Thus, despite an impressive classification of the lipid composition of membranes from animals, plants and microorganisms, and a rapidly improving knowledge of the proteins associated with membranes, the excellent work of Schmitt and Bear in the late 1930s and, more recently, Finean has attracted few fellow workers to this important field.

However, the behavior and properties of biologically-important lipids have been extensively studied, notably by Luzzati and colleagues and Chapman and colleagues, and speculations have been made on their structural role in membranes. A fairly detailed understanding of the behavior of lipids and more recently lipid–protein complexes, together with the demonstration by Wilkins and co-workers that meaningful X-ray scattering data may be obtained from unoriented membrane preparations, has resulted in an upsurge of applications of X-ray diffraction methods to membrane and other lipoprotein systems. It is the aim of this chapter to review the more recent advances resulting from the application of this particular technique to each of the above-mentioned fields, in doing so realizing the benefits which in general accumulate from a more multi-disciplined approach to complex biological problems of which membrane structure and function are archetypal.

II. X-ray Diffraction

Since a vast literature exists on the theory and experimental methods of X-ray diffraction and scattering, for the purposes of this chapter only a brief description is warranted.

The fundamental concept of X-ray diffraction relates to the fact that ordered solids, crystals for example, by virtue of the spacial arrangement of their component atoms or molecules, enables them to act as three-dimensional diffraction gratings for incident X-rays, one condition for diffraction requiring that the wavelength of the X-rays (~ 1 Å) is of the same order of magnitude as the interatomic or intermolecular separations. Thus diffraction is a simple interference phenomenon based upon the constructive and destructive interference of secondary X-ray wave fronts produced by the lattice defining the ordered solid. The well-known Bragg equation $n\lambda = 2d \sin\theta$ defines the geometry of diffraction necessary for constructive interference. The equation relates the distance separating equivalent lattice planes, d, to the angular relationship between the incident X-ray beam and the orientation of the lattice planes, 2θ, λ being the X-ray wavelength and n is an integer. Measurement of the angle between the incident and diffracted beams, 2θ, on the X-ray photograph enables the calculation of the distances d separating the equivalent lattice planes to be made.

The fact that the lattice dimension, d, and the angle of diffraction, 2θ, are inversely related means that for macromolecular systems such as membranes the diffraction or scattering effects occur at very low diffraction angles. The special techniques necessary to record diffraction effects at small angles have been summarized by Guinier and Fournet (1955) and Beeman *et al.* (1957). Examples of camera geometries useful in this field are described by Franks (1958), Elliot (1965) and Kratky (1963).

The positioning of the constituent atoms or molecules within the lattice defines the intensity distribution of the diffraction pattern. The intensity of (I) of the diffracted wavefront from each set of lattice planes is related to the amplitude of the wave F by the usual relationship

$$I = F^2$$

Thus from the measured intensities the calculation of structure amplitude $|F|$, but not the phase, corresponding to each diffracted wave is straight forward. The Fourier relationship relating electron density distribution of the diffracting object and its diffraction pattern is given by

$$\rho(x) = \int F(s) \exp(-2\pi i s \cdot x) \, dv$$

where $\rho(x)$ defines the electron density vectorially in real space, and $F(s)$ is the structure amplitude of the diffracted wave defined vectorially in diffracted or reciprocal space. This equation can be solved correctly only if we are able to determine the correct phasing for each structure amplitude F. The methods used to overcome or circumvent this problem of phase assignment are beyond the scope of this chapter. Suffice it to say that a number of techniques are being used sucessfully in the application of diffraction methods to the problem of lipid and membrane structure. The solution of the Fourier equation gives $\rho(x)$, the electron density distribution, which defines the number of electrons, and hence atoms, at different points in the repeating unit of the diffracting object. This electron density distribution may then be quantitatively interpreted in terms of the positioning of the component atoms or molecules and the spacial relationship between them.

These basic concepts form the starting point for the more advanced treatments of X-ray diffraction by ordered and disordered solids, fluids, etc., given by James (1963), Guinier (1963) and Hosemann and Bagchi (1962). For the literature dealing specifically with the theory applied to layered membrane-like structures reference is made to Blaurock and Worthington (1966), Burge and Draper (1967), Worthington (1969a) and Wilkins *et al.* (1971). The interpretation of diffraction photographs of lipid–water systems in terms of the lattice type and the dimensions (repeat distance, d; lipid thickness d_1; surface area per molecule, S; etc.) of each phase has been described in detail by Luzzati (1968).

III. Membrane Lipids

As indicated above the isolation and identification of individual lipid classes from a wide variety of biological sources is well documented and has become standard laboratory procedure. Furthermore, advances in gas–liquid chromatography have facilitated the detailed characterization of the fatty acid compositions of individual lipid species again from a wide variety of membranes (for reviews see van Deenen, 1965; and Rouser *et al.*, 1968). In

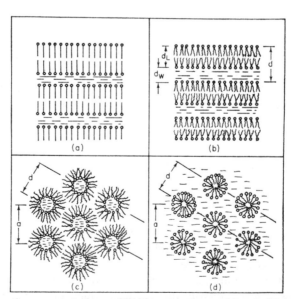

FIG. 1. Schematic representations of lipid–water phases. (a) gel; (b) lamellar liquid–crystalline; (c) hexagonal type II; (d) hexagonal type I. For further definition of these and other liquid–crystalline phases see Luzzati (1968).

addition, developments in lipid organic chemistry have provided synthetic analogues of isolated phospholipid classes with the added advantage of being able to control their fatty acid composition.

Particularly over the last ten years isolated and synthetic phospholipids have been the subject of extensive biophysical studies aimed at elucidating their structural and functional role in biological membranes. One aspect of this approach, concentrated on by Luzzati and co-workers, has been to determine the phase behavior of biologically important lipids in water. In order to determine the structure of the individual phase extensive use has been made of low-angle and wide-angle X-ray diffraction and for phospholipids in water the following structural classes have been described: (i) gel, (ii) lamellar, (iii) hexagonal, (iv) cubic, (v) micellar. Schematic representations of examples of these structural organizations are shown in Fig. 1. For further details on the derivation of these and other structures, together with an excellent review of the literature up to 1968, the reader is referred to Luzzati (1968). In this section, the phase behavior of the different classes of membrane lipids will be described with emphasis placed upon the structural parameters revealed by X-ray diffraction studies. Reference will be made to some of the early studies particularly where they are pertinent to more recent developments

A. PHOSPHOLIPIDS

1. *Phosphatidylcholines*

The zwitterionic phosphatidylcholine or lecithin (see Table I) is perhaps the most widely occurring phospholipid in animal cell membranes. It is present to a much lesser extent in higher plant cells but its occurrence in membranes of bacteria and other microorganisms is minimal and infrequent. The phase behavior of both naturally occurring and synthetic lecithins in water has been studied extensively by X-ray diffraction to a stage where it is possible to see how the hydrocarbon chain distribution influences phase behavior.

(a) *Dipalmitoyl phosphatidylcholine*. This synthetic lecithin has been studied extensively by Chapman *et al.* (1967) in both the dry state and in the presence of water. Luzzati and co-workers (1968) have concentrated on defining the phases present in the dry state and in the presence of less than $\sim 5\%$ water.

The phase diagram for the 1,2-dipalmitoyl-L-phosphatidylcholine/water system is shown in Fig. 2a. It is seen that on addition of water the transition temperature, T_c, of the phospholipid is lowered to a certain limiting value (T_c^*). This transition temperature is the minimum temperature required for the water to penetrate between the layers of the lipid molecules. Above the T_c line, the phosphatidylcholine/water system exists in the liquid–crystalline

TABLE I
Molecular formulae of phospholipids

(a)

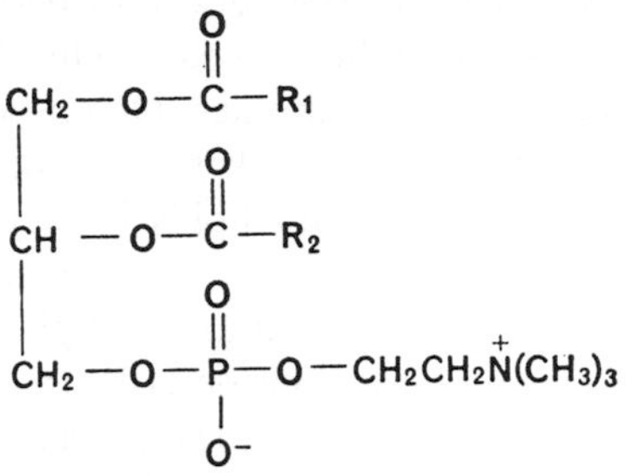

$$CH_2-O-\overset{\displaystyle O}{\overset{\|}{C}}-R_1$$

$$CH-O-\overset{\displaystyle O}{\overset{\|}{C}}-R_2$$

$$CH_2-O-\overset{\displaystyle O}{\underset{\underset{\displaystyle O^-}{|}}{\overset{\|}{P}}}-O-CH_2CH_2\overset{+}{N}(CH_3)_3$$

Phosphatidylcholine
(lecithin)

(b)

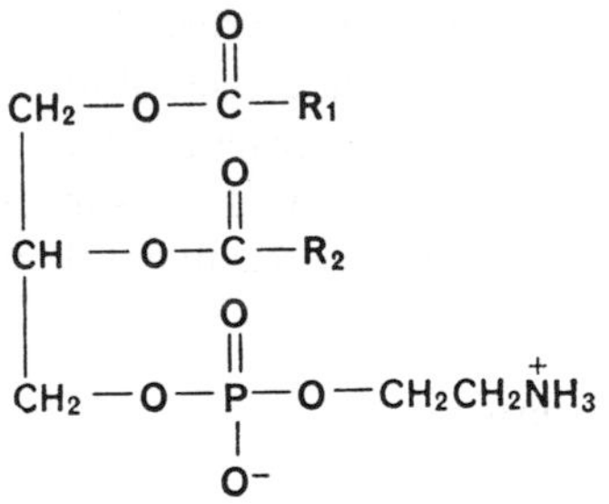

$$CH_2-O-\overset{\displaystyle O}{\overset{\|}{C}}-R_1$$

$$CH-O-\overset{\displaystyle O}{\overset{\|}{C}}-R_2$$

$$CH_2-O-\overset{\displaystyle O}{\underset{\underset{\displaystyle O^-}{|}}{\overset{\|}{P}}}-O-CH_2CH_2\overset{+}{N}H_3$$

Phosphatidylethanolamine

(c)

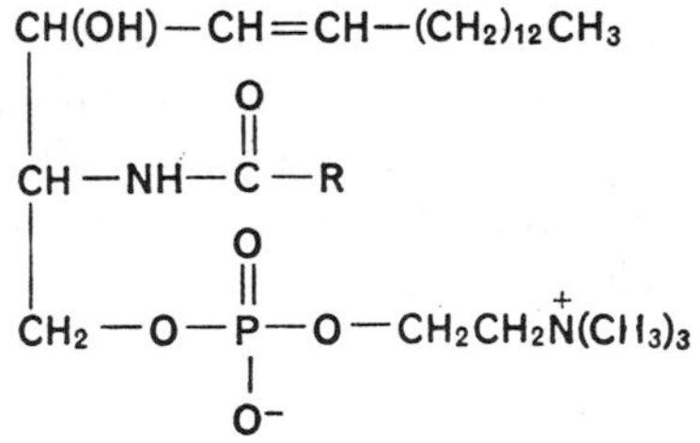

$$CH(OH)-CH=CH-(CH_2)_{12}CH_3$$

$$CH-NH-\overset{\displaystyle O}{\overset{\|}{C}}-R$$

$$CH_2-O-\overset{\displaystyle O}{\underset{\underset{\displaystyle O^-}{|}}{\overset{\|}{P}}}-O-CH_2CH_2\overset{+}{N}(CH_3)_3$$

Sphingomyelin

(d)

$$CH_2-O-R_1$$

$$CH_2-O-R_2$$

$$CH_2-O-\overset{\displaystyle O}{\underset{\underset{\displaystyle O^-}{|}}{\overset{\|}{P}}}-O-CH_2CH_2\overset{+}{N}(CH_3)_3$$

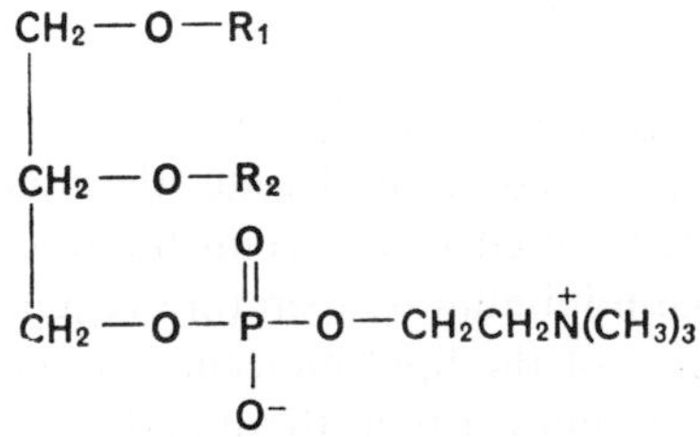

Plasmalogen

TABLE I cont.

(e) Phosphatidylserine

(f) Diphosphatidylglycerol (cardiolipin)

lamellar phase consisting of bimolecular layers of lipid molecules separated by layers of water. The hydrocarbon chains of the lecithin molecules are in a fluid state and the hydrophilic groups occupy the surface separating the lipid and water layers. The composition of the system at maximum hydration is ~ 40 wt % water, similar to that determined for the egg-yolk lecithin/water system (see p. 11). On addition of more than 40 wt % water the system exists as two phases: the liquid–crystalline lamellar phase and water.

As shown in Fig. 2b, at 50°C the lamellar repeat distance, d, increases until $c \approx 0.85$, where there is a sharp drop of 5 Å corresponding to the phase transition. This is followed by an increase in the diffraction spacing until $c \approx 0.6$, after which d remains constant at 56 Å. This transition at $c = 0.85$ is matched by a change in the wide-angle diffraction pattern from a sharp diffraction line at 4·2 Å above $c = 0.85$ to the broad diffraction line at 4·6 Å, characteristic of the liquid–crystalline phase, below $c = 0.85$.

On the other hand, at 25°C the repeat distance, d, increases as the water content increases, and reaches a limit at $c = 0.73$ (see Fig. 2b) there being no sharp change in d corresponding to a phase transition. The presence of a

G. G. SHIPLEY

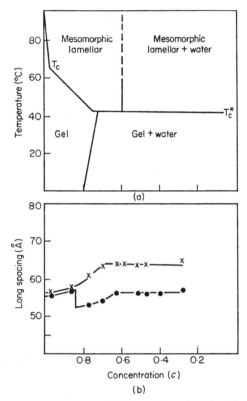

FIG. 2. The phase behavior of 1,2 dipalmitoyl-L-phosphatidylcholine. (a) phase
diagram; (b) variation of long spacing, d, with lipid concentration, c, at 25°C ($\times$)
and 50°C ($\bullet$) (from Chapman *et al.*, 1967).

single wide-angle diffraction spacing of 4·19 Å, indicates that the system is in
a gel phase. The thickness of the lipid layer is calculated to be 46 Å with an
area, S, occupied by a polar group at the phospholipid/water interface of
48 Å². Utilizing this information it can be shown that the hydrocarbon chains
of the lecithin molecules are packed in a two-dimensional hexagonal lattice
with the chain axes inclined at 58° to the lipid/water interface. The structure
of this phase, the gel, is lamellar and has been observed with other lipid/water
mixtures, e.g. potassium soaps and monoglycerides.

At very low water contents and in the dry state dipalmitoyl lecithin exhibits
a complex phase behavior as a function of temperature. In addition to the
lamellar phases referred to above, at higher temperatures structures based
upon body centered cubic and two-dimensional centered rectangular lattices
(Chapman *et al.*, 1967; Luzzati *et al.*, 1968) have been reported, with probably
a transition between the two as small quantities of water are added.

A somewhat different approach to the study of hydrated dipalmitoyl lecithin has been adopted by Levine *et al.* (1968). These authors utilize the building up of multilayers of dipalmitoyl lecithin on solid surfaces (e.g. Teflon, glass, etc.) using the surface deposition technique originally described by Blodgett and Langmuir (1937). These multilayer preparations have the advantage of being preferentially oriented with respect to the surface. When examined as a function of relative humidity and water content X-ray diffraction showed that the repeat distance of the multilayer changed from 73 Å with the specimen under a thin layer of water to 58 Å when dry. Almost no change in the spacing occurred on desorption in the range of relative humidities from 92–20%. A spacing of 4·2 Å was observed corresponding to a sharp

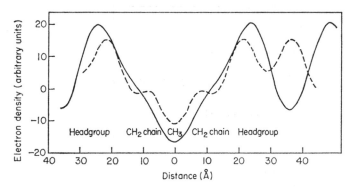

Fig. 3. Fourier synthesis or electron density profile of dipalmitoyl-DL-phosphatidyl-choline obtained from dry (- - - - -) and hydrated (————) Langmuir–Blodgett multilayers (from Levine *et al.*, 1968).

reflexion at right angles to the lamellar reflexions. This reflexion is characteristic of hexagonally packed hydrocarbon chains, indicating a high degree of order in the chains in the multilayer. The orientation of the reflexion indicates that the lecithin chains must lie almost perpendicular to the planes of the lamellae in contrast to the gel phase reported above.

Converting the observed four diffracted intensities into structure amplitudes and using the phase assignments $+, +, -, +$ for $h = 1$ to 4, Fourier syntheses were calculated giving electron density profiles normal to the plane of the lamellae. As shown in Fig. 3, the profiles obtained from the dry material and when under water are characterized by two electron dense peaks, separated by 44 Å and 50 Å respectively, on either side of a deep central well of low electron density. It is reasonable to assume that the peaks correspond to the phosphate containing headgroup region and to identify the trough in terms of the less electron dense hydrocarbon region of the phospholipid. The two

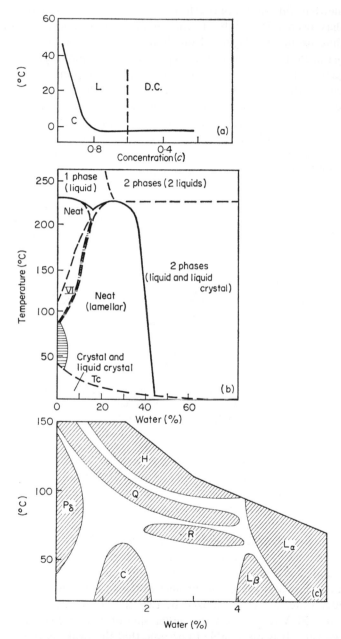

FIG. 4. Comparison of the phase diagrams of egg-yolk phosphatidycholine in water obtained by (a) Reiss-Husson, 1967; (b) Small, 1967; (c) Luzzati *et al.*, 1968. For explanation of the different regions see text.

profiles show that centers of the headgroup regions move apart on adsorption of water and that their half widths increase from 8 Å when dry to 14 Å when the multilayer is under water. The authors suggest that this change of half width may arise partly as a consequence of change of resolution of X-ray data, but that it may be evidence in favor of the idea that the headgroup of the lecithin molecule, the phosphorylcholine group, is curled up in the dry state and becomes fully extended in the presence of excess water.

(b) *Egg-yolk phosphatidycholine.* The isolated phosphatidylcholine which has been the subject of the most detailed biophysical studies is that derived

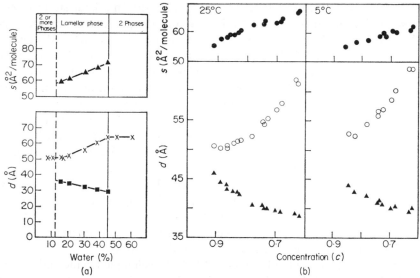

FIG. 5. Comparison of the structural parameters, d, d_1, and S of egg-yolk phosphatidylcholine determined by (a) Small (1967); (b) Reiss-Husson (1967).

from hen egg-yolk, and as with nearly all naturally occurring lipids egg-yolk lecithin has a heterogeneous distribution of fatty acids. Many authors have described the phase behavior of egg-yolk lecithin and for comparison the phase diagrams derived by Reiss-Husson (1967), Small (1967) and Luzzati *et al.* (1968) are shown in Fig. 4. The general features of the phase diagram shown in Fig. 4a indicate the similarity to dipalmitoyl lecithin described above, i.e. the presence of a transition from a gel base below the transition temperature T_c to the liquid crystalline lamellar phase above T_c. Again, above T_c two regions are identifiable; the single swelling lamellar phases up to $c \simeq 0.6$ and a two-phase system at higher water contents. The main difference with respect to dipalmitoyl lecithin is accounted for by a shift of the phase diagram down the temperature axis. This effect can be caused either by a decrease in the number of carbon atoms per fatty acid chain or by an increase in their unsaturation, the latter being the major cause in the case of egg-yolk lecithin.

Small (1967) was able to identify additional phases, particularly in the low water, high temperature part of the phase diagram (see Fig. 4b). These include a viscous isotropic (or cubic) phase which transformed on increasing the temperature into a phase with a texture which, when viewed under the polariz-ing microscope, was characteristic of another neat phase, probably hexagonal. At lower temperatures a poorly defined phase giving a complex diffraction pattern was identified (see hatched-area in Fig. 4b). In a later study Luzzati et al. (1968) further investigated this region of the phase diagram, and were able to assign the low temperature phase to two-dimensional centered rectangular lattice (Pδ in Fig. 4c). Again, this study confirmed the presence of both the cubic phase (Q in Fig. 4c) and identified the high temperature phase as hexagonal (H in Fig. 4c) although some differences in the temperature limits of these phases are apparent when the two studies are compared.

The swelling behavior of the lamellar liquid–crystalline phase of egg-yolk lecithin as detected by the change in repeat distance (d) is shown in Fig. 5 and compares the dimensions obtained by Small (1967) and Reiss-Husson (1967). The limiting repeat distance, d, at 24°C shown in Fig. 5a is 64 Å, with a limiting value of S, the surface area occupied per lipid molecule, of 72 Å^2. Reiss-Husson's studies at 25°C and 5°C gave limiting values of $d = 62$ Å and 64 Å with S changing from 68 Å^2 to 62 Å^2, respectively, on decreasing the temperature. A further report by Lecuyer and Dervichian (1969) and work in this laboratory are in good agreement with these values.

Again making use of the Langmuir–Blodgett technique Levine and Wilkins (1971) have studied the swelling of oriented egg-yolk lecithin multilayers. At 14% water content (g of water per 100 g of specimen) corresponding to 57% relative humidity the multilayer period had a value $d = 49 \cdot 7$ Å with a calculated area per molecule $S = 58 \cdot 9$ Å^2. At 100% relative humidity, giving a water content of 21%, values of $d = 51 \cdot 5$ Å and $S = 62 \cdot 7$ Å^2 are obtained, in good agreement with the values reported above for the unoriented preparations. With a critical assessment of the phasing problem, Fourier syntheses were calculated for egg-yolk lecithin in the two hydration states and the electron density profiles obtained are shown in Fig. 6. The authors interpret these electron density profiles as showing that a large proportion of the terminal methyl groups are localized near the center of the hydrocarbon region at 14% water content consistent with the presence of a deep trough of low electron-density in the center of the hydrocarbon region. The trough is similar to that shown by the electron density profiles of dipalmitoyl lecithin (see

Fig. 6). However, at 21% water content, this trough in the electron density profile is broadened, indicating an increase in the positional disorder of the terminal methyl groups located near the center of the bilayer.

It would appear from this and other studies that the trough in the middle of the syntheses does represent a region of low electron density in the middle of the hydrocarbon region. Again the peaks correspond to the electron dense phosphate groups. The peak-to-peak distance across the hydrocarbon region and the syntheses for egg lecithin bilayers decreases from 39·6 Å at 14 % water content to 36·8 Å at 21% water content ·A similar reduction is obtained when calculations of the lipid thickness are made using the lipid–water composition (see above).

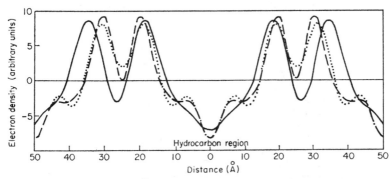

FIG. 6. Fourier syntheses of egg-yolk phosphatidylcholine bilayers in a direction perpendicular to the plane of the bilayer; —·—·— at 57% relative humidity ($d=50$ Å, 14% water) using four diffraction orders; ...... at 57% relative humidity using six diffraction orders; ——— at 100% relative humidity ($d=52·5$ Å, 21% water using four diffraction orders (from Levine and Wilkins, 1971).

In summary, Levine and Wilkins claim that in lecithin bilayers the hydrocarbon chains undergo large molecular motions. They argue in favor of considerable orientation of the chains, a localization of the terminal methyl groups near the center of the bilayer and suggest that increasing hydration results in both chain disorder and a more diffuse distribution of the methyl groups at the center of the bilayer.

(c) *The effect of sonication on egg-yolk lecithin dispersions.* When, for example, egg-yolk lecithin is swollen in excess water the resulting dispersion consists of large multi-bilayered aggregates, approximately spherical in shape, but extremely inhomogeneous with respect to the number of bilayers per aggregate. This, of course, results in a size inhomogeneity which for many biophysical studies is inconvenient. The use of sonication to produce a more homogeneous preparation of egg-yolk lecithin in water has been extensively described, and, depending upon many factors including power of sonication, time of sonication, volume of lecithin dispersion, subsequent

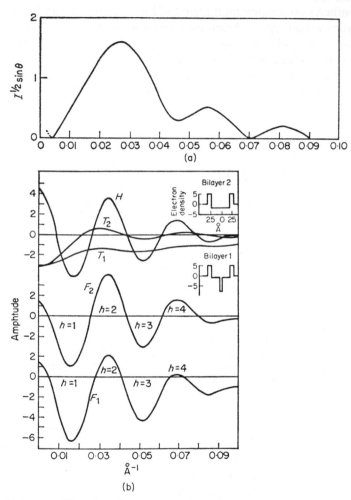

Fig. 7. (a) X-ray diffraction data from sonicated dispersion of egg-yolk phosphatidylcholine. (b) Fourier transforms calculated from simple step function bilayer models (Wilkins *et al.*, 1971). For an explanation of symbols see text.

column chromatography, uniform size single-bilayer shelled vesicles may be produced.

Early X-ray scattering studies (Chapman *et al.*, 1968) of sonicated egg-yolk lecithin dispersions noted the removal of the sharp Bragg diffraction lines originating from the multilamellar aggregates, these being replaced by a broad diffraction band centered around the position of the original first order diffraction maximum. This was interpreted in terms of removal of the multilayer structure, the consequent size reduction of the aggregate giving rise to a line broadening effect. Although line broadening is possible, particularly at early stages of sonication, the work of Wilkins *et al.* (1971) showing the presence of three broad maxima in the scattering curve of egg-yolk lecithin (see Fig. 7) allows a better interpretation of X-ray scattering data.

These authors confirm the disappearance of sharp Bragg diffraction lines and show that they are replaced by continuous diffraction or scattering. For egg yolk lecithin (15 weight % in water) three broad intensity maxima are observed with the minima between the peaks tending to zero intensity. Using $\sin^2 \theta$ as the intensity correction factor from a random dispersion of sheets the amplitude curve $I^{0.5} \sin \theta$ versus $\sin \theta$ shown in Fig. 7a is derived. This amplitude curve shows a strong maximum corresponding to $D \simeq 36$ Å and the weaker bands at $D/2$ and $D/3$. If the lipid bilayer has a symmetry plane through its center this amplitude curve should *reach* zero where the phase of the amplitude changes from $+$ to $-$. Although the intensity does not quite reach zero between the maximum it is felt by the authors that this could be attributed to either a small variation in the thickness of the bilayer or to bilayer curvature in the small vesicles. In any case the Bragg reflexions from lecithin with water multilayers 10 Å thick (see above, Levine and Wilkins, 1971) conform to this amplitude curve and add weight to the argument that the vesicles still contain the lipid arranged in a bimolecular array.

Thus the first scattering maximum may be interpreted in terms of interference within the lipid bilayer. To further define the origin of this interference the authors calculated the transforms from different model electron density profiles across the bilayer as shown in Fig. 7b. It is assumed that the polar head group (phosphorylcholine) layers have an electron density greater than that of water and the hydrocarbon part of the structure ($-CH_2-$ and terminal CH_3 groups) has an electron density less than that of water. Therefore, diffraction occurs from the two electron dense layers, their interference giving an amplitude $H = K \cos 2\pi\theta D/\lambda$ where D is the separation between the centers of the two layers. The hydrocarbon layer of lower electron density gives rise to the negative amplitude T. In the case of a hydrocarbon layer of uniform electron density, the amplitude curve T_2 results whereas with a narrow central region of even lower electron density corresponding to a localized layer of opposed $-CH_3$ groups, the amplitude curve T_1 is calculated. The combined amplitude $F = H + T$ produces the Fourier transforms shown in Fig. 7b. These transforms show a series of bands with a $\cos 2\pi\theta D/\lambda$ modulation and a maximum $|F|$ at approximately $\theta = h\lambda/2D$

when $h = 1, 2, 3$, etc. It will be noted that the model with a localized terminal methyl layer, i.e. T_1, produces bands with h even which are much stronger than those with h odd.

The positions of the maxima, particularly $h = 1$, are sensitive to the finite width of the head group layers and the authors suggest that the "thickness" of the bilayer D should be derived from the positions of the weaker, $h = 2, 3$, etc. bands. The authors have utilized this theory of scattering from fairly simple phospholipid bilayer dispersions to interpret similar X-ray scattering observed from isolated membrane dispersions (see section V).

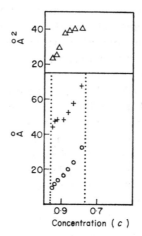

FIG. 8. Dimensions of the hexagonal type II phase of egg-yolk phosphatidylethanol-amine at 55°C. $+$, distance d between cylinder axes; $\bigcirc$, diameter of the water cylinder; $\triangle$, surface area per molecule, S (from Luzzati, 1968).

2. *Phosphatidylethanolamines*

This class of phospholipids (for molecular formula see Table I) also occurs extensively in both animal and plant cells, and in addition would seem to play an important structural role in cell membranes of bacteria and other microorganisms where it is often the major lipid present. However, relatively little information on the structure and phase behavior of phosphatidyl-ethanolamines is currently available.

Early work (Chapman *et al.*, 1966) examined an homologous series of synthetic phosphatidylethanolamines and showed that in the crystalline state the bimolecular thickness obeyed the usual rules of chain length dependence. At higher temperatures a transition occurred into a liquid–crystalline phase, exhibiting a broad diffuse X-ray diffraction line at 4·6 Å but giving only one diffraction line in the low-angle region. Thus, it was not possible on the basis of the X-ray evidence to assign one of the established structures to this high temperature phase. Most of the more recent work has concentrated on the phase behavior of phosphatidylethanolamines isolated from various natural sources.

(a) *Egg-yolk phosphatidylethanolamine.* The phase behavior of egg-yolk phosphatidylethanolamines (Reiss-Husson, 1967) is complex in that at 25°C and 35°C two phases, lamellar and hexagonal II, co-exist. At a higher temperature 55°C only the hexagonal phase is present the distance between the cylinder axes, d, increasing with increasing water content. When the lipid concentration c is less than 0·80 a two phase system is observed, in this case the hexagonal lipid–water phase plus excess water. This limited swelling behavior appears similar to that observed for the lamellar phase of lecithins reported above. The lattice dimensions characterizing the hexagonal phase are shown in Fig. 8.

On decreasing the temperature to 35° or 25°C an additional lamellar phase is observed with the relative proportion of the lamellar phase increasing as the temperature is lowered. A transition, presumably to the gel phase, is obtained at approximately 15°C.

(b) *Beef liver phosphatidylethanolamine.* Phosphatidylethanolamine extracted from beef liver (R. M. Williams, private communication) shows similar behavior, except that the hexagonal phase alone occurs at a lower temperature, 23°C. The swelling behavior of the hexagonal phase shown in Fig. 9 almost matches that exhibited by egg-yolk phosphatidylethanolamine at 55°C,

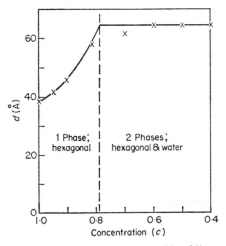

FIG. 9. Dimensions of the hexagonal type II phase of beef-liver phosphatidylethanolamine at 23°C. ×, distance between cylinder axes.

presumably this temperature dependence being related to the fatty acid compositions of the two extracted phosphatidylethanolamines.

(c) *Pseudomonas fluorescens phosphatidylethanolamine.* As part of a recent study (Cullen *et al.*, 1971) on the effects of temperature on the composition and properties of the lipids of a cold-tolerant microorganism *Pseudomonas fluorescens*, the phase behavior of the major lipid, phosphatidylethanolamine, was studied. In this study the organism was grown at two temperatures, 22° and 5°C, the phosphatidylethanolamine isolated and its fatty acid composition determined. As expected lowering the growth temperature resulted in an increase in the unsaturation of the fatty acid composition. The phase behavior of the phosphatidylethanolamines differing in fatty acid composition were studied as a function of water content and the partial phase diagrams obtained

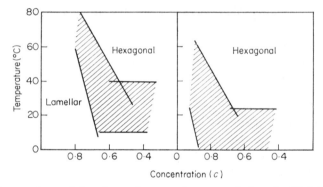

FIG. 10. Comparison of the phase diagrams in water of phosphatidylethanolamine from *Pseudomonas fluorescens* grown at 22° and 5°C. The hatched area indicates the mixed phase region.

are shown in Fig. 10. Although the phase boundaries are not accurately defined and the presence of more than one phase makes a quantitative interpretation difficult, a few significant points emerge. At lower water contents the existence of a lamellar phase only is demonstrated. On increasing the water content, a two-phase system is obtained, lamellar and hexagonal. Both of these phases are capable of only a limited uptake of water, three phases being present above a total of approximately 35% water. On increasing the temperature the lamellar phase gradually converts to the hexagonal phase, in both cases at greater than 40°C only the hexagonal phase is present. In the case of the organism grown at 22°C, there is evidence for a transition to a gel phase at about 10°C.

Obvious similarities exist between the two-phase diagrams with evidence for a shift of the phase diagram down the temperature axis for the phosphatidylethanolamine from the organism grown at 5°C. In fact when the lattice

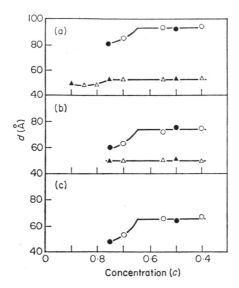

FIG. 11. The dimensions of the lamellar and hexagonal phases of phosphatidyl-ethanolamine (*P. fluorescens*)-water compared at fixed temperature differences above the growth temperature: (a) at growth temperature $\triangle T=0°C$; (b) $\triangle T=23°C$; (c) $\triangle T=43°C$. ($\triangle, d^{22}$ lamellar; $\bigcirc, d^{22}$ hexagonal; $\blacktriangle, d^5$ lamellar; $\bullet, d^5$ hexagonal.) The superscripts refer to growth temperature (from Cullen *et al.*, 1971).

dimensions of the phases are compared at fixed temperature differences from the growth temperature (Fig. 11), an almost exact identity is observed for both phases when present as a mixed phase system, and for the single hexagonal phase at $\Delta T=43°C$. This may suggest a feedback-regulatory mechanism that controls the lipid composition such that particular physical properties of the membrane containing this lipid are maintained when the environmental temperature is altered.

(d) *Pig erythrocyte phosphatidylethanolamine/blowfly larvae phosphatidyl-ethanolamine.* Rand *et al.* (1971) have examined the phase behavior of two pure phosphatidylethanolamines isolated from porcine erythrocytes (PPE) and from the larvae of the blowfly (FPE). Except for the PPE at low water content, which is a mixture of lamellar and hexagonal phases, at 20°C, as indicated in Fig. 12, a lamellar phase only is present for both phosphatidyl-ethanolamines. The structural parameters defining the lamellar phases at 20°C are shown in Fig. 13. At neutral pH both PPE and FPE incorporate water in the concentration range $c=0.90$ to 0.70. At higher water contents a two-phase system with excess water is obtained. The limiting dimensions for PPE are 52·1 Å, with a limiting surface area per molecule of 60 Å², and 50·0 Å for FPE. At higher temperatures both lipids exhibit the mixed lamellar/

alcohol sphingosine. The polar headgroup of this class of compounds is, however, identical to that of the phosphatidylcholines described above.

The only report on the phase behavior of sphingomyelin is by Reiss-Husson (1967) who noted that mixtures of beef brain sphingomyelin water possessed hydrocarbon chains that were crystalline at 25°C. On raising the temperature to 40°C a liquid crystalline lamellar phase is formed with a swelling behavior similar to that of the phosphatidylcholines described above. As shown in Fig. 14, over the concentration range $c = 0.8$ to 0.6 the lamellar spacing d increases from 62 Å to 78 Å while the lipid thickness decreases to approximately 48 Å. The surface area per molecules at maximum hydration is 55 Å. At concentrations c less than 0.6 a two-phase system is formed. The behavior of sphingomyelin is similar to lecithin with the exception that the lipid thickness is larger and the surface area per molecule smaller in the case of sphingomyelin, indicating a more saturated hydrocarbon chain composition.

4. *Plasmalogens*

Not all phospholipids have their hydrocarbon residues linked exclusively by an ester bond to glycerol. Plasmalogens possess a vinyl ether linkage and this occurs always at the 1-position (see Table 1). The bases esterified to the phosphate are usually choline or ethanolamine but a serine derivative has been described. Most animal tissues contain rather small amounts of plasmalogen but they are abundant in nervous tissue and heart muscle mitochondria.

In a recent paper (Gottleib and Eanes, 1972) the influence of electrolytes

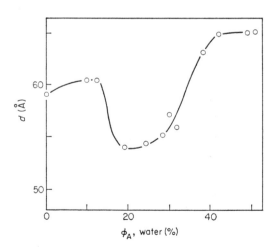

FIG. 15. Variation of long spacing d as a function of the volume concentration ϕ_A for the plasmalogen 1 octadec-9-enyl, 2 hexadecyl glycerophosphocholine (from Gottlieb and Eanes, 1972).

on the lamellar phase of a synthetic plasmalogen, 1 octadec-9-enyl, 2 hexadecyl glyceryl-phosphorylcholine has been described. As shown in Fig. 15, the swelling behavior of the lamellar phase of the plasmalogen at a water content greater than 18% is essentially similar to that of egg-yolk lecithin described above. The limit of swelling is approximately 40% volume per cent water. In the hydration range 0–13% water only limited swelling is observed, the wide angle diffraction pattern indicating the presence of a crystalline or gel phase. The sharp decrease in diffraction spacing d in the concentration range 13–18% water clearly corresponds to the phase transition into the liquid crystalline lamellar phase.

The authors show that in water the bilayer thickness, d_1, decreases with increasing interbilayer separation, d_A, until a limiting thickness is reached; d_1 then remains constant up to the maximum interbilayer separation attainable, i.e. maximum swelling. This behavior differs from that of egg-yolk lecithin (Small, 1967; Reiss-Husson, 1967) where d_1 decreased continuously with d_A up to the point of maximum swelling. No explanation for this difference is given. An interpretation in terms of a modification of a theory of lipid bilayer thickness developed by Parsegian (1967) suggests that when the interbilayer separation is large enough to accommodate two fully extended phosphorylcholine groups from two opposing bilayers no further changes in the surface area per molecule at the interface, S, and consequently d_1, should be observed. Bond-length data on the phosphorylcholine headgroup suggest that the limiting value of d_A should occur at an interbilayer distance of about 23 Å, in reasonable agreement with the experimental value of 21 Å. It would appear that these data could be interpreted in terms of a gradual unfolding of the phosphorylcholine group from a parallel to perpendicular orientation with respect to the plane of the lipid bilayer.

The authors then show interestingly that the swelling behavior of the plasmalogen, and the dependence of d_1 and d_A, is modified in the presence of IN solutions of various electrolytes (LiCl, NaCl, KCl, CsCl, HCl, CaCl$_2$, Na$_2$SO$_4$). It would appear that there is an ion specificity of the ability to incorporate water, the maximum amount of water incorporated increasing for the alkali chlorides in a simple order of ionic radii from Li$^+$ to Cs$^+$. Depending on the particular electrolyte used these may decrease, increase or have little effect on the thicknesses of the lipid bilayer in plasmalogen mesophases. As the authors point out, clearly much more work needs to be done in this area of phospholipid hydration and the effect of ions and electrolyte solutions before adequate explanations can be given.

5. *Phosphatidylserines*

This class of phospholipids with a widespread distribution in cell membranes of animals, plants and microorganisms, although rarely occurring at levels greater than 10%, is of interest since the molecules bear a net negative charge (see Table I).

A brief report of previously unpublished data on the complex behavior of a series of synthetic 1,2-diacyl phosphatidylserines both as the free acids and the mono-potassium salts in the dry state is included in a review by Williams and Chapman (1970). The behavior of phosphatidylserines in water has been examined by Shipley and Williams (unpublished results) for both the mono-potassium salt of 1,2-diacyl-L-phosphatidylserine, and also for the isolated phosphatidylserine from ox-brain. The swelling behavior of these two

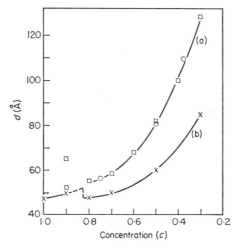

FIG. 16. Lamellar repeat distance d as a function of lipid concentration c at 22°C: (a) ox-brain phosphatidylserine, ○ and □ (symbols represent data from two separate experiments); (b) dilauroyl-L-phosphatidylserine, ×.

phosphatidylserines at 25°C is compared in Fig. 16. For the synthetic phosphatidylserine a transition from a gel phase to a liquid–crystalline phase is observed between $c=0.9$ and 0.8, the liquid crystalline phase then incorporating all the water added between the lipid bilayers at least to a concentration $c=0.3$. The ox-brain phosphatidylserine shows a similar behavior, having a single lamellar swelling phase over the concentration range $c=0.90$ to 0.30 the repeat distance increasing from 52 Å to 128 Å. The thickness of the lipid bilayer decreases to approximately 40 Å, the limiting surface area per molecule, S, being approximately 65 Å².

The differences between the isolated and synthetic phosphatidylserines can be partially explained by differences in their average chain length, i.e. one expects the hydrated dimensions of the shorter chain molecules to fall systematically below those of the phosphatidylserine with a longer average chain length. However, the difference increases with increasing hydration and one must suspect either inaccuracies in the water content of the synthetic phosphatidylserine or perhaps some molecular degradation.

The explanation of the continuous swelling behavior of these lipids normally invokes repulsion of the charged lipid bilayers as a major factor (see Gulik-Krzywicki *et al.*, 1969b; Shipley *et al.*, 1969a), the swelling effect being reversible on addition of excess monovalent or divalent cations or charged proteins. This will be discussed in more detail later.

6. *Cardiolipin*

The glycerol phospholipid frequently termed cardiolipin is a double unit in that it has two molecules of phosphatidic acid esterified to a central glycerol and is more appropriately designated as diphosphatidyl glycerol (see Table I). This phospholipid occurs in large amounts in bacterial membranes, mitochondria and chloroplasts, all of which contain respiratory assemblies.

The effect of Ca^{2+} on the phase behavior of beef heart cardiolipin has been studied in detail by Rand and SenGupta (1972). In the absence of added Ca^{2+} a lamellar phase only is present at lipid concentrations $c=0.39$ to 0.53 with the lipid thickness d_1 varying from 34·5 to 37·1 Å. At higher lipid contents $c=0.85$ to 0.95 a single hexagonal phase (type II) exists, the distance between the cylinder axes decreasing from 52 Å to 40 Å on increasing c. At intermediate concentrations the lamellar and hexagonal phases co-exist. This information is summarized in Table II.

TABLE II
Phases formed by cardiolipin in water

Dry wt % lipid	Phase	Dimensions	
39·3–53·4	Lamellar	d_1	34·5–37·1 Å
53·4–85·7	Lamellar and hexagonal	—	
85·7–95·0	Hexagonal	Distance between cylinder axes	52·2–40·5 Å
		Diameter of water cylinders	23·8–11·0 Å
		Polar to polar distance	38·4–29·5 Å
		S	34·7–24·0 Å²

(from Rand and SenGupta, 1972)

At all concentrations of $CaCl_2$ in the range 0·001 to 1·0 M cardiolipin precipitates as a pure hexagonal phase. This structure is stable over the temperature range 4 to 40°C with the usual temperature dependence of the lattice parameters. The Ca^{2+}/cardiolipin^{2-} mole ratio is one, indicating the formation of stoichiometric combination of the ions in the precipitate. The internal packing within the hexagonal structure is determined from the

TABLE III

Composition and structural dimensions of the hexagonal phases of cardiolipin complexes with divalent metal ions

Divalent ion	$T(°C)$	Dry wt and vol % of lipid in precipitate	Distance between cylinder axes (Å)	Diameter of water cylinder (Å)	Polar to polar distance (Å)	S, surface area available to polar groups at water interface (Å^2)
Ca^{2+}	4	92·6	58·0	16·6	41·4	29·4
	20	92·6	52·9	15·0	37·9	27·4
	40	92·6	49·8	14·2	35·6	29·2
Mg^{2+}	4	85·7	66·0	26·1	39·9	33·1
	20	85·7	61·0	24·2	37·8	36·1
	40	85·7	57·5	22·8	34·7	38·0
Ba^{2+}	40	84·5	59·4	24·6	34·8	38·9

(from Rand and SenGupta, 1972)

composition of the hexagonal phase and, as with all diacyl phospholipids, is type II with water cylinders in a hydrocarbon matrix. The composition and structural dimensions of the phase are given in Table III for the three temperatures, 4, 20 and 40° including the distance between cylinder axes, the diameter of the water cylinder, the polar to polar distance of the cardiolipin through the hydrocarbon matrix and the surface area S available to each half cardiolipin molecule on the water cylinder. The dimensions of the hexagonal phases are consistent with hexagonal phases observed with other diacylphospholipids when allowance is made for the double unit structure.

Precipitation of cardiolipin by $MgCl_2$ also produces hexagonal phases. However, at low temperatures and for the precipitates formed at low $MgCl_2$ concentrations, a lamellar phase co-exists with the hexagonal. The water content of the hexagonal phase and consequently the diameter of the water cylinder are larger than those of the corresponding Ca^{2+}-cardiolipin precipitate but the polar to polar distances, indicating twice the average length of cardiolipin molecule, are identical. In the case of $BaCl_2$ precipitates of cardiolipin, in the concentration range of 0·01 to 0·7 M $BaCl_2$ lamellar and hexagonal phases co-exist, the relative quantity of the former decreasing as the temperature is raised until a pure hexagonal structure exists at 40°C. The composition and structural dimensions of the hexagonal phase are similar to the Mg^{2+}-cardiolipin precipitates.

B. GLYCOLIPIDS

This section will review the hitherto neglected field of the phase behavior of sugar-containing lipids and for a review of their occurrence in animal cells and organelles see Rouser et al. (1968). However, it is in plant cells and cell organelles that the sugar-containing glycolipids and sulpholipids are of prime importance where one imagines they have an important role to play in membrane structure.

1. Cerebrosides

In animals the galactose or glucose ceramides known as cerebrosides (Table IV) occur extensively in the brain and rather less extensively in the lungs, kidney, liver, spleen, etc.

The phase behavior of a mixture of beef brain cerebrosides (Reiss-Husson, 1967) is of interest in that for lipid concentrations close to $c=0·70$ the temperature must be raised to approximately 70°C in order to effect the transition from a phase in which the chains are crystalline to the liquid–crystalline lamellar phase. The structural parameters of cerebroside–water mixtures at 77°C are shown in Fig. 17. At high lipid contents, greater then $c=0·9$, a crystalline phase exists. In the concentration range $c=0·9$ to 0·8

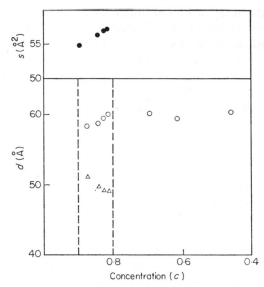

FIG. 17. Structural parameters of the lamellar phase of cerebrosides at 77°C as a function of lipid concentration c (from Reiss-Husson, 1967). Symbols as in Fig. 14.

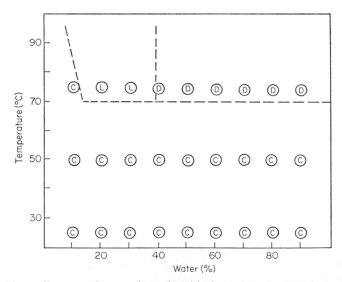

FIG. 18. Phase diagram of natural cerebroside in water. C, crystal or gel phase; L, lamellar phase; D, lamellar phase plus excess water (from Abrahamsson et al., 1972).

swelling occurs with d increasing minimally from about 58 to 60 Å with a concomitant decrease in d_1 from 51 Å to 49 Å. At concentrations less than $c = 0.8$ a two-phase zone exists, the lamellar phase in excess water.

Recently Abrahamsson et al. (1972) have examined the phase behavior of a number of glycosphingolipids in both the dry state and in the presence of water. Considering mainly the hydrated phases, for bovine cerebroside with long-chain base consisting of about 95% sphingosine and 5% dihydro-sphingosine with approximately equal amounts of normal fatty acids (mainly tetracosanoic acid) and 2-hydroxy fatty acids (mainly 2-hydroxytetracosanoic acid and 2-hydroxyoctadecanoic acid), the phase diagram shown in Fig. 18 was obtained. The lamellar liquid crystalline phase is observed only when the temperature exceeds 70°C as shown in Fig. 18. At a composition of 80% cerebroside the thickness of the bimolecular lipid layer (d_1) is 47·5 Å and the surface area per molecule in contact with water is 55·8 Å². The thickness of the unit layer in the lamellar mesophase when no water is present is too small (40 Å) in comparison with this value (47·5 Å) to correspond to the same type of structure and an interdigitating chain structure is envisaged (see Luzzati, 1968). A two phase liquid–crystalline dispersion is formed at water contents greater than 40 weight % compared with a value of 20 weight % described above (Reiss-Husson, 1967). The structural dimensions of the lamellar mesophase formed by the two cerebrosides are in good agreement.

The aqueous systems of homogeneous cerebroside with a C_{18} fatty acid was also studied by Abrahamsson et al. (1972) but no interaction with water was observed at any composition. As shown above, for the natural cerebroside no interaction with water takes place below about 70°C, this high temperature requirement presumably being determined by its lipid composition. For the synthetic cerebroside this transition temperature is not obtained below 100°C and consequently no hydration occurs.

2. Sulfatides

The sulfatides or sulfated cerebrosides (see Table IV) have a similar distribution and occurrence to the cerebrosides. The phase behavior has been determined for a sulfatide from human brain with a long-chain base composition similar to the beef brain cerebroside described above and with a fatty acid composition of about two-thirds 2-hydroxy fatty acids (mainly 2-hydroxytetracosanoic acid and 2-hydroxtetracosenoic acid) and one-third normal acids (mainly tetracosenoic acid).

The phase diagram of the aqueous system of natural sodium sulfatide (Abrahamsson et al., 1972) is shown in Fig. 19. On crystallization from aqueous mesophases a metastable gel is obtained. On losing water from the lattice a crystal form with vertical chains arranged according to the hexagonal chain packing is obtained. The repeat distance of this crystal form is 68·5 Å and the cross-section area per hydrocarbon chain is 19·5 Å², consistent with the chains of the molecule pointing in the same direction with respect to the polar end groups. In the gel state water layers of thickness up to 44 Å between the polar end groups have been observed to be stable for several days.

The lamellar liquid–crystalline phase was found also in this system. At 75°C and 10 wt % water the thickness of the bimolecular lipid layer of this

TABLE IV

Molecular formulae of glycolipids

(a) $CHOH-CH=CH-(CH_2)_{12}CH_3$

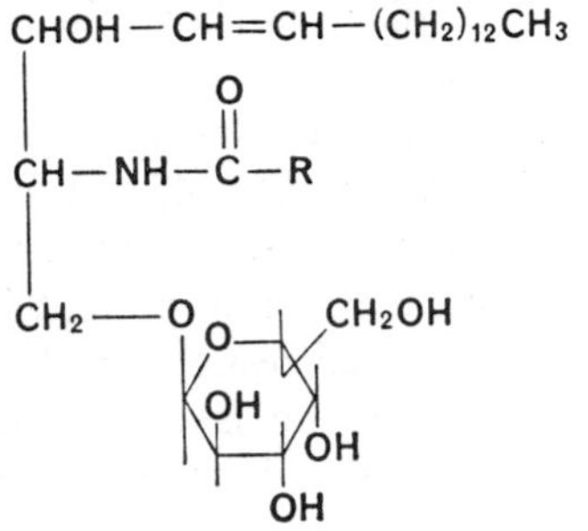

           O
           ‖
     $CH-NH-C-R$

                                     Cerebroside

     CH_2-O CH_2OH

         OH

           OH   OH

(b) $CHOH-CH=CH-(CH_2)_{12}CH_3$

           O
           ‖
     $CH-NH-C-R$

                                     Sulfatide

     CH_2-O CH_2OH

         OH

           OH

           $OSO_3^- Na^+$

           O
           ‖
(c) $CH_2-O-C-R_1$

           O
           ‖
     $CH-O-C-R_2$

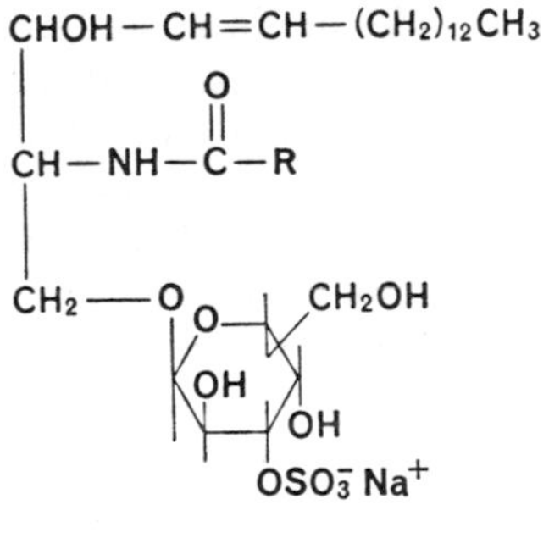

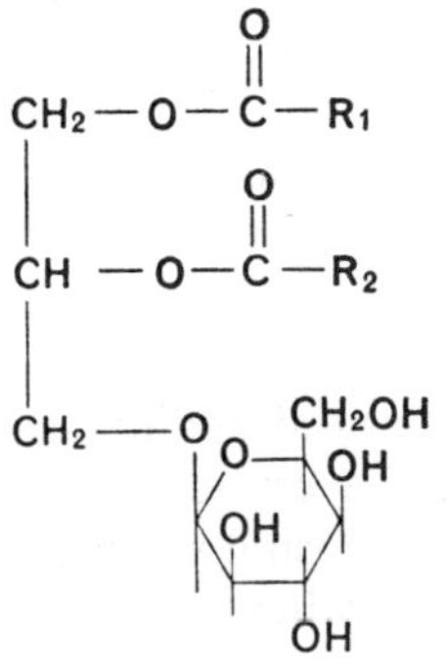

                                   Monogalactosyl diglyceride

     CH_2-O CH_2OH

           OH

           OH

           OH

TABLE IV *cont.*

(d)

O
‖
CH₂—O—C—R₁

O
‖
CH —O—C—R₂

CH₂—O

Digalactosyl diglyceride

phase is 52·2 Å, and the surface area per molecule at the water interface is 56·3 Å². The lipid layer thickness decreases with increasing temperature and/or increasing water content, a property typical of this phase.

At intermediate lipid compositions a cubic phase, I, with diffraction spacings in the following ration $\sqrt{2}/\sqrt{3}/\sqrt{8}/\sqrt{10}$, and a diffuse band near 4·5 Å attributed to liquid-like hydrocarbon chains, is observed. If the diffraction pattern is indexed as a primitive lattice, the indices are (110), (111), (220) and

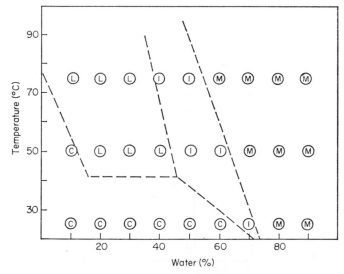

FIG. 19. Phase diagram of sodium sulfatide in water. *C,* crystal or gel phase; *L,* lamellar phase; *I,* isotropic cubic phase; *M,* micellar phase (from Abrahamsson *et al.,* 1972).

(310), and the unit cell dimension $a = 121 \cdot 3$ Å. There are too few observed lines to determine the space group unambiguously.

With an excess of water micellar solutions are formed which give a diffuse X-ray scattering in the range corresponding to 52–98 Å which is almost independent of the concentration. Such a maximum in the scattering curve is typical for micellar particles in which the polar groups form a shell with a considerably higher average electron density between the core of hydrocarbon chains and the water phase.

3. *Monogalactosyl Diglyceride*

Monogalactosyl diglycerides are particularly abundant in plant leaves and algae where they occur mainly in chloroplasts. They usually contain a high proportion of polyunsaturated fatty acids.

The phase behavior in water of monogalactosyl diglyceride (MGDG) from pelargonium leaves has recently been studied (Shipley *et al.*, 1973) in the concentration range $c = 1 \cdot 0$ to $0 \cdot 5$, and over the temperature range 0–80°C. The phase diagram shown in Fig. 20 indicates the presence of an hexagonal

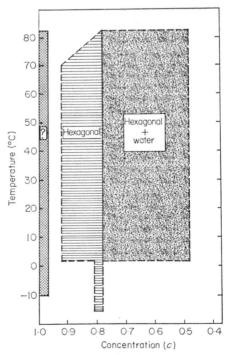

FIG. 20. Phase diagram of monogalactosyl diglyceride in water (from Shipley *et al.*, 1973).

phase over the concentration range $c=0.9$ to 0.78. At concentrations $c<0.78$ a two-phase system, hexagonal lipid–water in excess water, is present. X-ray diffraction in the wide-angle region confirms that the phase is liquid–crystalline and the thermal behavior of a 1 : 1 (w/w) MGDG/H_2O sample studied by differential scanning calorimetry indicates that there is a liquid–crystalline/gel transition in the region of $-30°C$.

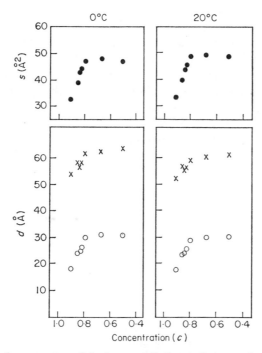

FIG. 21. Structural parameters of the hexagonal phase of monogalactosyl diglyceride as a function of lipid concentration, c, at $0°C$ and $20°C$. $\times$, distance between cylinder axes; $\bigcirc$, diameter of water cylinders; $\bullet$, surface area per molecule (from Shipley et al., 1973).

Fig. 21 illustrates the behavior of various lattice parameters at two temperatures $0°$ and $20°C$ as a function of lipid concentration. At $0°C$ the distance between the cylinder axes d increases to a limiting value of approximately 62.5 Å at an estimated concentration $c=0.78$, the lipid being fully hydrated at this point. At this limiting dimension the surface area per molecule at the lipid–water interface approaches 47 Å². On increasing the temperature to $20°C$ a small decrease in the limiting dimension d is observed, from 62.5 Å to 60.5 Å with a concomitant small increase in the surface area S to 48.7 Å².

4. *Digalactosyl Diglyceride*

Digalactosyl diglycerides are usually found together with MGDG in the chloroplasts of higher plants and algae. They are less abundant than MGDG but possess a similar high proportion of polyunsaturated fatty acids particularly linolenic acid.

Digalactosyl diglyceride (DGDG) again from pelargonium leaves (Shipley *et al.*, 1973) was examined over the concentration range $c = 1.0$ to 0.6 and over the temperature range -10 to $+80°C$. The phase behavior, illustrated in Fig. 22, indicates the presence of a liquid–crystalline lamellar lipid–water phase over the whole range of concentration and temperature studied. Differential scanning calorimetry shows a transition, probably the gel/liquid–crystalline, to be in the region of $-50°C$. For concentrations $c < 0.78$ the DGDG exists as a lamellar phase in excess water the system having reached full hydration at $c < 0.78$ whilst for $c > 0.78$ the lamellar phase expands with increasing water content or decreasing lipid concentration c.

Fig. 23 shows the structural parameters for DGDG at $0°$ and $20°C$ as a function of concentration. At $0°C$ the primary spacing d_{100} increases from

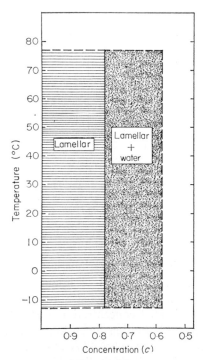

FIG. 22. Phase diagram of digalactosyl diglyceride in water (from Shipley *et al.*, 1973).

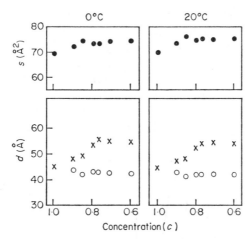

FIG. 23. Structural parameters of the lamellar phase of digalactosyl diglyceride as a function of lipid concentration, c, at 0°C and 20°C. ×, repeat distance; ○, lipid thickness; ●, surface area per molecule (from Shipley et al., 1973).

45 Å at $c=1.0$ to a limiting value of 55 Å at $c=0.78$ whilst the lipid thickness d_1 approaches a limiting value in the region $0.9 < c < 0.85$. At this limiting dimension the mean surface area per molecule at the lipid–water interface is 74 Å². On raising the temperature to 20°C a small decrease in the limiting dimension is seen; d decreases from 55 Å to 54 Å and d_1 from 42.3 Å to 41.6 Å, with a concomitant small increase in the surface area, S, to 75 Å².

C. MIXED LIPIDS

Although the behavior of single species of phospholipids in water has been reasonably well documented, relatively little attention has been paid to the phase behavior of mixed biological lipids. Ternary phase diagrams of systems containing long-chain fatty acids, alcohols, amines, etc., have been studied extensively by Ekwall and co-workers, (see Fontell et al., 1968). Similarly the phase interrelationships of the major lipid components of bile (lecithin, cholesterol, bile salts), has been determined by Small (1971).

Calorimetric methods have been used to investigate the role of the hydrocarbon chains in determining molecular mixing of lecithins (Phillips et al., 1972), the effect of phosphatidylethanolamine and dicetyl phosphoric acid on the phase transitions of dipalmitoyl lecithin (Abramson, 1970) and the molecular mixing of the component lipids of Mycoplasma laidlawii (Chapman and Urbina, 1971).

X-ray diffraction has been used to show that even at very small weight concentrations the incoporation of charged lipids, e.g. phosphatidylserine

(Shipley *et al.*, 1969a); cetyl trimethyl ammonium bromide, sodium oleate and sodium stearate (Gulik-Krzywicki *et al.*, 1969b) into egg-yolk lecithin, induces a continuously swelling lamellar phase analogous to that of a charged phospholipid alone, e.g. phosphatidylserine (see above). Clearly the behavior is charge dependent and repulsion between two charged "planar" surfaces can account for the swelling phenomenon. The addition of electrolytes reduces the swelling effect and the presence of equimolar amounts of oppositely charged lipids in egg-yolk lecithin completely removes the effect.

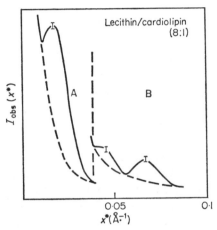

FIG. 24. Observed diffracted intensities $I_{obs}(x)$ as a function of x^* ($=2 \sin \theta/\lambda$) from dispersions of dipalmitoyl lecithin/cardiolipin. Regions A and B represent a change in intensity by a factor of 6·8 (Lesslauer *et al.*, 1971).

Mixtures of pig-liver lecithin and beef-heart cardiolipin have been shown by Rand (1971) to form lamellar phases at all compositions in the range 90 : 10 to 47 : 53 (lecithin : cardiolipin), the thickness of the biomolecular layer in the lamellar phase, d_1, varying between 39 and 41 Å, with a corresponding surface area per molecule of approximately 60 Å^2. Lesslauer *et al.* (1971) formed vesicles of dipalmitoyl lecithin–cardiolipin mixtures (8 : 1) by sonication and observed X-ray scattering consisting of three broad maxima, shown in Fig. 24, similar to that produced by sonicated egg-yolk lecithin dispersions (Wilkins *et al.*, 1971). A novel approach involving the calculation of a generalized Patterson function from the essentially single unit cell diffraction data and its deconvolution to produce the corresponding electron density distribution (Hosemann and Bagchi, 1962) were used together with conventional Fourier methods to produce the electron density profile shown in Fig. 25. As with egg-yolk lecithin the profile shows two peaks separated by about 40 Å identified as the phosphorus-containing headgroup regions of the bimolecular leaflet. Lesslauer *et al.* (1971) have utilized this technique to localize the lipid and membrane binding dye 1-anilo-8-napthalene-sulphonate (ANS) in phospholipid dispersions and lipid multilayers.

A mixed lipid system which has been studied extensively by X-ray diffraction methods is lecithin–cholesterol–water. The five key papers describing the phases present and their dimensions (Bourges *et al.*, 1967; Ladbrooke *et al.*, 1968; Rand and Luzzati, 1968; Lecuyer and Dervichian, 1969; Levine and Wilkins, 1971) have been reviewed recently Phillips (1972).

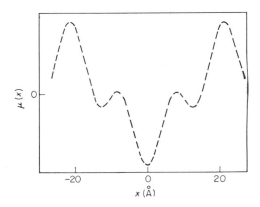

FIG. 25. Profile of the relative electron density $\mu(x)$ for the dipalmitoyl lecithin/cardiolipin bimolecular leaflet (from Lesslauer *et al.*, 1971).

D. EXTRACTED MEMBRANE LIPIDS

Luzzati and co-workers have examined the phase behavior of the total lipids extracted from four natural membrane systems. The behavior of the lipid extracts from human brain and beef-heart mitochondria have been discussed in detail by Luzzati (1968) and will be briefly summarized here whereas more detailed descriptions are given for erythrocyte membrane lipids and the lipids of maize chloroplasts.

1. *Mitochondria Lipids*

Gulik-Krzywicki *et al.* (1967) investigated a lipid extract from beef-heart mitochondria containing approximately 34% lecithin, 29% phosphatidylethanolamine, 10% phosphatidylinositol, 20% cardiolipids, 2% cholesterol and 5% neutral lipids. The phase behavior in water is shown in Fig. 26 and four phases were identified. At low water contents the hexagonal H_{II} phase is present which converts to a lamellar phase L_α on addition of more water. As shown in Fig. 27b at 25°C the dimension of the lamellar phase increases continuously with added water, the lipid thickness and surface area remaining

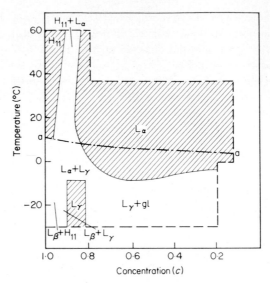

FIG. 26. Phase diagram of the mitochondria lipid–water system (from Gulik-Krzywicki *et al.*, 1967).

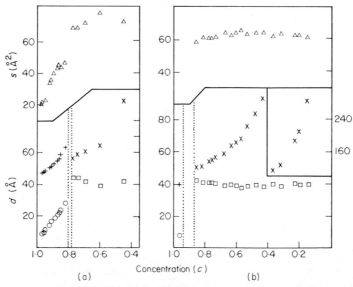

FIG. 27. Structural dimensions of lipid–water systems (a) human brain, 37°C (Luzzati and Husson, 1962); (b) beef-heart mitochondria, 25°C (Gulik-Krzywicki *et al.*, 1967). △, surface area per molecule; ×, lamellar repeat distance; □, thickness of lipid leaflet; +, distance between cylinder axes; ○, diameter of the water cylinder.

constant at approximately 40 Å and 64 Å², respectively. Over the range $c = 0.85$ to 0.14 the thickness of the water layer increases from 7 to 250 Å, thus showing a similar behavior to the charged phospholipids, e.g. phosphatidylserine, phosphatidylinositol (see above). Below the line a–a in Fig. 26 two lamellar phases L_γ and L_β of the gel type, characterized by a sharp diffraction line at 4·2 Å, are observed. The structures of these two phases have been described meticulously by Gulik-Krzywicki *et al.* (1967) and Luzzati (1968).

2. *Brain Lipids*

An early study (Luzzati and Husson, 1962) of human brain lipids containing approximately 52% phosphatidylethanolamine, 35% lecithin and 13%

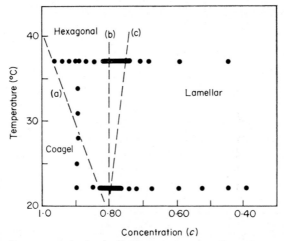

FIG. 28. Phase diagram of the brain lipid–water system (from Luzzati and Husson, 1962).

phosphatidylinositol, again shows (Fig. 28) a hexagonal phase at low water contents which transforms at $c \simeq 0.80$ to the expanding lamellar phase similar to that described for mitochondrial lipids. At low contents a gel type structure is present. The dimensions of the hexagonal and lamellar phases at 37°C are shown in Fig. 27a, for the lamellar phase the limiting value for the thickness of the lipid bilayer, d_1 and the surface area S being 40 Å and 70 Å², respectively.

3. *Erythrocyte Lipids*

Rand and Luzzati (1968) have reported the phase behavior of the total extracted lipids of human erythrocytes. The total lipid contains 26% cholesterol, the phospholipids being distributed between lecithin 36%, phosphatidylethanolamine 29%, sphingomyelin 23%, phosphatidylserine 10%

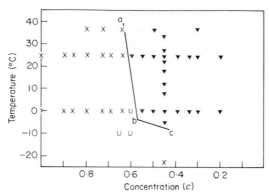

FIG. 29. Phase diagram of the human erythrocyte lipid–water system. ×, crystalline cholesterol plus another phase; u, unstable sample; ▼, single lamellar phase (from Rand and Luzzati, 1968).

and lysolecithin 2%. As shown in Fig. 29, at low water contents some cholesterol apparently separates from the lipid phase thus preventing a detailed structural analysis, whereas at higher water contents a single lamellar phase is present. The lamellar phase exists at temperatures between -5 and $+40°C$ and its dimensions at $0°C$ are shown in Fig. 30. Clearly the lamellar phase is of the continuously expanding type with d increasing up to 230 Å while the limiting values of d_1 and S are 45 Å and 77 Å², respectively.

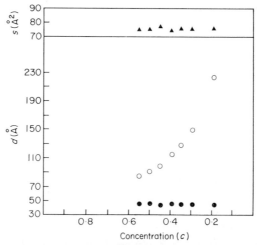

F,G. 30. Dimensions of the lamellar phase of human erythrocyte lipid at $0°C$: ○, repeat distance d; ●, lipid thickness d_1; ▲ surface area per molecule S. The surface area represented is that occupied by one phospholipid molecule plus 0·72 of a cholesterol molecule (from Rand and Luzzati, 1968).

From predictions of the approximate position of cholesterol within the lipid bilayer (see Fig. 31) and its known stoichiometric composition theoretical electron density distributions through the lipid bilayer can be derived in terms of the sum of a number of Gaussian functions. The best agreement between the amplitudes calculated from this model and the observed amplitudes is for the electron density profile shown at the bottom of Fig. 31. This best-fit model has part of the cholesterol in contact with the polar headgroup but much of it in contact with the phospholipid hydrocarbon chains. A detailed discussion of the interaction of cholesterol with phospholipids is given by Phillips (1972).

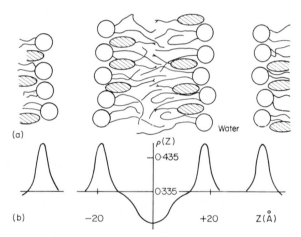

FIG. 31. (a) Schematic representation of the structure of the lamellar phase of human erythrocyte lipids. The circles represent the polar groups of the phospholipid molecules. The curved lines represent the paraffin chains. The hatched group represents the steroid nucleus of the cholesterol molecule. (b) The electron density distribution through the lipid leaflet, indicating the absolute levels of electron densities and the dimensions, z, from the center of the leaflet (from Rand and Luzzati, 1968).

4. Chloroplast Lipids

Rivas and Luzzati (1969) have separated the lipids from maize chloroplasts into two fractions, a polar lipid fraction containing phosphatidylglycerol, lecithin and phosphatidylinositol, and a galactolipid fraction containing monogalactosyl diglyceride, digalactosyl diglyceride and sulfolipid. The phase diagrams of the two fractions are compared in Fig. 32. For the mixed polar lipids at low-water content and high temperatures the hexagonal phase type II is present but at all other compositions the continuously expanding lamellar phase exists (Fig. 32a). As shown in Fig. 33a the limiting lipid

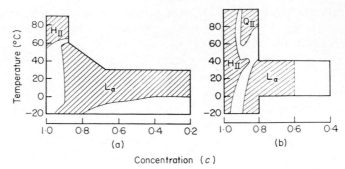

FIG. 32. Phase diagrams of (a) polar lipids, (b) galactolipids, from maize chloroplasts (from Rivas and Luzzati, 1969).

thickness, d_1, and the surface area, S, have values of 39 Å and 56 Å^2, respectively, again showing similar behavior to the extracted lipids described above.

However, the galactolipid fraction shows a more extensive hexagonal phase at low-water contents as well as an isotropic cubic phase (Q_{II} in Fig. 32b) at higher temperatures. Furthermore, the lamellar phase shows only limited uptake of water, similar to lecithin and the pure digalactosyl diglyceride described earlier, a two-phase system lamellar phase in excess water being present at lipid concentrations less than $c \simeq 0.65$. The dimensions of this lamellar phase are shown in Fig. 33b.

With the benefit of all this information certain principles which appear to govern the phase behavior of biological (and other) lipids in water become

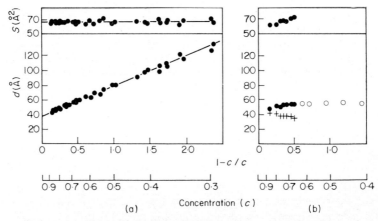

FIG. 33. Structural dimensions of the lamellar phases of maize chloroplast lipids. (a) polar lipids; (b) galactolipids. d, repeat distance; S, surface area per molecule. The abscissa is plotted as $1 - c/c$ where c is the weight concentration of lipid (from Rivas and Luzzati, 1969).

apparent. These principles and their possible relevance to membrane structure are considered in the discussion.

IV. Lipid–Protein Complexes

A. PHOSPHOLIPID–PROTEIN COMPLEXES

As indicated in section III we have now fairly extensive information regarding the phase behavior of biologically important lipids in water and some of the factors governing phase behavior are becoming apparent. For example, the effects of hydrocarbon chain length and chain unsaturation are predictable; the presence of charge on diacyl lipids tends to induce the continuously expanding lamellar phase; the presence of charged lipids in lipid mixtures produces expanding lamellar phases; the relative volumes of the headgroup and hydrocarbon chain moieties may dictate which of the two types of both hexagonal and cubic structures (see Luzzati, 1968) a particular lipid will adopt in the presence of water. Furthermore, X-ray diffraction methods have enabled us to determine quantitatively the parameters defining

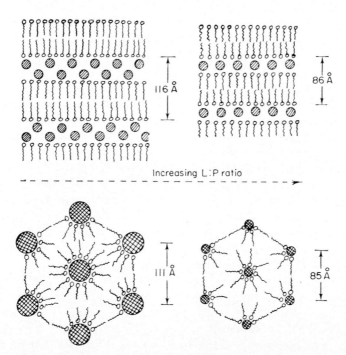

FIG. 34. Schematic representation of the lamellar and hexagonal phases formed between cytochrome c and phospholipids.

each phase. The approaches adopted have been invaluable to the understanding of X-ray diffraction from the more complex membrane systems (see below). However, an obvious intermediate step taken by Shipley *et al.* (1969a,b) and Gulik-Krzywicki *et al.* (1969a) was to apply these methods to lipid–protein complexes formed between the now well-characterized phospholipids and water soluble proteins for which X-ray crystallographic studies had provided excellent, detailed structural information.

In a study mainly restricted to the interaction of cytochrome *c* with phospholipids, structural parameters were derived by X-ray diffraction methods for lipoprotein complexes precipitated in the aqueous phase (Shipley *et al.*, 1969a) and also for the lipoprotein complexes soluble in hydrocarbon solvents (Shipley *et al.*, 1969b). Using a mixture of egg-yolk phosphatidylcholine/ox-brain phosphatidylserine (2 : 1 w/w), precipitated complexes with ferricytochrome *c* were shown to be lamellar structures of two basic types (see Fig. 34) with dimensions of either 87 Å and 116 Å depending on the lipid : cytochrome *c* ratio. These dimensions were consistent with the incorporation of either one or two layers of cytochrome *c* molecules, respectively, between a phospholipid bilayer. In both cases the complexes were thought to be stabilized by ionic/electrostatic interactions between the basic groups on the protein and the acidic groups on the phosphatidylserine. As shown in Fig. 34, although the affect of low ionic strength (NaCl) is as yet unclear the presence of a high salt medium (100 mM NaCl) predictably disrupts the complex. Occasionally a hexagonal phase with a cylinder to cylinder axia $a = 85$ Å was obtained.

Using phosphatidylserine alone as the interacting lipid an equally complex polymorphic behavior was observed (Shipley, Atkinson, Leslie and Davis, unpublished). Again two lamellar phases with similar dimensions ($d = 117$ Å and $d = 85$ Å) are obtained, the formation of which are dependent on both the composition and the presence of added salt. In the absence of added NaCl a lipoprotein complex (molar ratio phosphatidylserine : cytochrome *c* 10 : 1) gave a hexagonal diffraction pattern with a cylinder axis to cylinder axis dimension $a = 111$ Å. Although it has not been possible to determine the molecular organization within the two hexagonal phases it is likely that the phospholipid will form the continuous phase as indicated schematically in Fig. 34. The sequence of phases observed, hexagonal ($a = 111$ Å) → lamellar ($d = 117$ Å) → lamellar ($d = 85$ Å) → lamellar ($d = 96$ Å), as the lipid to protein ratio increases suggests that the specific structure formed may be governed by geometrical factors involving the packing of cytochrome *c* molecules at the lipid interface.

In a more extensive study Gulik-Krzywicki *et al.* (1969a) have investigated the interactions between either ferricytochrome *c* or lysozyme with phosphatidylinositol, cardiolipin and mitochondria lipids, using both X-ray diffraction

TABLE V
Ferricytochrome c-phosphatidyl inositol–water

Phase	T (°C)	d (Å)	c_p/c_1	c_0	d_p (Å)	d_1 (Å)	d_0 (Å)
I	20	78·5	0·78	0·18	23·0	39·8	15·7
	20	68·0	0·78	0	24·8	43·2	0
II	20	112·0	1·49	0·23	42·9	39·6	29·5
	20	91·0	1·49	0	47·0	43·8	0
	20	—	—	0·2–	—	39·5	—
				0·8			

(from Gulik-Krzywicki *et al.*, 1969a)

and circular dichroism methods. As shown in Table V the ferricytochrome
c–phosphatidylinositol system gives two lamellar structures with, at maximum
hydration, $d=78·5$ Å and $d=112$ Å, again the interpretation being in terms
of arrays of hydrated proteins, either one or two molecules thick, intercalated
between the lipid bilayers. From the molecular composition parameters
describing the thickness of the lipid, protein and aqueous layers are made.
Similarly with cardiolipin, two lamellar phases with $d=108$ Å and $d=75$ Å
were obtained at different lipid to protein weight ratios. The authors describe
three phases following the interaction of cytochrome c with beef-heart
mitochondria lipids. Two of these are lamellar, the dimensions of the hydrated
complexes being $d=86·5$ Å and $d=116$ Å, and the other has an hexagonal
structure with $a=83$ Å. Clearly these are extremely similar to the complexes
described above for phosphatidylserine/phosphatidylcholine mixtures.

The interaction of the same lipids with lysozyme produces an interesting
variation. Cardiolipin–lysozyme mixtures give two lamellar phases (see
Table VI) for which the contributions of the three components to the overall
lattice thickness are computed. Interestingly, in one of the lamellar phases,

TABLE VI
Lysozyme–cardiolipin–water

Phase	T (°C)	d (Å)	c_p/c_1	c_0	d_p (Å)	d_1 (Å)	d_0 (Å)
III	25	92·0	1·13	0·28	29·0	34·8	28·2
	25	73·0	1·13	0	34·5	38·5	0
IV	25	77·0	1·69	0·17	35·0	27·0	15·0
	25	67·0	1·69	0	36·9	30·1	0
V	35	85·2	1·69	0·17	38·8	29·9	16·5
	25	—	—	0·2–	—	35·0	—
				0·8			
	35	—	—	0·2–	—	34·5	—
				0·8			

(from Gulik-Krzywicki *et al.*, 1969a)

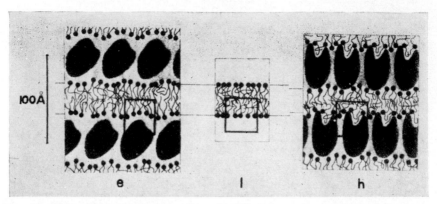

FIG. 35. Schematic representation of two lamellar phases formed between lysozyme and phospholipids. Note that the thickness of the lipid leaflets is the same in phase *e* and in the lipid–water phase *l*, and that it decreases in phase *h* (from Shechter *et al.*, 1971).

III, d_1 is the same as for the lamellar lipid–water phase alone whereas in the other case IV, the lipid layer thickness is decreased considerably when compared to the lipid–water phase, as shown in Fig. 35. Further structural and fluorescence spectroscopy studies (Gulik-Krzywicki *et al.*, 1970; Shechter *et al.*, 1971) indicate differences in the type of interaction responsible for complex stability. In essence these studies suggest that for weak electrostatic interactions no decrease in d_1 occurs whereas when hydrophobic interactions occur, involving the apolar peaks of both lipid and protein, the observed shrinkage in d_1 results. For the system lysozyme–phosphatidylinositol in which both temperature and pH were raised a number of phases (see Table VII) based upon lamellar hexagonal and two-dimensional square lattice were obtained. The two hydrated phases with $d = 83$ Å and 106 Å appear similar to those described above. For further information on the structure and interactions of phospholipid–cytochrome *c* complexes reference is made to Quinn and Dawson (1969a,b, 1970), Horne and Watkins (1970), Fromherz (1970), Kimelberg *et al.* (1970) Kimelberg and Lee (1970), and Steinemann and Lauger (1971).

A similar approach has been adopted by Rand (1971) in a study of the interaction of pig-liver phosphatidylcholine/beef-heart cardiolipin mixtures with bovine serum albumin at pH 3·33 where the protein is expanded or unfolded. For molar ratios of lecithin/cardiolipin in the range 93 : 7 to 50 : 50 the precipitated complexes are single lamellar phases with *d* changing from 90 Å to 68 Å. The compositions vary systematically, the weight per cent lipid remaining constant at about 41%, the protein increasing from 10 to 30% and water decreasing from 50 to 30%. The reduction in *d* from 90 Å to 68 Å is correlated with the increasing cardiolipin/serum albumin mole ratio giving

TABLE VII
Lysozyme–phosphatidyl inositol–water

pH	T (°C)	c_p/c_1	c_0	Lattice type	Lattice dimensions (Å)
4	0–35	1·09	0·27	Two-dimensional square	63·5
4	0–35	1·09	0	Two-dimensional square	57·0
5	25	0·50	—	Lamellar	74·0
5·5–7	0–35	1·96	0·23	Two-dimensional hexagonal	80–82·5
7	0	—	—	Two-dimensional hexagonal	96
7	30	—	—	Lamellar	106
7	40	—	—	Lamellar	83
8	25	0·97	—	Lamellar	106
8	40	—	—	Lamellar	83

(from Gulik-Krzywicki et al., 1969a)

rise to a decreasing net positive charge on the complex. When the net charge approached zero the structure is condensed and the area available per protein molecule in the plane of the layer indicates tight packing. Rand concludes that since the two components have opposite charges electrostatic interactions are responsible for the initial interactions but that subsequent complex stabilization may be achieved by shorter range polar and hydrophobic interactions between the lipid and protein, a view already expressed by Shipley et al. (1969b) and Gulik-Kryzwicki et al. (1969a).

For a discussion of the structure of this type of lipoprotein complex when extracted into hydrocarbon solvents (Das et al., 1965) the reader is referred to Shipley et al. (1969b) and Lesslauer et al. (1970).

B. SERUM LIPOPROTEINS

Although not directly related to membrane structure small-angle X-ray scattering studies of water-soluble serum lipoproteins will in the future provide information on the mutual arrangement of lipids and proteins in these systems. So far preliminary X-ray scattering data have been reported for human serum high-density lipoproteins (Shipley et al., 1972), low-density lipoproteins (Mateu et al., 1971) and abnormal low-density lipoproteins (Hamilton et al., 1971). X-ray scattering curves of the type shown in Fig. 36 are obtained which give information on the size, shape, molecular weight, etc., of the lipoprotein. Furthermore, the scattering curve may be interpreted in terms of step electron density models where different regions of electron

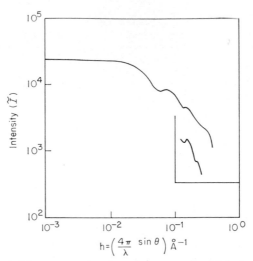

FIG. 36. Small-angle X-ray scattering from human serum high density lipoproteins, HDL$_2$.

density correspond to the location of different molecular components, leading eventually, when sufficient X-ray scattering data are available, to the calculation of electron density profiles across the lipoprotein particle. However, although many further studies in this field are predicted, any structural relationship between serum lipoproteins and membrane lipoproteins has yet to be established.

V. Membranes

For the purposes of this chapter an arbitrary division of membranes into three classes is made. The naturally occurring multilayered membranes myelin, retinal rods, and chloroplasts are considered in the first part of this section. However, most membranes, e.g. plasma membranes, do not occur in such an organized multimembrane array and the intermembrane diffraction data derived from the "artificial" stacking of such membranes has not been too informative. The significant advances in this field have come from the demonstration by Wilkins *et al.* (1971) that interpretable data could be derived from unstacked membranes. Structural information from these "single" membranes is given in the second section. Recent X-ray diffraction studies of membranes associated with viruses are considered in the last section.

A. MULTI-MEMBRANES

1. *Myelin*

The molecular organization of nerve myelin (see Fig. 37) has been studied extensively by optical birefringence, X-ray diffraction, and electron micro-

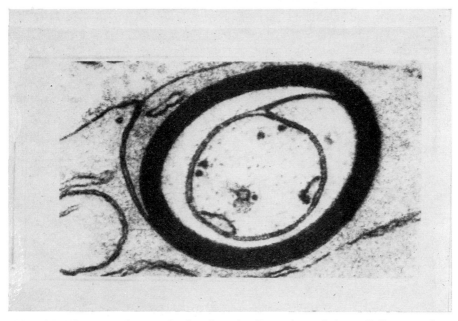

FIG. 37. Electron micrograph showing the formation of the radially repeating structure of myelin.

scopy. On the basis of these physical studies, together with a knowledge of its chemical composition, nerve myelin has been shown to consist of concentric layers of lipoprotein spirally wrapped around the nerve axon. The first X-ray diffraction pattern of intact or live myelinated nerve (Schmitt *et al.*, 1935, 1941) showed a series of discrete low-angle X-ray reflexions which are related to the radial packing of the concentric lipoprotein layers. The diffraction pattern consisted of the first five orders of a radial repeat distance of 170–185 Å depending on the variety of nerve, and an interpretation was given based on a double lipid bilayer with linear protein molecules located at the lipid interfaces. The center-to-center layer distance is the radial repeating unit and each layer consists of two Schwann cell membranes.

In an excellent series of papers Finean and co-workers combined X-ray diffraction and electron microscopy to examine both peripheral and central nervous system myelin from a variety of sources, deriving additional structural information from the effects of dehydration, heating, cooling, freezing, thawing, etc. This work has been extensively reviewed elsewhere (Finean, 1966; Stoeckenius and Engelman, 1969).

In studying the effects of swelling and shrinking of the nerve myelin sheath Finean noted that the lipoprotein layers were separated by increasingly thick layers of water whereas the X-ray diffraction patterns and electron

micrographs indicated that the lipoprotein layer itself remained virtually
unaltered. If this is correct, as Finean and Burge (1963) showed, then as the
one dimensional unit cell expands or contracts, each point on the lattice
explores a finite length of the Fourier transform of the membrane structural
unit. Thus, if the observed diffracted intensities are plotted against the
reciprocal spacings of the changing lattice points, the intensity profile should

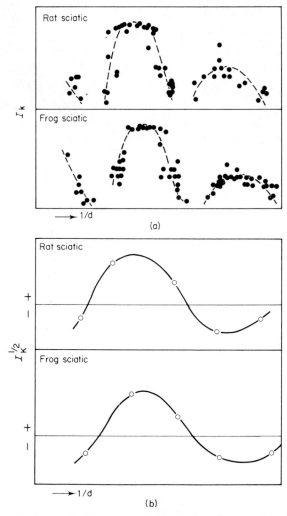

Fig. 38. (a) Experimental values of diffraction intensities for swollen preparations
of rat and frog sciatic nerve myelin. (b) Fourier transforms calculated from the
diffraction data (from Finean and Burge, 1963).

be enveloped by a smooth curve. The change in the diffracted intensities (and their position $1/d$) from rat and frog sciatic nerve myelin preparations induced by swelling in hypotonic diluted Ringer solutions is shown in Fig. 38a and clearly the intensity profile is enveloped by a smooth curve. The curves tend to reach zero at three points on the ordinate corresponding to positions at which the Fourier transform changes sign. The Fourier transforms calculated for the two myelin preparations are shown in Fig. 38b. Positioning the first five orders of diffraction from the unmodified myelin unit on the Fourier transform allows the allocation of signs of the structure amplitudes corresponding to these reflexions, in this case, $-++--$. With this information the Fourier summations were calculated giving electron density profiles for rat optic and sciatic nerve. These are shown in Fig. 39 together with a molecular interpretation. The profiles show peaks to either side of the deep trough which are about 50 Å apart, and it was suggested that these might correspond to the separation between phosphate groups at the surfaces of a lipid bilayer, although it was emphasized that the peak positions could be

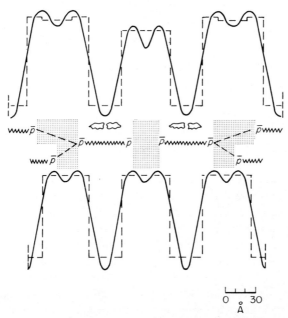

FIG. 39. Electron density profiles for myelin layers in rat sciatic (upper) and rat optic (lower) nerves. The continuous curve was calculated directly from the measured diffraction intensities without applying any corrections. Essential correction factors made the curves approximate more closely to the simple step functions indicated with an interrupted line. Cholesterol molecules and phospholipid molecules are indicated schematically but approximately to scale. Dotted regions indicate the probable locations of non-lipid components (from Finean, 1969).

affected by intensity corrections. In terms of molecular structure it was pointed out that profiles of this type could be accounted for by a lipoprotein-sandwich type of structure and that they could distinguish myelin from rat optic nerve (periodicity 160 Å) and rat sciatic nerve (periodicity 180 Å) in terms of the thicknesses of non-lipid layers in the region where the outer surfaces of glial or Schwann-cell membranes came together during myelin formation. This important advance in phase determination has been utilized by many other workers as indicated below.

The electron density profile shown in Fig. 39 was calculated using five orders ($h = 1-5$) of the lamellar diffraction. Blaurock and Worthington (1969) obtained diffraction patterns from a variety of peripheral nerve and central nervous system myelin in which discrete diffraction lines were observed out to $h = 11$. These diffraction orders with $h > 5$ only have small intensities, partly due to the cylindrical layering involved and partly because of the intrinsic membrane structure (see below). The authors note differences in the radial repeat distances for peripheral nerve myelin (171–182 Å) and central nervous system myelin (153–159 Å). They show, too, that peripheral nerve and central nervous system myelin exhibit different intensity distributions for $h = 1$ to 5, but that these are independent of the myelin source, in this case, frog, rat and chicken. These authors (Worthington and Blaurock, 1969a) then characterized the swelling behavior of peripheral frog sciatic nerve myelin in distilled water, Ringer's solution and sucrose solutions. In sucrose, for example, low-angle diffraction patterns were obtained with thirteen ($h = 1-13$) diffraction orders of a repeating unit $d = 388$ Å. A Fourier transform was obtained very similar to that obtained by Finean and Burge (1963) and an identical sequence of signs for the structure amplitudes for native myelin was determined.

In a series of papers (Worthington and Blaurock, 1968, 1969a,b; Worthington, 1969b; Worthington and King, 1971) Worthington has adopted the approach of proposing a model for the myelin membrane and comparing its diffraction pattern with that obtained experimentally. The model based on a simple step function is defined by a number of parameters determining in the one dimensional case, the limits of regions of differing electron density, these regions being identified with different parts of the double membrane system. These parameters are then adjusted and refined until optimal agreement is obtained between intensities predicted by the model and those determined experimentally. In this procedure it is, of course, a prerequisite that the number of diffraction parameters exceed the number of model parameters being refined.

In their first model (Fig. 40A) they distinguished a myelin layer thickness, w, from the periodicity d, which was considered to include a contribution from "intermembrane fluid". The myelin membrane was assumed to consist of two closely apposed lipoprotein leaflets with identical symmetrical profiles. A value for w was obtained directly from the zero positions along the Fourier transform which with this particular model are spaced regularly with respect to the reciprocal spacing l/d. The original model thus had 5 independent parameters, l, v, P, L and F and optimal agreement with the frog sciatic

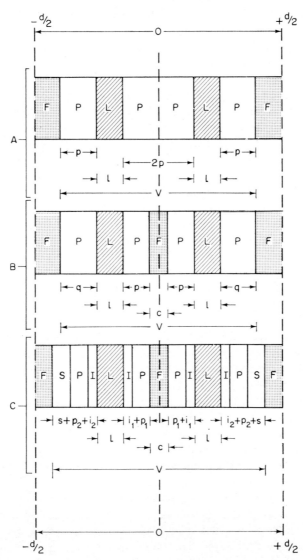

FIG. 40. Centrosymmetrical electron density strip models for myelin. For explanation, see text.

nerve data (d=171 Å) occurs when v=156 Å, l=19 Å and the electron density levels for lipid hydrocarbon and non-lipid were 0·25 electrons/Å³ and 0·38 electrons/Å³, respectively. The various intact peripheral nerves studied showed a variation in d of 171–182 Å, and the model parameters show some variation, the largest change occurring in v. The width of the fluid channel (d–v) showed only a small variation of 16·0–18·5 Å. The intact myelins from central nervous system showed a variation in d of 153–159 Å. The lipid thickness l remained fairly constant as in the case of the peripheral nerves, but v showed variation in the range 144–150·5 Å. The width of the fluid channel (d–v) varied minimally in the range 6·5–9 Å.

As stated earlier, the original X-ray diffraction gave only five orders of lamellar diffraction (h=5), thus restricting the number of refinable parameters to a maximum of five, but improvements in techniques, etc., have enabled higher orders now up to eighteen (h=18), to be measured thus allowing more complex models to be refined. Worthington (1969b) then proposed a seven-parameter model which included an independent fluid layer between the inner cytoplasmic surfaces within the membrane pair as well as allowing the layers of higher electron density on either side of the region of low electron density to become independent in width. This model is shown in Fig. 40B. The number of operational parameters can be reduced to five v, c, p, l and α where α defines the relationship between the electron densities P, F, and L. Using seven orders of diffraction (h=7) the refined model parameters shown in Table VIII were obtained. The low-density regions l within the plasma membrane components are 22–23 Å wide for the three varieties of sciatic nerve; these values are somewhat larger than those derived previously with the earlier symmetric membrane model. There is, however, a cytoplasmic fluid layer of 6–10 Å and an extracellular fluid layer of 15–17 Å. The outer high-density layers of the plasma membrane components are not symmetric: the outer layer which faces the extracellular fluid exceeds the inner layer by about 3–5 Å in each of the three sciatic nerves.

This model was also tested for central nervous system myelin, frog optic nerve and frog spinal cord, and good agreement values (R) were obtained between the observed and calculated intensities, partially due, no doubt, to the fact that the number of diffraction orders was only slightly greater than the number of operational parameters. The model suggested that low-density regions within the plasma membrane components are 23–24 Å wide for the two nerves of the central nervous system, these values being again somewhat larger than those derived using the earlier symmetric membrane model. The cytoplasmic fluid layer is 6–8 Å thick and this layer has about the same width as the extracellular fluid layer. The outer high-density layers of the plasma membrane components are not symmetric, the outer layer being wider than the inner layer. In both cases (see Table VIII) the outer layer exceeds the

TABLE VIII

Model parameters and the corresponding R values for three live peripheral nerve myelins ($h=7$) and two central nervous system myelins ($h=6$)

	d(Å)	v(Å)	$d-v$(Å)	c(Å)	p(Å)	l(Å)	q(Å)	α	$R\%$
Frog sciatic nerve	171	156	15	6	$24\frac{3}{4}$	$22\frac{1}{3}$	28	0·30	8·2
Rat sciatic nerve	176	159	17	10	24	22	$28\frac{1}{2}$	0·34	9·2
Chicken sciatic nerve	182	165	17	8	$26\frac{1}{4}$	23	29	0·30	5·2
Frog optic nerve	154	148	6	6	$21\frac{1}{4}$	24	$25\frac{3}{4}$	0·36	2·0
Frog spinal cord	153	147	6	8	$20\frac{1}{3}$	23	26	0·25	1·4

(from Worthington, 1969b)

inner layer by 5–6 Å, a slightly greater difference than that shown by the peripheral nerves.

Recently Worthington and King (1971) with eighteen orders of diffraction ($h = 18$) available have developed a twelve-parameter model (see Fig. 40) in which the following notation is used: d, radial repeat distance; v, thickness of the membrane pair; c, thickness of the cytoplasmic fluid layer composed of electron density F; l, thickness of the layer composed of electron density L; i_1 and i_2, thicknesses of the layers composed of electron density I; p_1 and p_2, thicknesses of the layers composed of electron density P; s, thickness of the layer composed of electron density S. The twelve parameters are the five electron densities (F, P, I, L, S) and the seven layer thickness ($c, p_1, p_2, i_1, i_2, l, s$) but, because the fluid density F is known and because X-ray intensities are on a relative scale, there are only ten operational parameters. The best agreement with the intensity data for frog sciatic nerve was obtained using the following model parameters: $P = 0.37$ electrons per Å^3, $I = 0.36$ electrons per Å^3, $L = 0.27$ electrons per Å^3, $S = 0.35$ electrons per Å^3, $c = 10$ Å, $p_1 = 13$ Å, $i_1 = 10$ Å, $l = 21$ Å, $i_2 = 9$ Å, $p_2 = 12$ Å, $s = 10$ Å. A diagram of the twelve-parameter electron density model is shown in Fig. 40C. In summary, these results indicate a thickness for the single membrane of 75 Å, which is similar to that derived using the simpler models (see above). The single membrane is asymmetric with the thickness of the central low electron density region again being unexpectedly narrow (as in the three earlier models). The actual value of l is 21 Å and this region is thought to contain the lipid hydrocarbon chains. The thicknesses of electron density layers I and P have the property $i_1 \simeq i_2$, and $p_1 \simeq p_2$, and therefore the five layers (p_1, i_1, l, i_2, p_2) form an approximately centrosymmetrical unit. The asymmetry of the nerve membrane unit arises primarily through the presence of an electron dense layer 10 Å thick which faces the extracellular fluid space. The observation of a symmetrical unit within the nerve membrane suggests that these five layers may constitute a symmetric lipid bilayer with the protein component confined to only one layer, the 10 Å layer.

The phases derived by Worthington for the first twelve reflexions using

TABLE IX

Model phases for the first twelve diffraction orders

Model type	$h=1$	2	3	4	5	6	7	8	9	10	11	12
Five-parameter	−	+	+	−	−	+	+	−	−	+	+	+
Seven-parameter	−	+	+	−	−	+	+	−	+	−	−	−
Eight-parameter	−	+	+	−	−	+	+	−	+	−	−	−
Twelve-parameter	−	+	+	−	−	+	+	−	+	−	−	−

(from Worthington and King, 1971)

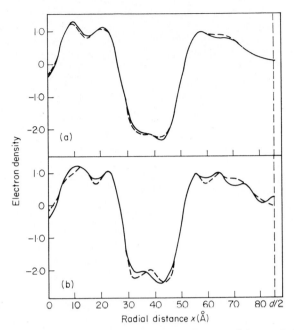

FIG. 41. Electron density profiles for the myelin layers of frog sciatic nerve computed using (a) the first twelve reflexions; (b) the first eighteen reflexions. The continuous and discontinuous curves refer to the Fourier synthesis using the observed and calculated structure amplitudes F, respectively (from Worthington and King, 1971).

four different models are shown in Table IX. The only difference is that the sign combinations of reflexions $h=9$ to 12 are reversed in the three later models. Fourier syntheses were calculated using the diffraction data out to the $h=12$ and $h=18$ reflexions (Worthington and King, 1971) and these are shown in Fig. 41. The continuous curves are the profiles for the myelin of frog sciatic nerve using the phases from the twelve parameter model and correspond to effective resolutions of 7 Å and 4·8 Å, respectively. The discontinuous curves are the Fourier syntheses calculated using the theoretical diffraction amplitudes derived from the model together with the model phases. Clearly only small differences exist between the observed and calculated Fourier syntheses suggesting that the twelve-parameter model is a reasonable choice for nerve, and therefore the parameters of this electron density model closely resembles the molecular parameters of live nerve.

Worthington and Blaurock (1969a) and Blaurock (1971) have commented on additional factors which may affect the interpretation of X-ray data derived from swollen myelin preparations. Worthington and Blaurock (1969a) have suggested on the basis of refined electron density models that

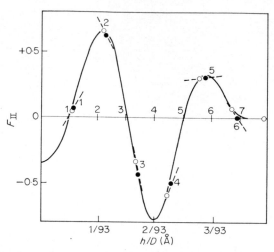

FIG. 42. The Fourier transform, F_{II}, derived from the cytoplasmic swelling of frog sciatic nerve myelin (from Blaurock, 1971).

the dimensions l and v of the membrane pair change with swelling and that the sampling of a unique Fourier transform, on which many phase assignments are made, may not be obtained. Blaurock (1971) has interpreted the Patterson functions of swollen myelin preparations at various salt concentrations and shows that, for example, in 1 mM $CaCl_2$ a small amplitude swelling at the cytoplasmic interface occurs, distinguishable from that occurring normally at the extracellular surface. The different Fourier transform derived from cytoplasmic swelling is shown in Fig. 42. When electron density profiles are calculated at the same resolution for membranes in various bathing media which promote the different types of swelling, the single membrane profiles obtained are near identical (see Fig. 43). This is determined by the existence of the two Fourier transforms for extracellular and cytoplasmic swelling but only one choice of signs $(-++--)$ gives this agreement in the electron density profile for the native membrane. For example, the $(+++++)$ phasing of Akers and Parsons (1970b) (see below) give well-defined differences in the single membrane profile.

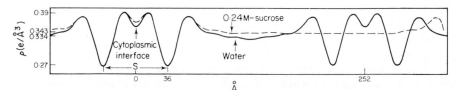

FIG. 43. Fourier synthesis showing the electron density profile of frog sciatic nerve myelin (from Blaurock, 1971).

Moretz *et al.* (1969a,b) have shown that OsO_4, $KMnO_4$ and aldehyde fixatives used in electron microscopy induce changes in both the periodicities and the intensities of myelin preparations, indicating that some structural rearrangement takes place on fixing. They also showed that post-fixation treatment with acetone and alcohol dehydrating agents could produce further changes in the diffraction of both polar and nonpolar (particularly cholesterol) lipids.

Accurate diffraction spacings and intensities for orders $h=1$ to 5 were obtained from frog sciatic nerve myelin by Akers and Parsons (1970a) using a modified Kratky slit-collimation camera with a proportional counting system. By examining the half-widths of the intensities and the effect of stretching the nerve on the relative intensities, the authors conclude that the restricted number (five) of major reflexions is not limited by disorder of the membrane but results from a specific electron density distribution across the myelin membrane pair. Akers and Parsons (1970b) then approached the problem of determining the phases by a heavy atom labelling method in which osmium tetroxide, platinic chloride or potassium permanganate is titrated into the frog sciatic nerve myelin structure. Changes in both the periodicities and the intensities of the diffraction pattern were observed. Using the osmium labelled data the Patterson function indicated that the metal goes mainly to a single site although on increased labelling either a second site is established or membrane reorganization occurs. The major point of this study was, however, to demonstrate a novel approach to phasing the myelin structure amplitudes. Computer analogue techniques were based on the addition of Gaussian distributions of electron density, representing the heavy atoms, to one or two sites on the electron density profile of the native membrane, the latter being calculated for all 32 (or 16 equivalent pairs) combinations of the phases. The intensity distribution corresponding to the modified electron density profile was calculated and compared with that of the non-labelled profile. The differences between the intensity distributions were then compared to those observed experimentally after labelling. The authors state that only one phase sequence $(+ + + + +)$, with appropriate parameters describing the Gaussian distribution function, gives an acceptable fit with the observed data. An extension of this approach was used to calculate the phase combination $(+ + +)$ for three additional reflexions, $h=6$, 8, and 11 and the Fourier transforms using 5 and 8 reflexions resulted in the electron density profiles shown in Fig. 44. Obvious differences are detectable in the profile (cf. Figs. 39, 41 and 43) particularly the high peaks at the two centers of symmetry. However, since objections to this interesting approach to the solution of the phase problem which produces a unique set of signs have been raised (Worthington, 1970), we must await further confirmation before a detailed analysis of the electron density profile is made.

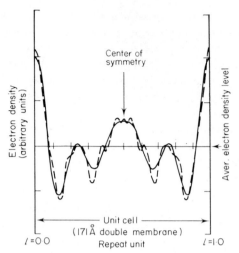

FIG. 44. Electron density profile for frog sciatic nerve myelin calculated using a different combination of phases (from Akers and Parsons, 1970).

TABLE X
Unit cell dimensions and structure factors for different myelins

h	RSC $a = 180$ Å		ROP $a = 156$ Å		FSC $a = 170$ Å	
	$F(h)$	$\delta F(h)$	$F(h)$	$\delta F(h)$	$F(h)$	$\delta F(h)$
1	$-0\cdot10$	$0\cdot02$	$-0\cdot08$	$0\cdot02$	$-0\cdot10$	$0\cdot04$
2	$1\cdot41$	$0\cdot02$	$1\cdot67$	$0\cdot05$	$1\cdot41$	$0\cdot02$
3	$0\cdot62$	$0\cdot02$	$0\cdot25$	$0\cdot04$	$0\cdot74$	$0\cdot04$
4	$-1\cdot73$	$0\cdot03$	$-1\cdot52$	$0\cdot12$	$-1\cdot51$	$0\cdot08$
5	$-0\cdot66$	$0\cdot05$	$-0\cdot20$	$0\cdot10$	$-0\cdot84$	$0\cdot06$
6	$0\cdot26$	$0\cdot07$	$-0\cdot22$	$0\cdot10$	$0\cdot27$	$0\cdot07$
7	$-0\cdot03$	$0\cdot03$	$0\cdot04$	$0\cdot04$	$-0\cdot14$	$0\cdot05$
8	$0\cdot28$	$0\cdot09$	$-0\cdot05$	$0\cdot05$	$0\cdot31$	$0\cdot08$
9	0	$0\cdot08$	$0\cdot05$	$0\cdot05$	0	$0\cdot07$
10	$0\cdot28$	$0\cdot12$	$0\cdot56$	$0\cdot18$	$0\cdot32$	$0\cdot09$
11	$0\cdot36$	$0\cdot10$	0	$0\cdot08$	$0\cdot50$	$0\cdot10$
12	$-0\cdot28$	$0\cdot10$	$0\cdot06$	$0\cdot06$	$-0\cdot11$	$0\cdot04$
13	$-0\cdot16$	$0\cdot04$	$0\cdot06$	$0\cdot06$	$-0\cdot26$	$0\cdot08$
14	$-0\cdot04$	$0\cdot04$	$-0\cdot06$	$0\cdot06$	0	$0\cdot08$
15	$-0\cdot27$	$0\cdot10$	0	$0\cdot09$	$-0\cdot16$	$0\cdot08$
16	$0\cdot04$	$0\cdot04$	—	—	$0\cdot04$	$0\cdot04$
17	$0\cdot04$	$0\cdot04$	—	—	0	$0\cdot09$
18	$-0\cdot04$	$0\cdot04$	—	—	—	—

(from Caspar and Kirschner, 1971)

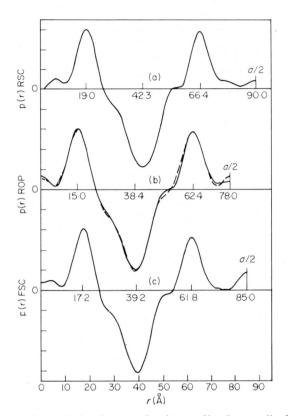

FIG. 45. A comparison of the electron density profiles for myelin from (a) rabbit sciatic nerve; (b) rabbit optic nerve; (c) frog sciatic nerve (from Caspar and Kirschner, 1971).

Recently Caspar and Kirschner (1971) have collected X-ray data from rabbit and frog sciatic nerves and rabbit optic nerve to the equivalent of a 10 Å spacing. Using a detailed crystallographic method the sign sequences shown in Table X were obtained for the three sets of data. The calculated electron density profiles, in each case with asymmetry about the membrane center, are shown in Fig. 45. The profiles based on diffraction data to a 10 Å spacing have an effective point resolution of about 7 Å, which, it is claimed, is adequate to define the location and dimensions of specific parts of the structure. The authors' schematic interpretation of the electron density profile in molecular terms is shown in Fig. 46.

The centers of the pair of high density peaks again represent the mean position of the lipid phosphate and sugar groups and the fact that the width of these peaks seems comparable with the dimensions of the polar groups,

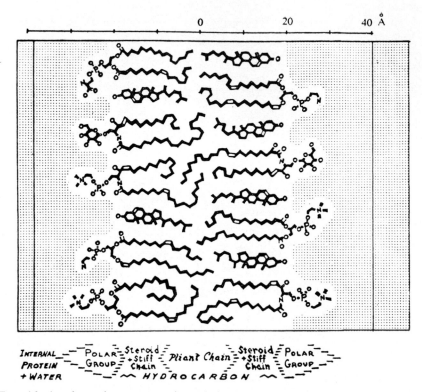

FIG. 46. A schematic representation of the nerve myelin structure. The electron density curves of rabbit optic (dotted) and sciatic myelin are shown above (from Caspar and Kirschner, 1971).

the position normal to the membrane surface is well determined. The low density trough is relatively broad, indicating that the ends of the hydrocarbon chains, which have the lowest electron density, are distributed within a layer about 15 Å thick. However, the most interesting feature of the high resolution profiles is the asymmetry of the profile on either side of the central hydrocarbon region. The asymmetry is identified by the authors with the steroid portion of cholesterol and suggests that cholesterol is present at a greater concentration on the external side of the hydrocarbon layer.

In an attempt to quantitatively locate the different lipid components, protein and water in myelin structure, the electron density profile on an absolute scale indicates that the middle of the membrane has a density in the range 0.27–0.29 electrons/Å^3, which is close to the density of 0.27 electrons/Å^3 for longer chain liquid paraffins and olefins. The upper limit for the estimated density suggests that protein occupies less than 10% of the space in this

central region. The asymmetric steps on either side of the central low
density region are in the same position as the symmetric steps identified with
steroid in lipid bilayers formed from phospholipids and cholesterol. Assuming
that all the steroid portion of the cholesterol is located in the two-step regions
each of width about 10 Å, and that the relative area occupied by the lipid is
the same as that in monolayers and bilayers, the electron densities indicate
an approximately equimolar ratio of cholesterol and polar lipid on the outer
side of the hydrocarbon layer, and a ratio of about 3 : 7 on the inner side. In
agreement with other methods, the density measurements indicate that the
oriented hydrocarbon is predominantly close packed in the steroid regions
and that the ends of the chains near the bilayer center are more disordered.
The hydrocarbon thickness is about 38 Å in rabbit myelin and 35 Å in a frog
myelin.

The hydrated lipid polar groups and protein are identified with the peaks
having electron densities of about 0.40 electron/$Å^3$. The authors consider
that protein and water are distributed in the spaces between the membrane
bilayers with perhaps the protein concentration higher near the lipid surface,
but the possibility of a small amount of protein extending across the lipid
bilayer is not excluded.

2. *Retinal Rods*

Early studies suggested that structural organization within the retinal rods
of frogs, for example, was giving rise to their observed optical birefringence.
Schmidt (1938) interpreted the birefringence in terms of the organization of
lipid within the rod, the rod consisting of a lipid bilayer with protein intercalated parallel to the plane of the discs. Subsequent electron microscopy
and X-ray diffraction studies showed that the stacking of the membrane
discs was regular with a well-defined disc-to-disc repeat. The first X-ray
diffraction study (Finean *et al.*, 1953) indicated a repeat distance of 370 Å
for wet, osmium tetroxide-fixed guinea-pig rods and at about the same time
electron microscopy was begining to indicate the presence of sub-unit or
globular substructure within the plane of the discs. Improved X-ray diffraction
patterns of retinal rods were obtained (for example, Robertson, 1966) but
usually only one or two orders of the disc-to-disc repeat distance were
recorded.

Since 1965 a systematic structural investigation of frog retinal rod photoreceptors has been made, principally by C. R. Worthington and J. K. Blasie,
and which now provides us with a fairly detailed molecular picture of this
particular membrane system. X-ray diffraction patterns were obtained by
Blasie *et al.* (1965) from preparations of isolated disc membranes in which
retinal rods from a number of retinae were sedimented such that the photoreceptor disc membranes were organized into a single rod at the bottom of

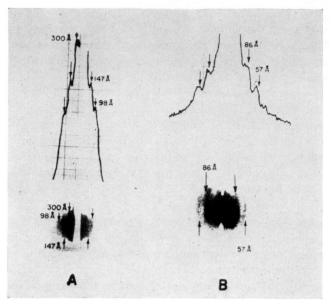

Fig. 47. Low-angle X-ray diffraction patterns and densitometer traces of frog retinal rod membranes. (A) X-ray beam parallel to the planes of the membranes, (B) X-ray beam normal to the planes of the membranes (from Blasie *et al.*, 1965).

an ultracentrifuge tube. It could be shown that the planes of the disc membranes were oriented at right angles to tube axis and X-ray diffraction patterns were recorded with the X-ray beam directed either parallel to the centrifuge tube axis or at right angles to this axis, providing information on either the subunit structure or the lamellar structure respectively. With the X-ray beam at right angles to the tube axis, a lamellar spacing, d, of about 300 Å was observed with three orders of this repeat distance being recorded (Fig. 47A). This value agrees fairly closely with the two orders ($h=2$) of a disc-to-disc repeat ($d=320$ Å) recorded from intact retinal rods (Robertson, 1966). The lamellar diffraction from retinal rods will be discussed in more detail below.

With the X-ray beam parallel to the tube axis, the X-ray pattern consisted of two diffuse reflexions at about $d\approx 80$ Å and $d\approx 50$ Å (Fig. 47B). For these X-ray reflexions the intensities and the ratio of the spacings varied with the hydration of the disc membrane preparations. For example, when the sample was air-dried, the ratio of X-ray spacings was $\sqrt{2}$ consistent with a square lattice of 70 Å by 70 Å. This, together with electron microscopic evidence, indicated a subunit structure within the plane of the membrane with particles 40–50 Å in diameter at the corners of this lattice. Similar interpretations in terms of subunits of particles could be made for untreated disc membrane preparations, from preparations treated with chemicals and from various preparations at different degrees of hydration. The particles were originally considered to be either rhodopsin molecules or lipid micelles.

Blasie and Worthington (1969) and Blasie *et al.* (1969) using improved X-ray data from wet specimens observed that the X-ray reflexions from fully hydrated disc membranes (Fig. 48) were more diffuse than the corresponding reflexions from the air-dried preparations and that the ratio of X-ray spacings changed from $\sqrt{2}$ to 1·5. This indicated that the particles had less order in the wet state and suggested an analysis of the X-ray data in terms of radial distribution of the particles using analogies with the theory of X-ray diffrac-

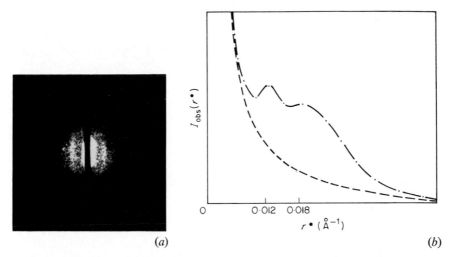

(a)   (b)

FIG. 48. X-ray diffraction pattern and densitometer trace of completely wet specimens of frog retinal rod membranes, obtained with the X-ray beam normal to the planes of the membranes (from Blasie *et al.*, 1969).

tion from gases and liquids. The subunits were considered as a planar two-dimensional liquid and the planar radial distribution function $2\pi r p'_m(r)$ was plotted against the radial distance r from any arbitrary subunit center as shown in Fig. 49a. The number of subunits at a distance from the center of any arbitrary subunit contained within a ring of radius r and thickness dr is given by $2\pi r p_m(r)\,dr$. The radial distance to the first peak gives the center-to-center distance between nearest neighbors and the area under the first peak determines the number of the nearest neighbors around any arbitrary subunit. The planar radial distribution function for fully hydrated preparations of isolated disc membranes at a temperature of 26°C had 3·0 nearest neighbors at a distance of 56 Å compared with the four nearest neighbors at a distance of 70 Å for the air-dried preparations described earlier.

In the same study the subunits in the disc membrane preparations were identified as rhodopsin, retinal covalently linked to the protein (or lipoprotein) opsin. The identification was based upon a study of the attachment of an antibody to the disc membrane subunits, the antibody being specific for the rhodopsin of frog retinal receptors, followed by critical comparisons of the low-angle X-ray data from both untreated and antibody-treated fully hydrated disc membrane preparations. Furthermore, the radial distribution functions (see above) indicated that the best model for the photoreceptor particle was a sphere of uniform density which had a diameter of about 42 Å. This was in reasonable agreement with other physico-chemical measurements on the isolated rhodopsin which suggested a spherical molecule of molecular weight 28,000 and diameter 46 Å.

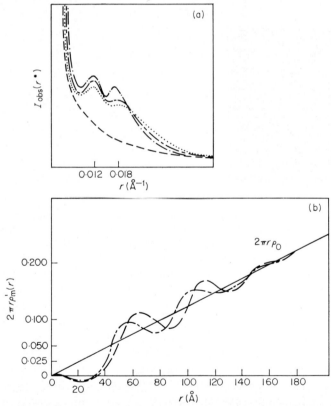

Fig. 49. (a) Microdensitometer tracings of X-ray diffraction patterns from frog retinal rod membranes at ...... 42·5°C; – – – – 26°C; –··–··– 4·5°C. (b) Corresponding planar radial distribution functions, symbols as in (a) (from Blasie and Worthington, 1969).

By recording X-ray diffraction patterns from fully hydrated preparations of disc membranes at three different temperatures, 4·5°C, 26·0°C, and 42·5°C, shown in Fig. 49a, an interpretation of the resulting radial distribution functions (see Fig. 49b) in terms of the distance and the number of the first and second nearest neighbors as a function of temperature was possible. The authors found that the higher the temperature, the smaller the distance is between the first neighbors and as a consequence there are less nearest neighbors; the lower the temperature the larger the distance between neighbors, with, in this case, an increase in the number of nearest neighbors. Essentially this means that as the temperature increases the rhodopsin molecules come closer together but, as their concentration density remains constant, they have fewer neighbors. Liquids behave similarly as a function of temperature, suggesting that the rhodopsin molecules within the disc membranes behave like a planar liquid, the rhodopsin molecules within the disc membranes possessing considerable freedom of movement which is temperature dependent.

A recent study by Blasie (1972a) aimed at locating the position of the photopigment rhodopsin molecules relative to the lipid hydrocarbon layer made use of X-ray diffraction data recorded as a function of the electron density of the surrounding medium. Using glucose, LiCl or CsCl in phosphate buffer as the bathing medium, the scaled integrated intensity data (see Fig. 50a) for the photopigment molecules was measured at different values of the electron density utilizing the, presumably unaffected, integrated intensity at $r^* \simeq 1/4·5 \text{ Å}^{-1}$, arising from the lipid hydrocarbon chains, for intensity scaling. As indicated in Fig. 50b increasing the electron density $\bar{\sigma}_M$ decreases the

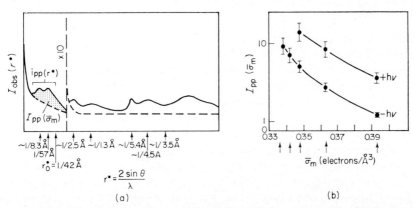

FIG. 50. (a) Diffracted intensity $I_{obs}(r^*)$ obtained from a wet pellet of frog retinal rod membranes. (b) Observed integrated intensity $I_{pp}(\sigma_M)$ diffracted by photopigment molecules in wet frog retinal rod membranes as a function of the electron density of the sedimentation medium $\bar{\sigma}_M$ and of bleaching the membrane (from Blasie, 1972a).

integrated diffracted intensity from the photopigment molecules relative to that from the lipid hydrocarbon chains. This behavior was observed for both unbleached and bleached (exposed to white light) membranes, the integrated intensity being greater for the bleached membranes compared with the unbleached membranes for all values of the electron density. Using a simple model consisting of spherical photopigment molecules of radius ~ 22 Å and uniform electron density located at the interface separating the aqueous surface layer and the lipid hydrocarbon core of the disc membrane, the photopigment molecules being located only on one side of the membrane, an embedding parameter R was calculated from the X-ray data. This parameter determines the extent to which the photopigment molecule is located in both the aqueous surface layer and the lipid hydrocarbon core. Calculation of the embedding parameter for unbleached membranes indicates that photopigment

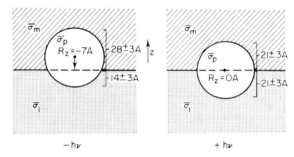

FIG. 51. The embedding of photopigment molecules in the aqueous and lipid phases of the retinal rod membrane before and after bleaching (from Blasie, 1972b).

molecules with a molecular diameter of ~ 42 Å protrude about 28 Å into the aqueous surface layer of the disc membrane and are embedded in the lipid hydrocarbon core of the disc membranes to an extent of about 14 Å, as shown in Fig. 51. After the photopigment is bleached, the projection of the photopigment molecules into the aqueous surface layer decreases to about 21 Å, while the extent of its embedding in the lipid hydrocarbon core increases to about 21 Å. Thus bleaching of the photopigment molecules results in their sinking into the lipid hydrocarbon core of the disc membrane by about 7 Å or approximately 16–17% of their molecular diameter.

Arguing that the surface of the photopigment molecule which protrudes into the aqueous surface layer should consist of primarily the polar residues of the protein, perhaps resulting a net electrical charge, Blasie (1972b) investigated the effect of ionic strength and pH on the X-ray diffraction pattern of frog retinal disc membranes. Again changes in the diffraction patterns, and consequently the radial distribution function, were observed for both bleached and unbleached membranes. The radial position of the first nearest neighbor maximum was shown to be sensitive to pH changes, with the average separation of nearest neighbors for the planar arrangement of photopigment molecules increasing with increasing pH, in the range pH 6–8, for both bleached and unbleached photopigment. At a given pH the average separation of nearest neighbors is less when the photopigment is bleached. Assuming that the lipids in the membrane are unaffected by variations in the ionic strength from 0·115 to 1·840 M NaCl, an increase in the ionic strength at pH 7 decreases the average separation of the nearest neighbor photopigment molecules. These pH and ionic strength dependences of the average nearest neighbor are consistent with the photopigment molecule bearing a net negative electric charge and that this charge is reduced when the photopigments are bleached (bleaching apparently exposing polar groups with a positive charge). The fact that counterions in the aqueous phase can effectively shield this electric charge indicates that exposed polar groups are responsible for this charge. Blasie considers that reduction of the net charge by bleaching could result in a decrease in the area occupied by the polar groups on the surface of the photoequipment in the aqueous layer. This would increase the lipid solubility of the photopigment molecule and thus provide an explanation for the earlier studies demonstrating a sinking of the photopigment molecule into the lipid hydrocarbon core of the disc membrane on bleaching.

Improvements in both membrane isolation procedures and in X-ray diffraction techniques have led to great improvements in the lamellar diffraction patterns obtained from photoreceptor membranes in the outer rods of intact untreated retinae. In particular, the number of lamellar diffraction orders recorded has increased dramatically. For three varieties of frog, *Rana pipiens* ($d=316$ Å), *Rana catesbeiana* ($d=314$ Å), *Rana tempororia* ($d=275$–300 Å), 11, 11 and 7 orders respectively have been obtained from intact retinae (Gras and Worthington, 1969; Blaurock and Wilkins, 1969). Gras and Worthington (1969) have also recorded diffraction patterns from the retina of rats ($d=325$ Å) and cattle ($d=312$ Å). The X-ray diffraction patterns all refer to the disc-to-disc repeat in photoreceptors and provide information on the lamellar structure of the membranes, but since different methods of interpreting the X-ray data were used by the two groups they will be discussed separately.

The structure analysis of Gras and Worthington (1969) is based upon the X-ray pattern obtained from the frog (*Rana catesbeiana*) which shows the first eleven orders ($h=11$) of diffraction from a disc-to-disc repeat. A photograph of the X-ray pattern obtained from intact frog retina is shown in Fig. 52. Their interpretation is based upon building electron density strip models of the structure, calculating the theoretical diffraction pattern and then

D

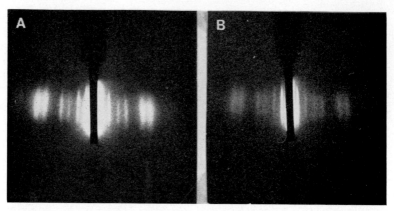

FIG. 52. Low-angle X-ray diffraction pattern from intact, untreated frog retinal photoreceptors showing a disc-to-disc repeat period of 314 Å (from Gras and Worthington, 1969).

comparing this with the observed X-ray intensities (see similar approach to myelin structure). Making the assumption that the double membrane unit is centrosymmetric and thus choosing a centrosymmetric model a total of seven parameters (five of which are operational) are used to define the model (see Fig. 53); P is the electron density of the outer high-density layer; L is the electron density of the central low-density layer; F is the electron density of the fluid medium; d is the radial repeat distance; v is the thickness of the membrane pair; c is the thickness of the cytoplasmic fluid layer; $d-v$ is the thickness of the extracellular fluid layer; l is the thickness of the central low electron-density region; p is the thickness of the cytoplasmic outer high-density layer; q is the thickness of the extracellular outer high-density layer.

The theoretical diffraction was computed when these five parameters were varied over a wide range and the theoretical diffraction was then compared with the observed diffraction. The best agreement with the observed diffraction for frog retinal photoreceptors is obtained using the parameters $c=5$ Å, $v=154$ Å, $p=40$ Å, $l=15$ Å, $q=18.5$ Å, and as $d=314$ Å then $d-v=160$ Å. Thus each disc membrane is 74·5 Å thick and the thickness of the central low-density region is 16 Å. An important feature of the model is that the thickness of the cytoplasmic outer layer is comparatively large with $p=40$ Å and the authors are tempted to assign the rhodopsin molecules of diameter 42 Å to this particular layer.

This interpretation would suggest that the photoreceptor membrane has a lipid bilayer structure but with an extremely narrow central region the hydrocarbon chains. The disc membrane subunits are thought to be the rhodopsin molecules and as indicated by the model shown in Fig. 53 they

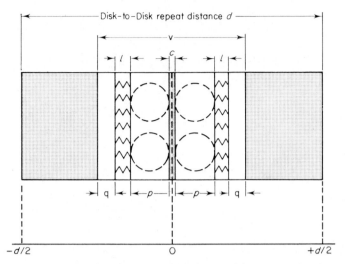

FIG. 53. Centrosymmetrical step electron density model for retinal rod membranes. The model is centrosymmetrical and has repeat distance d and disc thickness v. The disc contains two asymmetric membrane units with a fluid layer of width c between the units. The clear regions refer to electron density P, the zigzag lines refer to the electron density l, and the dotted regions refer to the fluid electron density F. The dotted circles refer to the photopigment molecules (from Worthington, 1971).

occur on only the cytoplasmic side of the photoreceptor membrane. This would mean that the rhodopsin molecules are located on the inside of the disc in retinal rods outer segments.

Blaurock and Wilkins (1969) have obtained seven orders of diffraction ($h=7$) and continuous diffraction at higher angles from the retinal rods of intact frog (*Rana tempororia*) retina. Using the corrected intensity data, a Patterson–Fourier synthesis showed large peaks, which represent vectors between regions of different electron density in the lamellae, at 40 Å and 85 Å. The peak at 85 Å is thought to result from a distance between the centers of adjacent membranes, their correlation giving the 85 Å peak in the Patterson function. The peak at 40 Å is interpreted as arising from the vectorial separation of layers of high electron density on each side of a typical lipid bilayer.

Now, making the assumption that the discs are symmetrical with a mirror plane through the center of each disc, Fourier syntheses were calculated using the observed amplitudes together with all combinations of the phases (+ or −) which were consistent with the Patterson function. The only Fourier synthesis consistent with all the available evidence is shown in Fig. 54, together with the authors' interpretation of the electron density profile. In the middle of each membrane unit a deep trough of low electron density, corresponds to

G. G. SHIPLEY

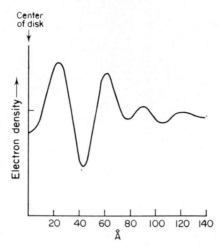

FIG. 54. Electron density profile for frog photoreceptor membranes (from Blaurock and Wilkins, 1969).

the lipid hydrocarbon layer, with peaks of high density on each side which correspond to the lipid headgroups and protein. Initially, the authors interpreted the separation of the peak maxima of 40 Å as being greater than the separation of the lipid headgroups in the bilayer and thus assigned to protein a fairly narrow layer outside the lipid bilayer. However, in a recent paper Blaurock and Wilkins (1972) have shown that when an array of disc membranes is swollen in various concentrations of glucose–Ringer solution the electron density profile for a single membrane of the disc remains unaltered, this observation being consistent with a correct selection of the phases of the structure amplitudes. Furthermore, the electron density profile is reinterpreted in terms of a lipid bilayer in which the separation of the lipid headgroups is 40 Å and a significant part of the membrane protein does penetrate into the bilayer. Again equatorial reflexions are observed at ~55 Å and considered to result from the arrangement of rhodopsin molecules in the membrane but at present it does not seem possible to infer an asymmetric distribution of the rhodopsin molecules in the retinal membrane at least from these electron density profiles.

3. Chloroplasts

Chloroplasts, the organelles in which photosynthesis takes place in eucaryotic plant cells, are surrounded by a double-membraned envelope with other membranes in the interior forming a highly laminated internal structure. In procaryotic bacterial or algae cells the chloroplasts do not exist as separate organelles and the laminated membrane structure is distributed throughout the cell.

Kreutz has investigated this multimembrane system using X-ray diffraction methods and by orienting the chloroplast preparations with respect to the X-ray beam, has derived information on both the lamellar layer structure and also the structure within the plane of the membrane. The corrected lamellar diffraction from preparations of different *Chlorella* and *Antirrhinum* chloroplasts (Kreutz and Menke, 1962) shows lamellar repeat distances in the range 170–250 Å. As an example the meridional and equatorial diffraction from oriented chloroplasts of *Antirrhinum majus* is shown in Fig. 55. Both model building and the use of the characteristic function of Porod (1951) were used to determine the phases, and the electron density profile shown in Fig. 56a was derived. On the basis of the model calculations and the electron density profile the molecular interpretation shown in Fig. 56b given by Kreutz (1966). On the outside of the thylakoid there are two protein layers 30–40 Å thick which sandwich a layer 70–80 Å thick containing the lipid and chlorophyll. The molecular arrangement in which a "reverse" lipid bilayer is present with the headgroups at the center explains the peak of high electron density at the mirror plane of the profile and Kreutz suggests that a mono-

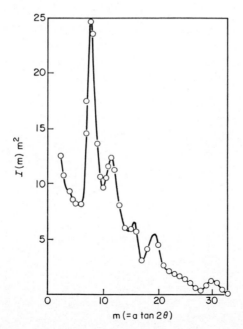

FIG. 55. The corrected X-ray diffraction pattern from *Antirrhinum majus* chloroplasts (from Kreutz, 1963).

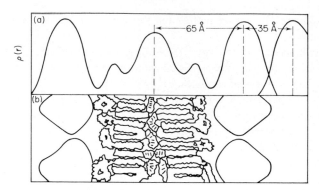

FIG. 56. (a) Electron density profile for chloroplast membranes, and (b) a molecular interpretation (from Kreutz, 1966).

molecular layer of porphyrin rings, components of the chlorophyll molecules, is arranged between each lipid and protein layer. The long hydrocarbon phytyl chain of the chlorophyll is organized within the "reverse" lipid bilayer. Although the limits of resolution of the electron density profiles obtained by Kreutz (1963a, 1964) do not justify this detailed molecular positioning, if the sign determination is correct then new concepts of lipid-protein organization may be necessary to explain the observed profile. In passing, one may note similar features in the electron density profile obtained by Akers and Parsons (1971) for myelin (see Fig. 44).

By subtracting out the X-ray diffraction due to the lamellar structure, Kreutz (1963b) was able to show the presence of diffraction or scattering

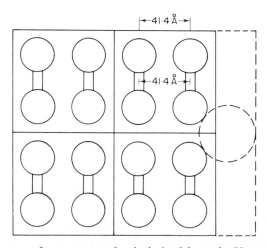

FIG. 57. Substructure of quantasome that is derived from the X-ray diffraction data.

arising from structure within the plane of the membrane. The observed diffraction was consistent with that arising from a two dimensional lattice of crystallites either of the fluid type (see p. 65) or an exact lattice with periodic interruptions. Adopting the second alternative on the basis of electron microscope evidence, the diffraction data are consistent with a square planar arrangement of protein particles. Utilizing both the planar diffraction and parameters derived by X-ray scattering, an isolated protein preparation (Hosemann and Kreutz, 1966), each crystallite, identified with the protein part of the quantasomes, consists of four protein subunits of diameter ~27 Å with the arrangement shown in Fig. 57.*

B. SINGLE MEMBRANE

1. *Erythrocyte Membranes*

X-ray diffraction from erythrocyte membranes was first described by Finean *et al.* (1966, 1968) using centrifuged pellets of erythrocyte ghosts. Following controlled dehydration, low-angle X-ray diffraction patterns were produced from the artificially stacked membranes. The diffraction was extremely weak but as water was removed, at a point corresponding to a water content of about 10–20%, three or four lamellar reflexions (see Fig. 58) could be identified corresponding to a repeat distance of 100–120 Å. This diffraction pattern is produced by the stacked membranes containing essential water of hydration, the diffraction arising from a lamellar system where 100–120 Å represents the thickness of one membrane. An interpretation in terms of a lipid bilayer with non-lipid components located at the surface was considered to be quantitatively consistent with the observed intensity distribution in the diffraction pattern. Further dehydration of the pellet presumably resulting in membrane degradation produced changes in the diffraction pattern as shown in Fig. 58.

Recently Wilkins *et al.* (1971) have obtained diffraction from dispersions of haemoglobin-free erythrocyte ghosts containing about 5% solids. The corrected scattering shown in Fig. 59 consists of a broad band with a maximum at 45 Å, a weak band at 15 Å, and in over-exposed films, possibly a band at ~22 Å. In interpreting this type of diffraction data from dispersions of membranes Wilkins *et al.* (1971) noted that the $|F|$ curves, where $|F| = I^{0.5}$ sin θ, closely resembled the F curve obtained for lipid bilayers (see section on egg-yolk lecithin), i.e. a main band and weaker sub-multiple bands. Thus scattering curves of this type were taken to indicate the presence of extensive, though not necessarily continuous, phospholipid bilayer-type structures in membranes. The location of protein with respect to the lipid bilayer remains unclear. The presence of cholesterol in both lipid dispersions and membranes, causing an increased ordering of the hydrocarbon chains, seems to correlate

---

* X-ray diffraction studies of chloroplast membranes have been reviewed extensively by W. Kreutz (1970) in *Advances in Botanical Research* 3, 53–169.

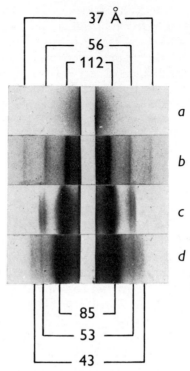

FIG. 58. Low-angle X-ray diffraction patterns recorded during dehydration of a sample of rat erythrocyte ghosts (from Finean *et al.*, 1966).

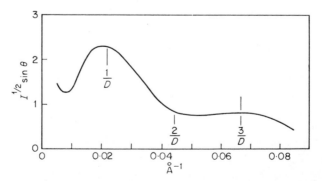

FIG. 59. The corrected continuous diffraction from rat erythrocyte membranes (from Wilkins *et al.*, 1971).

with the observation that the third maximum ($h=3$) is stronger then the second ($h=2$). In the absence of cholesterol the second maximum is stronger than the third.

Clearly, erythrocytes give diffraction data consistent with the presence of extensive phospholipid bilayer structure in the membrane, the high cholesterol content of erythrocytes producing the strong third maximum. The relationship between the diffraction patterns of the sedimented pellets and the dispersions has not yet been discussed.

2. Mitochondrial Membranes

Diffraction patterns obtained from hydrated pellets of both the inner and outer mitochondrial membranes from rat liver, show two or three orders of a lamellar repeat of 118 Å at water contents of 20–30% (Thompson et al., 1968). The lamellar phase is thought to represent the intact membrane which retains water essential to its structural integrity. Subsequent changes in the diffraction patterns on further dehydration probably represent molecular reorganization within the membrane. These authors could find no X-ray evidence for a pronounced subunit structure within the plane of the membrane in contrast to, for example, chloroplast or retinal rod membranes (see above).

3. Brush-border Membranes

Again using the sedimentation-dehydration technique Limbrick and Finean (1970) obtained a series of X-ray diffraction patterns from a brush border membrane preparation of guinea-pig intestinal epithelium. At a water content of 20–30% the second to eighth orders of a fundamental repeat period of 300 Å were obtained where the repeat corresponds to a double membrane unit (cf. erythrocyte membranes). Reasonable agreement with the observed intensity data is obtained when a symmetric step-function is used as a model. Changes in the diffraction pattern are observed on further dehydration of the membrane preparation.

4. Mycoplasma laidlawii Membranes

The ease of isolation of pure plasma membranes of the microorganism Mycoplasma laidlawii (a pleuropneumonia organism) together with the ability to vary their lipid composition has made this system the subject of many physical studies. Early studies on membrane solubilization and reaggregation (Razin et al., 1965) were followed by calorimetric studies of lipid phase changes in lipid extracts, membranes and whole organisms (Steim et al., 1969). Recently Engelman (1970) has further investigated these phase transitions using X-ray diffraction methods, and finds that on going through the phase transition changes occur in the low-angle and wide-angle diffraction from both palmitic acid-enriched and erucic acid-enriched membrane pellets.

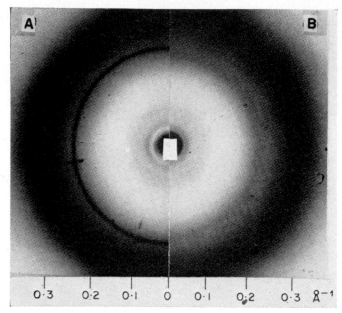

FIG. 60. X-ray diffraction patterns from isolated erucic acid-enriched membranes from *Mycoplasma laidlawii*. (A) membranes at 10°C, (B) membranes at 40°C (from Engelman, 1970).

Specifically, in the wide-angle region there is a gradual shift from the broad diffuse ring at 4·6 Å to an intense sharp ring at 4·2 Å as the temperature is lowered (see Fig. 60). These transitions are interpreted in terms of the lipid component of the membrane and represent a conversion from a relatively disordered hydrocarbon chain packing to one in which the chains are close packed probably in an ordered hexagonal array. The same diffraction behavior is observed for dispersions of the isolated lipids as well as from intact organisms. In each case the transitions occur over a relatively wide temperature range (20–50°C). Similar behavior has been reported recently (Esfahani et al., 1971) for the lipids in membranes of *Escherichia coli*.

Low-angle X-ray scattering from dispersions of *M. laidlawii* membranes was obtained below and above the phase transition (Wilkins et al., 1971; Engelman, 1971) as shown in Fig. 61a. For erucate enriched membranes below the phase transition, three bands are observed with maxima near 52 Å, 26 Å, and 17 Å, with the second and third bands being of comparable intensity. The erucate-enriched membranes give equivalent Bragg spacings which are larger than those for palmitate-enriched (multiples of ~47 Å) and oleate-enriched (multiples of ~46 Å) membranes, presumably reflecting chain length differences in the major fatty acids. The fact that the third band ($h = 3$)

is as intense as the second ($h=2$) is interpreted as resulting from a localization of the CH_3 groups at the center of the bilayer (cf. the similar effect induced by cholesterol in erythrocyte membranes and lipid dispersions). The dimensions of the lipid bilayer thickness, determined from the positions of the maxima, are consistent with the orientation of ordered hydrocarbon chains perpendicular to the membrane surface. Above the phase transition each of the three bands moves to a smaller spacing (multiples of 41 Å) and the third band becomes much weaker than the second (see Fig. 61b). The basic form of the diffraction remains the same with a strong main band and weaker bands at sub-multiples of the main band spacing. As is well recognized for lipid systems, as the hydrocarbon chains become more fluid the dimensions of the bilayer will decrease, consistent with the observed shifts in the maxima, and the CH_3 groups become less well localized. As indicated above this should increase the intensity of the third maximum compared to the second and, indeed, as shown in Fig. 61, this does occur.

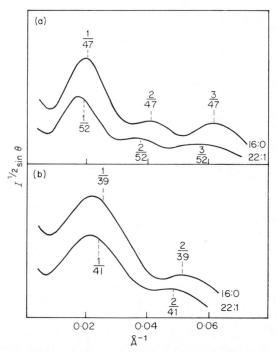

FIG. 61. Low-angle X-ray diffraction from *Mycoplasma laidlawii* membranes (a) below the thermal phase transition: upper curve, palmitate-enriched membranes at 10°C; lower curve, erucate-enriched membranes at 10°C. (b) above the thermal phase transition: upper curve, palmitate-enriched membranes at 43°C; lower surve, erucate-enriched membranes at 37°C (from Engelman, 1971).

5. *Halobacterium halobium Membranes*

Certain halophilic bacteria can only sustain growth in, for example, high concentrations of NaCl (Brown, 1965) and when the salt is removed the cell membrane is disrupted into fragments. One of the fragments, characterized by a deep purple color, has been isolated from *Halobacterium halobium* by Oesterhelt and Stoeckenius (1971) and the purple color shown to be due to retinal bound to an opsin-like protein (see retinal rod membranes). Low-angle X-ray diffraction patterns (Blaurock and Stoeckenius, 1971) from the purple membrane dispersed in water consist of both diffuse scattering and a number of sharp diffraction lines as shown in Fig. 62. The sharp diffraction lines in the range $d = 54$ Å to 7 Å are all indexed on a planar hexagonal lattice with the center-to-center distance of 63 Å. Diffraction from oriented membrane preparations show that this hexagonal array extends in the plane of the membrane.

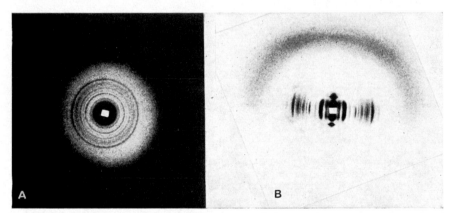

FIG. 62. Low-angle X-ray diffraction patterns of the purple membrane fraction of *Halobacterium halobium*: (A) for an unoriented dispersion of the membrane sheets in water; (B) for a dispersion dried on a smooth horizontal surface (from Blaurock and Stoeckenius, 1971).

Oriented preparations of highly hydrated membranes gave the continuous diffracted intensity along the meridian shown in Fig. 63 and on drying the diffuse scattering is replaced by a series of sharp diffraction lines of periodicity 49 Å. With only minor changes in the equatorial reflexions from the hexagonal lattice the membrane structure seems, unusually, unaffected by drying. The diffraction profile from the wet, oriented membranes showed two broad bands at 43 Å and 21 Å and is consistent with diffraction from a bilayer profile. The electron density profile shown in Fig. 64 was calculated using the phases of a symmetrical bilayer and shows the usual separation of the electron dense layers by 40 Å but, strikingly, the electron density at the center of the bilayer is greater than that of water. This suggests to the authors that substantial amounts of protein are located at the center of the membrane. It is hoped that this excellent diffraction data will eventually yield information on the precise location of this rather unique membrane protein.

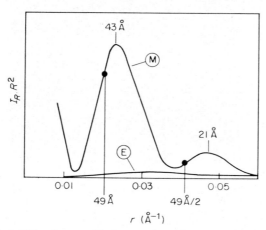

FIG. 63. Corrected diffraction profile from oriented wet purple membrane preparation of *H. halobium* (from Blaurock and Stoeckenius, 1971).

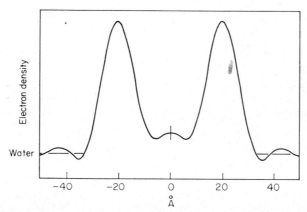

FIG. 64. Electron density profile across the bilayer of the purple membrane fraction of *H. halobium* (from Blaurock and Stoeckenius, 1971).

C. VIRUS-ASSOCIATED MEMBRANES

An interesting new approach to the problem of membrane structure has been through the investigation of the lipoprotein layers which envelope certain viruses. For example, the marine bacteriophage PM2, in which it is

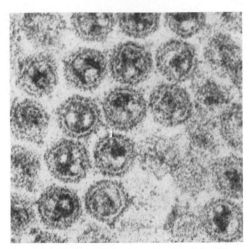

FIG. 65. Electron micrograph of bacteriophage PM2, pelleted, embedded, sectioned and stained with uranyl acetate and lead citrate (from Harrison *et al.*, 1971).

suggested that protein and lipid are actually assembled around the nucleic acid, negatively stained electron micrographs of particles give a membrane-like appearance near the virus surface as shown in Fig. 65. Harrison *et al.* (1971a) have obtained excellent X-ray diffraction patterns from centrifuged pellets of the spherical virus (Fig. 66), the low angle region consisting of a number of sharp rings characteristic of diffraction from a spherically sym-metric object extending to about 25 Å. Phase assignment was accomplished using the diffraction data from the viruses in media of different electron density and the Fourier syntheses shown in Fig. 67 were calculated. The most prominent feature of the electron density curves is the deep minimum centered at $r=220$ Å with the two peaks on either side separated by about 40 Å and this part of the profile clearly resembles the profile of both phospholipid bilayers and membranes described earlier. The external region between the outermost peak and the surface of the particle (at $r=300$ Å) is probably occupied by protein subunits packed such that the characteristic icosahedral surface lattice is produced. The interior of the particle necessitates incorporation of protein with the DNA in order to account for the high electron density in the center of the particle.

Similar X-ray diffraction patterns have been obtained for the Sindbis virus grown in hamster kidney cells which acquires a lipoprotein coat as it perme-ates through the infected cell surface (Harrison *et al.*, 1971b). The electron density profile for Sindbis virus particle of radius of 350 Å contains the same dominant feature shown by PM2 consisting of a deep minimum at a radius of 232 Å with the two peaks on either side separated by 48 Å. This feature is

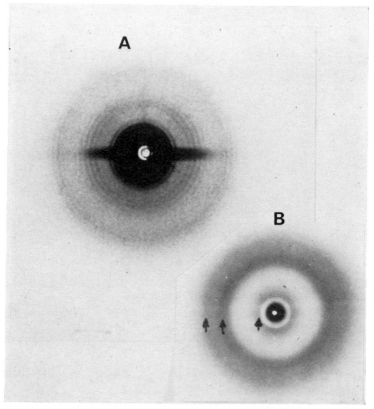

FIG. 66. X-ray diffraction patterns of PM2; (A) low-angle region, (B) wide-angle region (from Harrison *et al.*, 1971a).

attributed to the lipid component arranged in a bilayer structure. Again RNA and protein are thought to contribute to the core and glycoprotein is located on the outside of the virus particle.

VI. Discussion and Conclusions

In the last ten years significant advances have been made in our understanding of the structural organization of lipids, lipoproteins and membranes, although it is salutary to look back occasionally at the work of Schmitt and his colleagues performed in the 1930s. As far as biological lipids are concerned, we have a library of information on which to discuss their phase behavior. This information enables us to predict, not always correctly, the major features of the phase diagram to be adopted by a lipid of known structure.

G. G. SHIPLEY

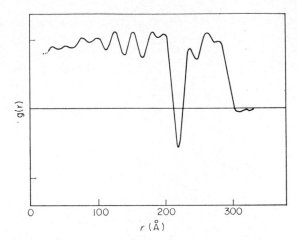

FIG. 67. Electron density profile across the PM2 particle, where *r* is a radial distance across the particle (from Harrison *et al.*, 1971a).

For example, it is possible to argue that the following points have been established by the studies of phase behavior.

(i) Decreasing the hydrocarbon chain length tends to shift the phase diagram down the temperature axis.

(ii) Increasing the hydrocarbon chain unsaturation also tends to shift the phase diagram down the temperature axis.

(iii) Diacyl lipids with a net charge form the continuously expanding lamellar phase.

(iv) The incorporation of charged lipids into lipids normally exhibiting limited swelling enables them to incorporate more water and act like the charged lipid.

(v) All the mixed lipids extracted from membranes contain charged lipids and at high water contents produce expanding lamellar phases (see Section III (D)).

(vi) In general, the high lipid-high temperature region of the phase diagram is the most complex in terms of the number of phases present.

(vii) Occasionally a "stable" mixed lamellar-hexagonal phase exists over a wide range of temperature and composition. Increasing the temperature gradually converts the lamellar phase to hexagonal II.

(viii) Luzzati (1968) has emphasized the formation of the hexagonal and cubic type II phases by diacyl lipids whilst monoacyl lipids tend to produce the inverse hexagonal and cubic type I phases, the explanation being in terms of the relative bulkiness of the hydrophilic and paraffin moieties of the lipid molecule. One awaits with interest the phase diagrams of lipids such as gangliosides and phosphate inosites with the combination of highly charged, bulky hydrophilic groups and a paraffin region comprising two hydrocarbon chains.

The definition of the bimolecular leaflet of phospholipids in both multilayer lamellar structures and single bilayers is of prime importance to our understanding of diffraction from membranes. The lipid bilayer is now defined in terms of its electron density profile which consists of a trough of low electron density, identified with the hydrocarbon chain region, separating two peaks of high electron density corresponding to the lipid headgroups. Since these features are recognized in the electron density profiles of certain membranes it is reasonable to assume that the lipid bimolecular leaflet contributes significantly to the time-average structure of these membranes. The importance of the other molecular organizations, for example, the hexagonal or cubic phases, in terms of membrane structure or function has yet to be established.

The phospholipid-protein complexes described in section IV shows the same kind of complex polymorphism as lipid–water systems with both lamellar and hexagonal phases defined. The initial interaction responsible for complex formation is almost certainly electrostatic but with the two components suitable oriented there is a finite possibility that apolar or hydrophobic interactions may then occur. The correlation between a structural change and changes in various spectroscopic parameters has offered operational definitions of the two types of complexes, one stabilized solely by electrostatic interactions and another in which hydrophobic interactions occur. Sufficient expertise is now available to shift the emphasis to the interaction of phospholipids with more membrane-relevant proteins. The current activity in the field of membrane protein isolation and characterization suggests that structural information on membrane lipid–membrane protein complexes should be available in the near future.

In terms of membrane structure, considering first the two natural multi-membranes, myelin and retinal rods, the advances made have tended to be in two different directions. In the case of myelin improved X-ray data and the use of more sophisticated diffraction theory has led to a higher resolution picture of the molecular organization perpendicular to the plane of the membrane. Features consistent with a bimolecular leaflet of phospholipid are recognizable but do not necessarily exclude other organizations. However, no consistent viewpoint on the localization of the protein is apparent from the different studies of myelin. In the case of retinal rods, the molecular organization of the photopigment molecules within the plane of the membrane seems well established thus enabling the first steps to be taken which will eventually relate membrane structural changes to a primary functional activity.

The relationship between the X-ray diffraction of phospholipid dispersions

E

and membrane dispersions recognized by Wilkins *et al.* (1971) will enable plasma and other single membrane systems to be the subject of diffraction studies in the future. Such future studies should be addressed to the problems concerning the extent of bilayer structure in a membrane and a precise location of its protein component. In view of the encouraging results with retinal rods some diffraction studies aimed at probing the structure-function relationship of other membranes seem long overdue. The structural investigation of highly organized but localized membrane regions with perhaps the responsibility for a specific membrane function is desirable but limited by the size of these regions. This program may involve the study of not only native membranes but also the reconstituted membrane lipid-protein complexes referred to above. That both approaches will benefit from the experience gained in the study of lipid phase behavior and its quantitation is unquestioned.

VII. Acknowledgements

The author wishes to thank Dr. D. M. Small and Dr. M. C. Carey for reading the manuscript of this chapter and for their helpful suggestions, Miss M. Skibbs for typing the manuscript, and former colleagues at Unilever Research Laboratory for friendly and stimulating collaboration.

References

Abrahamsson, S., Pascher, I., Larsson, K. and Karlsson, K-A. (1972). *Chem. Phys. Lipids* **8**, 152–179.

Abramson, M. B. (1970). *In* "Surface Chemistry of Biological Systems". (M. L. Blank, ed.), Vol. 7, pp. 37–53. Plenum Press, New York.

Akers, C. K. and Parsons, D. F. (1970a). *Biophys. J.* **10**, 101–115.

Akers, C. K. and Parsons, D. F. (1970b). *Biophys. J.* **10**, 116–136.

Beeman, W. W., Kaesberg, P., Anderegg, J. W. and Webb, M. B. (1957). *In* "Handbuch der Physik". Vol. 32, pp. 321–442.

Blasie, J. K. (1972a). *Biophys. J.* **12**, 191–204.

Blasie, J. K. (1972b). *Biophys. J.* **12**, 205–213.

Blasie, J. K. and Worthington, C. R. (1969). *J. molec. Biol.* **39**, 417–439.

Blasie, J. K., Dewey, M. M., Blaurock, A. E. and Worthington, C. R. (1965). *J. molec. Biol.* **14**, 143–152.

Blasie, J. K., Worthington, C. R. and Dewey, M. M. (1969). *J. molec. Biol.* **39**, 407–416.

Blaurock, A. E. (1971). *J. molec. Biol.* **56**, 35–52.

Blaurock, A. E. and Stoeckenius, W. (1971). *Nature New Biol.* **233**, 152–154.

Blaurock, A. E. and Wilkins, M. H. F. (1969). *Nature, Lond.* **223**, 906–909.

Blaurock, A. E. and Wilkins, M. H. F. (1972). *Nature, Lond.* **236**, 313–314.

Blaurock, A. E. and Worthington, C. R. (1966). *Biophys. J.* **6**, 305–312.

Blaurock, A. E. and Worthington, C. R. (1969). *Biochim. Biophys. Acta* **173**, 419–426.

Blodgett, K. B. and Langmuir, I. (1937). *Phys. Rev.* **51**, 964–982.
Bourgès, M., Small, D. M. and Dervichian, D. G. (1967). *Biochim. biophys. Acta* **137**, 157–167.
Brown, A. D. (1965). *J. molec. Biol.* **12**, 491–508.
Burge, R. E. and Draper, J. C. (1967). *Acta Cryst.* **22**, 6–13.
Caspar, D. L. D. and Kirschner, D. A. (1971). *Nature New Biol.* **231**, 46–52.
Chapman, D. and Urbina, J. (1971). *FEBS Lett.* **12**, 169–172.
Chapman, D., Byrne, P. and Shipley, G. G. (1966). *Proc. R. Soc., A,* **290**, 115–142.
Chapman, D., Williams, R. M. and Ladbrooke, B. D. (1967). *Chem. Phys. Lipids* **1**, 445–475.
Chapman, D., Fluck, D. J., Penkett, S. A. and Shipley, G. G. (1968). *Biochim. biophys. Acta* **163**, 255–261.
Cullen, J., Phillips, M. C. and Shipley, G. G. (1971). *Biochem. J.* **125**, 733–742.
Das, M. L., Haak, E. D. and Crane, F. L. (1965). *Biochemistry* **4**, 859–865.
Elliot, A. (1965). *J. scien. Instrum.* **42**, 312–316.
Engelman, D. M. (1970). *J. molec. Biol.,* **47**, 115–117.
Engelman, D. M. (1971). *J. molec. Biol.,* **58**, 153–165.
Esfahani, M., Limbrick, A. R., Knutton, S., Oka, T. and Wakil, S. J. (1971). *Proc. natn. Acad. Sci. U.S.A.,* **68**, 3180–3184.
Finean, J. B. (1966). *In* "Progress in Biophysics and Molecular Biology". (J. A. V. Butler and H. E. Huxley, eds.) Vol. 16, pp. 143–170. Pergamon Press, Oxford.
Finean, J. B. and Burge, R. E. (1963). *J. molec. Biol.* **7**, 672–682.
Finean, J. B., Sjostrand, F. S. and Steinmann, E. (1953). *Expl. Cell Res.* **5**, 557–559.
Finean, J. B., Coleman, R., Green, W. G. and Limbrick, A. R. (1966). *J. Cell Sci.* **1**, 287–296.
Finean, J. B., Coleman, R., Knutton, S., Limbrick, A. R. and Thompson, J. E. (1968). *J. gen. Physiol.* **51**, 19s.
Fontell, K., Mandell, L., Lehtinen, H. and Ekwall, P. (1968). *Acta Polytechnica Scand., Chem. and Metallurgy Ser.* No. 74.
Franks, A. (1958). *Br. J. appl. Phys.* **9**, 349–352.
Fromherz, P. (1970). *FEBS Lett.* **11**, 205–208.
Gottlieb, M. H. and Eanes, E. D. (1972). *Biophys. J.* **12**, 1533–1548.
Gras, W. J. and Worthington, C. R. (1969). *Proc. natn. Acad. Sci.* **63**, 233–238.
Guinier, A. (1963). "X-ray Diffraction in Crystals, Imperfect Crystals, and Amorphous Bodies". W. H. Freeman, San Francisco.
Guinier, A. and Fournet, G. (1955). "Small-angle Scattering of X-rays". John Wiley and Sons, New York.
Gulik-Krzywicki, T., Rivas, E. and Luzzati, V. (1967). *J. molec. Biol.* **27**, 303–322.
Gulik-Krzywicki, T., Shechter, E., Luzzati, V. and Faure, M. (1969a). *Nature, Lond.* **223**, 1116–1121.
Gulik-Krzywicki, T., Tardieu, A. and Luzzati, V. (1969b). *Mol. Cryst. & Liquid Crystals* **8**, 285–291.
Gulik-Krzywicki, T., Shechter, E., Iwatsubo, M., Ranck, J. L. and Luzzati, V. (1970). *Biochim. biophys. Acta* **219**, 1–10.
Hamilton, R. L., Havel, R. J., Kane, J. P., Blaurock, A. E. and Sata, T. (1971). *Science, N.Y.* **172**, 475–478.
Harrison, S. C., Caspar, D. L. D., Camerini-Otero, R. D. and Franklin, R. M. (1971a). *Nature New Biol.* **229**, 197–201.
Harrison, S. C., David, A. Jumblatt, J. and Darnell, J. E. (1971b). *J. molec. Biol.* **60**, 523–528.

Horne, R. W. and Watkins, J. S. (1970). *Micron* **1**, 394–404.

Hosemann, R. and Bagchi, S. N. (1962). "Direct Analysis of Diffraction by Matter". North Holland, Amsterdam.

Hosemann, R. and Kreutz, W. (1966). *Naturwissenschaften* **53**, 298–304.

James, R. W. (1963). "The Optical Principles of the Diffraction of X-rays". Bell, London.

Kimelberg, H. K. and Lee C. P. (1970). *J. Membrane Biol.* **2**, 252–262.

Kimelberg, H. K., Lee, C. P., Claude, A. and Mrena, E., (1970). *J. Membrane Biol.* **2**, 235–251.

Kratky, O. (1963). *In* "Progress in Biophysics and Molecular Biology". (J. A. V. Butler, H. E. Huxley and R. E. Zirkle, eds.) Vol. 13, pp. 105–173. Pergamon Press, Oxford.

Kreutz, W. (1963a). *Z. Naturf.* **18b**, 1098–1104.

Kreutz, W. (1963b). *Z. Naturf.* **18b**, 567–571.

Kreutz, W. (1964). *Z. Naturf.* **19b**, 441–446.

Kreutz, W. (1966). *In* "Biochemistry of Chloroplasts". (T. W. Goodwin, ed.) Vol. 1, pp. 83–88.

Kreutz, W. and Menke, W. (1962). *Z. Naturf.* **17b**, 675–683.

Ladbrooke, B. D., Williams, R. M. and Chapman, D. (1968). *Biochim. biophys. Acta* **150**, 333–340.

Lecuyer, H. and Dervichian, D. G. (1969). *J. molec. Biol.* **45**, 39–57.

Lesslauer, W., Wissler, F. C. and Parsons, D. F. (1970). *Biochim. biophys. Acta* **203**, 199–208.

Lesslauer, W., Cain, J. and Blasie, J. K. (1971). *Biochim. biophys. Acta* **241**, 547–566.

Levine, Y. K. and Wilkins, M. H. F. (1971). *Nature New Biol.* **230**, 69–72.

Levine, Y. K., Bailey, A. I. and Wilkins, M. H. F. (1968). *Nature, Lond.* **220**, 577–578.

Limbrick, A. R. and Finean, J. B. (1970). *J. Cell Sci.* **7**, 373–386.

Luzzati, V. (1968). *In* "Biological Membranes". (D. Chapman, ed.) pp. 71–123. Academic Press, New York.

Luzzati, V. and Husson, F. (1962). *J. Cell Biol.* **12**, 207–219.

Luzzati, V., Gulik-Krzywicki, T. and Tardieu, A. (1968). *Nature, Lond.* **218**, 1031–1034.

Mateu, L., Tardieu, A., Aggerbeck, L. and Scanu, A. M. (1971). *In* "Plasma Lipoproteins". (R. M. S. Smellie, ed.) pp. 87–88. Academic Press, London.

Moretz, R. C., Akers, C. K. and Parsons, D. F. (1969a). *Biochim. biophys. Acta* **193**, 1–11.

Moretz, R. C., Akers, C. K. and Parsons, D. F. (1969b). *Biochim. biophys. Acta* **193**, 12–21.

Oesterhelt, D. and Stoeckenius, W. (1971). *Nature New Biol.* **233**, 149–152.

Parsegian, V. A. (1967). *Science, N.Y.* **156**, 939–942.

Phillips, M. C. (1972). *In* "Progress in Surface and Membrane Science". (J. F. Danielli, M. D. Rosenberg and D. A. Cadenhead, eds.) Vol. 5, pp. 139–221. Academic Press, New York.

Phillips, M. C., Hauser, H. and Paltauf, F. (1972). *Chem. Phys. Lipids* **8**, 127–133.

Porod, G. (1951). *Kolloid-Z.* **124**, 83–114.

Quinn, P. J. and Dawson, R. M. C. (1969a). *Biochem. J.* **113**, 791–803.

Quinn, P. J. and Dawson, R. M. C. (1969b). *Biochem. J.* **115**, 65–75.

Quinn, P. J. and Dawson, R. M. C. (1970). *Biochem. J.* **116**, 671–680.

Rand, R. P. (1971). *Biochim. biophys. Acta* **241**, 823–834.

Rand, R. P. and SenGupta, S. (1972). *Biochim. biophys. Acta* **255**, 484–492.

Rand, R. P. and Luzzati, V. (1968). *Biophys. J.* **8**, 125–137.

Rand, R. P., Tinker, D. O. and Fast, P. G. (1971). *Chem. Phys. Lipids* **6**, 333–342.

Razin, S., Morowitz, H. J. and Terry, T. M. (1965). *Proc. natn. Acad. Sci. U.S.A.* **54**, 219–225.

Reiss-Husson, F. (1967). *J. molec. Biol.* **25**, 363–382.

Rivas, E. and Luzzati, V. (1969). *J. molec. Biol.* **41**, 261–275.

Robertson, J. D. (1966). *Ann N.Y. Acad. Sci.* **137**, 421–440.

Rouser, G., Nelson, G. J., Fleischer, S. and Simon, G. (1968). *In* "Biological Membranes". (D. Chapman, ed.) pp. 5–69. Academic Press, London.

Schmidt, W. J. (1938). *Kolloid-Z.* **85**, 137–148.

Schmitt, F. O., Bear, R. S. and Clark, G. L. (1935). *Radiol.* **25**, 131–151.

Schmitt, F. O., Bear, R. S. and Palmer, K. J. (1941). *J. cell. comp. Physiol.* **18**, 31–42.

Shechter, E., Gulik-Krzywicki, T., Azerad, R. and Gros, C. (1971). *Biochim. biophys. Acta* **241**, 431–442.

Shipley, G. G., Leslie, R. B. and Chapman, D. (1969a). *Nature, Lond.* **222**, 561–562.

Shipley, G. G., Leslie, R. B. and Chapman, D. (1969b). *Biochim. biophys. Acta* **173**, 1–10.

Shipley, G. G., Atkinson, D. and Scanu, A. M. (1972). *J. Supramolecular Struct.* **1**, 98–104.

Shipley, G. G., Green, J. P. and Nichols, B. W. (1973). In press.

Small, D. M. (1967). *J. Lipid Res.* **8**, 551–557.

Small, D. M. (1971). *In* "The Bile Acids". (P. P. Nair and D. Kritchevsky, eds.) pp. 249–356. Plenum Press, New York.

Steim, J. M., Tourtellotte, M. E., Reinert, J. C. McElhaney, R. N. and Rader, R. L. (1969). *Proc. natn. Acad. Sci. U.S.A.* **63**, 104–109.

Steinemann, A. and Lauger, P. (1971). *J. Membrane Biol.* **4**, 74–86.

Stoeckenius, W. and Engelman, D. M. (1969). *J. Cell Biol.* **42**, 613–646.

Thompson, J. E., Coleman, R. and Finean, J. B. (1968). *Biochim. biophys. Acta* **150**, 405–415.

Van Deenen, L. L. M. (1965). *In* "Progress in the Chemistry of Fats and Other Lipids". (R. T. Holman, ed.) Vol. 8, 1–127. Pergamon Press, Oxford.

Wilkins, M. H. F., Blaurock, A. E. and Engelman, D. M. (1971). *Nature New Biol.* **230**, 72–76.

Williams, R. M. and Chapman, D. (1970). *In* "Progress in the Chemistry of Fats and Other Lipids". (R. T. Holman, ed.) Vol. II, 1–79. Pergamon Press, Oxford.

Worthington, C. R. (1969a). *Biophys. J.* **9**, 222–234.

Worthington, C. R. (1969b). *Proc. natn. Acad. Sci. U.S.A.* **63**, 604–611.

Worthington, C. R. (1970). *Biophys. J.* **10**, 675–677.

Worthington, C. R. and Blaurock, A. E. (1968). *Nature, Lond.* **218**, 87–88.

Worthington, C. R. and Blaurock, A. E. (1969a). *Biochim. biophys. Acta* **173**, 427–435.

Worthington, C. R. and Blaurock, A. E. (1969b). *Biophys. J.* **9**, 970–990.

Worthington, C. R. and King, G. I. (1971). *Nature, Lond.* **234**, 143–145.

Chapter 2

Some Recent Studies of Lipids, Lipid–Cholesterol and Membrane Systems

D. CHAPMAN

Chemistry Department, Sheffield University, Sheffield, England

I. Introduction

The evidence that cell membranes contain appreciable regions of lipid in a bilayer form is now strong (see Chapter 1). The circumstantial evidence and the more direct evidence from X-ray methods all converge to support this conclusion. There are, however, still very many questions remaining to be resolved concerning the detailed lipid organization, the interaction of lipid with cholesterol and concerning the organization of protein before we can feel sure that we fully understand the detailed structure of cell membranes.

In this chapter I shall endeavour to summarize some recent studies where important information concerning these various questions has been obtained.

II. Studies of Lipids

Biochemical evidence has shown that many membranes contain a variety of lipid classes, whilst associated with each lipid class there is a distribution of fatty acids.

In recent years physical studies have been carried out on simple lipid systems in order to understand their behaviour. The majority of these studies have been carried out with lecithins but increasingly studies of other lipid classes, e.g. the sphingolipids, glycolipids, cerebrosides and gangliosides, are also being made.

We shall first briefly summarize these studies.

A. THERMAL TECHNIQUES

The lipids present in biological membranes, e.g. phospholipids and glycolipids, generally exhibit both thermotropic and lyotropic mesomorphism, and often form bilayer leaflet membrane structures in H_2O. Order → disorder (crystal → liquid crystal) phase transitions have been studied using both the techniques of differential thermal analysis and differential scanning calorimetry (Ladbrooke and Chapman, 1969). The temperatures at which the phase transitions occur are dependent upon the headgroup, the hydrocarbon chain length, and the degree and type of unsaturation present (Chapman, 1968). For the same headgroup and extent of hydration, lipids with more unsaturated chains have lower transition temperatures than more saturated ones (Ladbrooke et al., 1968), longer chains higher transition temperatures than shorter ones (Chapman et al., 1967), and cis-unsaturated chains lower transition temperatures than trans-unsaturated ones (Chapman et al., 1966).

Quantitative estimates of the heat involved in these melting processes have also been made (Phillips et al., 1969). Some of the appropriate thermodynamic data are shown in Table I. The change in entropy (ΔS) equals q_{rev}/T, where q_{rev} is the head absorbed or latent heat and T is the temperature (°K). For a first-order transition at constant pressure the free energy change is zero so that the latent heat becomes equal to the enthalpy change (ΔH) and hence $\Delta H = T\Delta S$. The entropies shown in Fig. 1 represent the changes in configurational freedom at the transition temperatures. These figures give a comparison of the relative amounts of disorder being introduced into the hydrocarbon chains as they undergo the transitions. The enthalpy values cannot be compared directly, since the transitions occur at different temperatures and the temperature dependence of ΔH is approximately 300 times greater than that of ΔS.

TABLE I

Thermodynamic data for the crystalline → liquid-crystalline transitions of 1,2-diacyl-L-phosphatidylcholines

Acyl chain length		Behenoyl $C^{\circ}22$	Stearoyl $C^{\circ}18$	Palmitoyl $C^{\circ}16$	Myristoyl $C^{\circ}14$	Oleoyl $C^{1}18$
At maximum hydration	Transition temperature $T_c(^{\circ}C)$	75	58	41	23	−22
	Enthalpy change ΔH kcal mol^{-1}	14·88	10·67	8·66	6·64	7·6
	Entropy change ΔS cal mol^{-1} deg^{-1}	42·8	32·4	27·6	22·4	30·3
Monohydrate α-form	T	—	78	65	51	—
	ΔH	—	5·25	4·55	3·66	—
	ΔS	—	15·0	13·5	11·3	—
Anhydrous crystalline β-form	T	120	115	—	—	—
	ΔH	17·3	13·8	—	—	—
	ΔS	44·0	35·6	—	—	—

1. *Liquid Crystalline Transitions of Lecithins*

(a) *Anhydrous lecithins.* X-ray diffraction has shown that anhydrous lecithins crystallize in the β form. These lecithins do not undergo a simple fusion process, they pass through a thermotropic mesophase before becoming an isotropic liquid (Williams and Chapman, 1970). The largest endothermic transition is from crystal to liquid crystal which, since it is chain length dependent, mainly involves a change in the hydrocarbon chains. The heat for this is about 95% of the total heat of fusion.

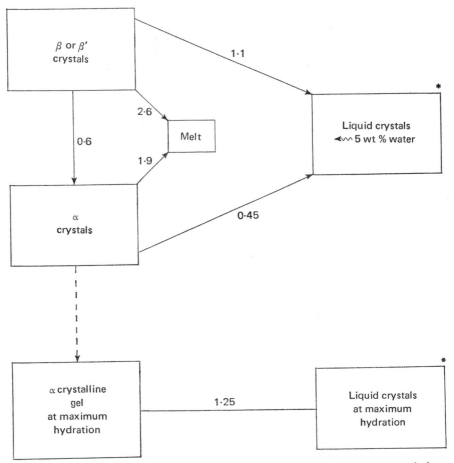

FIG. 1. Configurational entropies of transition (cal deg^{-1} mole^{-1}) per methylene group (Phillips *et al.*, 1969).

* The particular liquid-crystalline phases have not been specified. Heats, and hence entropy, changes per methylene group at lamellar to hexagonal, lamellar to cubic and cubic to hexagonal liquid-crystalline phase boundaries are <5% of those of the crystalline to liquid-crystalline transitions.

The total entropy per methylene group is the same for all long-chain compounds in the β crystalline form at their chain melting point. Now the entropy gain during the transition from β crystal to isotropic liquid for n-alkanes, triglycerides and fatty acids is 2·6 e.u. per CH_2 group (Fig. 1), while for the β crystal to liquid–crystal transition for lecithins the equivalent figure is 1·1 e.u. Thus, in the liquid-crystalline state, the chain fluidity is about half that found in liquid n-alkanes at the transition temperature. This may arise by inhibition of rotation about the carbon–carbon bonds. A study (Salsbury and Chapman, 1968) by nuclear magnetic resonance of chain motions in lecithin molecules in anhydrous liquid-crystalline phases indicates that considerable rotation about the carbon–carbon bonds occurs.

The reduction in chain motions, as compared with the melt of n-alkanes, could arise either because of the particular molecular structure of the lecithins or because of the influence of the ordered state of the mesophase. The β to α crystal transition for lecithins would have $\Delta S \sim 0.65$ e.u. per CH_2 group, which is close to the figures given earlier for rotational premelting in n-alkanes, etc., indicating that for this transition lecithins behave like other long-chain compounds.

(b) *Hydrated lecithins.* Lecithin monohydrates show more complex mesomorphic behaviour than the anhydrous compounds in that they form several liquid-crystalline phases (Chapman et al., 1967; Luzzati et al., 1968). However, once again the initial chain melting heat accounts for about 95% of the total heat of fusion. The α crystalline form of the monohydrate of lecithin has an entropy of transition to the liquid-crystalline state of 0·45 e.u. per methylene group. This compares with an entropy change of 1·9 e.u. $(CH_2)^{-1}$ for the transition α crystal to melt for n-alkanes, etc. The difference of 1·45 e.u. between these values compares favourably with the difference of 1·5 e.u. in the entropies of transition between β crystal to liquid crystal and β crystal to melt. Hence the chain motions in the monohydrate liquid crystals are essentially the same as those in anhydrous systems.

In excess water the lecithins undergo a transition (Chapman et al., 1967) from α gel to the neat or smectic phase (bimolecular lamellae). The temperature and heats of this transition are chain length dependent and it is at this transition that "chain melting" occurs. (There is also a small endothermic peak (Chapman et al., 1967) before the main transition which possibly arises from rearrangements within the polar group lattice.) DSC curves of distearoyl phosphatidylcholine indicate that 10 moles of water per mole of lecithin are bound, i.e. are unfreezable, at 0°C (Williams and Chapman, 1970). Addition of water to the anhydrous material results in hydration of the polar headgroups and this causes a lowering (Fig. 2) of the transition temperature. It is also apparent that the shape of the phase transition endotherm alters. The transition changes to a highly cooperative phenomenon on hydration,

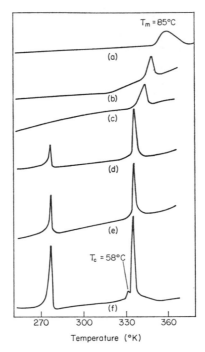

FIG. 2. Differential scanning calorimetry curves of distearoyl lecithin as a function of hydration. (a) anhydrous; (b) 10 wt % H_2O; (c) 20 wt % H_2O; (d) 25 wt % H_2O; (e) 30 wt % H_2O; (f) 40 wt % H_2O (Williams and Chapman, 1970).

consistent with laser–Raman evidence (Lippert and Peticolas, 1971). The entropy change at the main transition is 1·25 e.u. $(CH_2)^{-1}$ as compared with 0·45 e.u. $(CH_2)^{-1}$ for α crystal to liquid crystal. Since the hydrocarbon chain structure is identical for both the α crystal and α gel (Chapman *et al.*, 1967; Larsson, 1967) the difference of 0·8 e.u. must be accounted for by differences in the liquid-crystalline phases. We conclude that the hydrocarbon chain mobility in the fully hydrated liquid-crystalline system is much greater than that in the anhydrous or monohydrate liquid crystals. It has been noted (Luzzati, 1968) that the area occupied per molecule at a lipid/water interface increases as the water content of the system is increased at constant temperature. For dipalmitoyl lecithin the area occupied at the interface (Phillips and Chapman, 1968) increases from about 60 Å² to about 70 Å² as the water content increases from 20 to 40 wt %. This is in complete accord with the conclusion that the hydrocarbon chains in liquid crystalline phases have greater mobility in the presence of substantial amounts of water than in the anhydrous or monohydrate systems.

It is clear that for anhydrous lecithin liquid crystals at the transition temperature the configurational entropy per methylene group is 1·5 e.u. less than that for a liquid *n*-alkane at the melting point. However, in the fully hydrated systems, this difference is reduced by a factor of about 2 and is about 0·7 e.u. As rotation about a C–C bond gives rise to an entropy increase of about 2·2 e.u., an indication of the rotation in the two states is obtained.

As the lipid is heated above its transition temperature, a further gradual increase in disorder of the chains occurs. The degree of disorder of lipid chains therefore depends upon (a) whether the lipid is above or below its transition temperature, and (b) when in the liquid-crystalline phase, how high the temperature of the lipid is above this transition point. This is an important point which needs to be considered in intercomparison of the fluidity of homologous lipids.

(c) *Theoretical studies.* Theoretical studies describing the rotation of long chain paraffins in the α or hexagonal rotator form have been reported. Cooperation in these systems has been introduced using the Bethe (1935) or

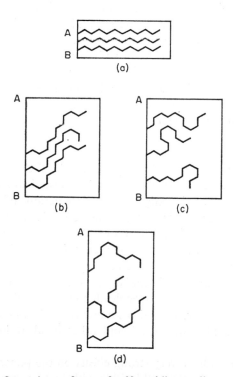

Fig. 3. Typical configurations of sets of self-avoiding walks on a two-dimensional hexagonal lattice with (a) Δ=1; (b) Δ=2; (c) Δ=3; and (d) Δ=4 (Whittington and Chapman, 1966).

Bragg–Williams (Bragg and Williams, 1934) approximations. Hoffman (1952) and Chapman and Whittington (1964) used potential energy diagrams to study the motion of the chains. In the latter study it was assumed that rotation of one molecule facilitated the rotation of other molecules in a gear-like system. A Monte-Carlo method has also been used to simulate the rotation of the chains in this form (Whittington and Chapman, 1965).

A very simple theoretical system has also been devised (Whittington and Chapman, 1966) to provide some insight into the motion of chains in the liquid-crystalline form. In the real system the twisting of CH_2 groups of one chain has to be concordant with the movements of CH_2 groups in adjacent chains. The model system to describe this is restricted to a set of simple chains in two dimensions with the chains confined to lie on a two-dimensional hexagonal lattice. The end-to-end distance of each chain (corresponding in a real system to the distance from the polar group to the methyl group) and other properties were determined as a function of the density of packing of the chains using two simple potential functions and using the Monte-Carlo computational method. Typical configurations of sets of simple chains on a two-dimensional hexagonal lattice are shown in Fig. 3. At the highest density only the fully extended configuration occurs. At lower densities other configurations are allowed. As the density of the chains decreases, the end-to-end distance of the chains suddenly falls, consistent with the occurrence of a cooperative phase transition. This model system shows the way in which cooperative movement of the adjacent chains may occur in a monolayer, bilayer or a membrane system.

2. Lipid–Metal Ion Systems

Electrostatic interactions of the polar head of the phospholipids have been shown to affect their thermotropic phase behaviour. Urbina and Chapman (1971) have investigated the interaction of different divalent cations on the endothermic phase transition of dipalmitoyl lecithin and ox-brain phosphatidylserine. From monolayer and lipsomes experiments, uranyl ions UO_2^{2+} are known to interact stoichiometrically with the polar groups of lecithin and to bind in the membrane surface of different biological membranes, presumably to phosphate groups. These authors confirmed that the strong binding of this ion is specifically to the phosphate groups by means of infrared spectroscopy. It is also known that phosphatidylserine forms a stoichiometric complex with Ca^{2+} and other divalent ions (Hauser et al., 1969).

These interactions have very strong effects on the phase transition of both lipids. A titration of dipalmitoyl lecithin with UO_2^{2+} shows that the original transition temperature is at 314·5 K (41·5°C) shifted in the 1 : 1 complex to 319 K (46°C) (see Fig. 4). With phosphatidylserine containing a net negative

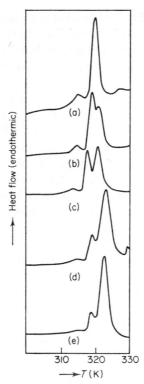

FIG. 4. Differential scanning calorimetry curves of dipalmitoyl lecithin water mixtures in the presence of (a) 0·0; (b) 0·1; (c) 0·2; (d) 0·4; (e) 0·8 equivalents of UO_2^{2+} per equivalent of phosphorus (Urbina and Chapman, 1971).

charge, the binding of Ca^{2+} at 1 : 1 proportion shifts the transition from 290 K observed with free phosphatidylserine to 295 K for the complex. This is an important point because it shows how the transition temperature and hence the degree of lipid chain fluidity and thus related permeability of a lipid bilayer forming part of a membrane could be affected by metal–ion interaction with the polar headgroup.

3. *Mixing Properties*

The mixing properties of various homologous lipids have been studied (Phillips *et al.*, 1970). Since biomembranes contain a varied population of chain lengths with different degrees of unsaturation and substitution together with (in general) a wide variety of polar headgroups, it is important to understand the phase behaviour of these mixed systems.

With widely dissimilar chain lengths, phase behaviour characteristic of a monotectic system (Phillips *et al.*, 1970; Bowden, 1954) is found, e.g. mixing of equimolar quantities of dioleyl lecithin with dibehenoyl lecithin results in transitions occurring at $T_c = -22°C$ and $T_m = 69°C$. With closer chain lengths, e.g. nC_{14} and nC_{18}, a mixed solid phase and solid + liquid-crystalline phase is present in the phase diagram. With $nC_{16} + nC_{18}$ or $nC_{14} + nC_{16}$ ideal mixing in both phases occurs.

Mixing of different polar headgroup containing lipids, e.g. cerebroside (a sugar lipid, $T_m = 65°C$) with egg lecithin ($T_m \simeq -5°C$), can result in a mixed gel–liquid crystal system with a lower ($\simeq 35°C$) T_{max} (Clowes *et al.*, 1971). The mean "fluidity" of the system is thus higher than that of the pure cerebroside but lower than that of the egg lecithin, at the same temperature. Similar behaviour is found in the system dimyristoyl lecithin–dimyristoyl phosphatidylentholamine (Fig. 5). Dimyristoyl lecithin has $T_c = 23°C$ (Fig. 5a) and for dimyristoyl phosphatidylethanolamine $T_c = 48°C$ (Fig. 5e). The pretransitional peak of the lecithin, thought to be due to a rearrangement of the polar headgroup, is removed on addition of small amounts of the dimyristoyl phosphatidylethanolamine (Fig. 5b) (Keough and Chapman, 1972).* At

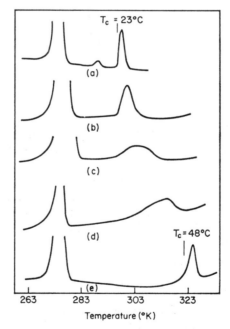

FIG. 5. Differential scanning calorimetry curves of dimyristoyl lecithin DML–dimyristoyl phosphatidylethanolamine DMPE. (a) DML 100 mol %; (b) DML 95 mol % + DMPE 5 mol %; (c) DML 70 mol % + DMPE 30 mol %; (d) DML 50 mol % + DMPE 50 mol %; (e) DMPE 100 mol % all in excess water (Keough and Chapman, 1972).

* See note 1, p. 144.

higher concentrations (Fig. 5c, d), the endotherm is very broad. This indicates (Phillips *et al.*, 1970) the presence of clusters of gel and liquid-crystalline lipid in the bilayer, and the reduction in enthalpy of the transition is characteristic of a lower cooperativity of the transition in the mixture compared to the individual components. Such broad transitions are typical of several biomembranes. The variation of enthalpy involved in the melting process as the concentration of the two lipids vary is shown in Fig. 6.

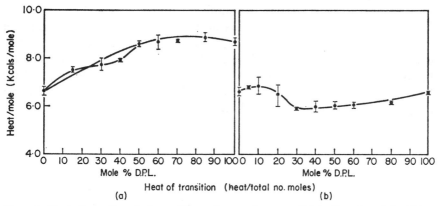

FIG. 6. Variation of enthalpy with various mixtures of (a) dimyristoyl lecithin–dipalmitoyl lecithin and (b) dimyristoyl lecithin–dimyristoyl phosphatidylethanolamine (Keough and Chapman, 1972).

4. *Interactions with Dye Molecules and Fluorescent Probes*

Träuble (1972) has studied the dye molecule bromothymolblue a pH indicator of a pK of 7·1. A marked decrease in the absorption band at $\lambda = 615$ nm is observed when lipid is added to an aqueous solution of bromothymolblue at pH 7. The extinction coefficient E_{615} of the membrane-bound species of bromothymolblue (BTB^- and/or BTB^0) is more than two orders of magnitude smaller than the extinction coefficient of bromothymolblue in an aqueous solution of pH 7 ($E_{615} = 1·4 \times 10^4$). The phase transition is therefore accompanied by a sharp decrease in the optical density. Titration experiments carried out at 25°C and 45°C indicate that this effect is mainly due to an increase in the number of binding places on the membrane ($n = 7·7$ at 25°C and $n = 20$ at 45°C, where $n =$ number of binding places per 100 lipid molecules). (The condition $pH \approx pK$ is critical for these measurements.) In the alkaline region at a $pH > 9·0$ the indicator exists exclusively in the double charged form (BTB^{--}) which shows no detectable interaction with the membranes. At low pH the indicator exists as BTB^- and/or BTB^0, depending upon the pH. Attachment of these species to the membrane causes, however, no spectral changes.

The binding of bromothymolblue, in contrast to that of ANS (see later), can cause detectable perturbations in the lipid structure leading to a shift in the transition temperature T_t. Träuble suggests that this indicates that BTB is buried deeper in the membrane phase than ANS. In order to study this effect in more detail, transition curves of dimyristoyl lecithin vesicles ($T_t=24°C$) were measured at different BTB concentrations at pH 7 using a light scattering method. An asymptotic value, $T_t \approx 20°C$, is reached for high BTB concentrations. Upon changing the pH from pH 7 to pH 9 the transition temperature, T_t, goes back from $T_t \approx 20°C$ to the ininitial value of $T_t \approx 24°C$, characteristic for the lipid without attached indicator molecules. A subsequent reduction in pH to pH 7 brings the value of T_t again to $T_t \approx 20°C$. These experiments demonstrate that in principle it is possible to trigger the phase transition (1) by the adsorption and desorption of ligands to the lipid structure, and (2) by changing the pH of the environment. The variation of the transition temperature with pH is shown in Fig. 7.

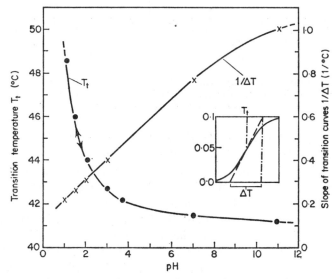

FIG. 7. The variation of transition temperature of dipalmitoyl lecithin with pH (after Träuble, 1972).

Several fluorescence probes are known which bind to lipid membranes and which are sensitive to the environmental polarity and/or constraint. The indicator ANS (1-anilino-8-naphthalenesulphonate) contains a hydrophobic ring system and a charged sulphonic acid group (pH > 3). ANS is soluble, both in organic solvents and in water. Due to its amphiphilic nature ANS is expected to bind in the semi-polar surface of a lipid structure and to orient with the sulphonic acid group in the aqueous phase. A strong inverse dependence of the quantum yield, Q, of ANS on the polarity of the solvent was established by Stryer (1965), Turner and Brand (1968). Addition of a lipid dispersion to an aqueous solution of ANS produces a strong increase in the ANS fluorescence. The quantum yield, Q, of ANS when bound to dipalmitoyl lecithin is $Q = 0.08 \pm 0.03$ compared to $Q = 0.004$ in water. The value of Q corresponds to an environmental dielectric constant of about $E = 35$, indicating that the molecule is bound within the polar groups of the lipid structure. This interpretation is supported by studies of several workers (Gulik-Kryzwicki, 1970; Vanderkooi and Martonosi, 1969; Brockelhurst *et al.*, 1970; Lesslauer *et al.*, 1971).

Thus ANS is expected to be sensitive against changes in the structure and arrangement of the polar headgroups at the lipid phase transition. A sharp increase in fluorescence intensity is observed in the temperature range between 40° and 42°C. Compared to the light-scattering and volumetric measurements the ANS transition curve is smeared out towards lower temperatures. The sharp fluorescence change between 40° and 42°C parallels the structural changes in the hydrocarbon chains, as they are observed in the light-scattering and volumetric measurements. Changes in the hydrocarbon chains are accompanied by changes in the packing of the polar headgroups. The "tail" in the ANS transition curve indicates that a gradual loosening of the polar groups precedes the main transition. Titrations of dipalmitoyl lecithin dispersions with ANS were carried out at temperatures below and above T_t. These experiments show that the increase in ANS fluorescence intensity at the phase transition is mainly due to an increase in the number of binding places for ANS on the membrane surface. The values of the number of binding places per 100 lipid molecules are: $n = 2.5$ at 25°C and $n = 6.7$ 45°C. The binding constants are $K = 0.06 \times 10^4$ at 25°C and $K = 0.5 \times 10^4$ at 45°C.

Several other adsorption indicators with different structural sensitivities can be used to study the lipid phase transition (TNS, Rhodamine 6G, N-phenyl-1-naphtylamine; cf. Radda, 1971). For example, the probe N-phenyl-1-naphtylamine senses primarily the environmental viscosity (Radda, 1971).

5. *Interactions with Drug Molecules*

Recently, thermal studies have been made using the lecithin–water system as a model for natural cell membrane systems. A number of drug molecules have now been examined. These include antibiotics (Pache and Chapman, 1972), analgesic and some antidepressant molecules (Chapman and Saville, 1972). This simple model system appears to have considerable potential for helping our understanding of the mode of action of these molecules. The effect of desiprimine on the transition temperature of dipalmitoyl lecithin is illustrated in Fig. 8. It can be seen that the shift in transition temperature is

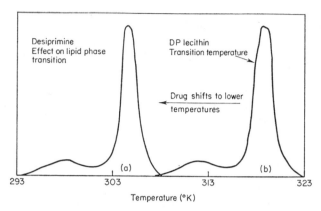

FIG. 8. The effect on the transition temperature of dipalmitoyl lecithin with increasing amounts of the antidepressant drug desiprimine (Chapman and Saville, 1972).

considerable, as much as 20°C, and is to *lower* temperatures. Thus the drug causes the lipid to be considerably more fluid and hence more permeable at a given temperature. Thus if we extrapolate this finding to the real situation and we think of a cell membrane of the central nervous system at 37°C, then this drug increases fluidity and hence permeability to water as well as for certain organic molecules. It may well be possible that the rotation and passage of carrier proteins may also be considerably increased. Thus a whole range of biochemical changes and process may be speeded up. The fact that the transition remains but shifts implies that the drug is acting upon the polar group of the lipid and is not intercalated among the lipid chains as occurs with cholesterol. (See also p. 129 and p. 130.)

6. *Monolayer Studies*

It is important to appreciate that there is a correlation between the bulk properties as indicated by the thermal studies and monolayer properties at the air–water interface (Chapman *et al.*, 1966).

Phospholipid molecules possess both hydrophobic and hydrophilic properties and are oriented at a water surface, forming monolayers. For many years, monolayer studies of phospholipids have been conducted. Usually this work has been performed with natural phospholipid mixtures and, in the vast majority of cases, with egg-yolk phosphatidylcholine. In recent years studies have been made with pure synthetic phospholipids (Van Deenen *et al.*, 1962). These have shown that, at room temperature, the saturated phospholipids exhibit monolayers which are more condensed than those of unsaturated phospholipids containing *cis* double bonds, i.e. the saturated phospholipids occupy less area at low surface pressure than the corresponding unsaturated compounds. The monolayer behaviour of the phospholipids reflects their transition temperatures (Chapman, 1966). Thus, at room temperature, a phospholipid which has a high transition temperature exhibits a condensed film; a phospholipid having a lower transition temperature exhibits an expanded film or greater area per molecule (Phillips and Chapman, 1968).

The influence of metal ions upon the transition temperature also finds its parallel in the effect on the monolayer properties. Where the effect is to cause an increase in transition temperature a more condensed film is obtained.

B. NUCLEAR MAGNETIC RESONANCE

1. Proton Wide-line and Pulsed NMR

Early wide-line studies indicated that NMR was a promising technique with which to study molecular mobility in lipid systems (Chapman and Salsbury, 1966).* Anhydrous lecithins at liquid nitrogen temperatures were shown to have chain proton linewidths of $\sim 1\cdot6$ mT† and this was gradually reduced on heating to $0\cdot8$ mT (distearoyl lecithin) (Salsbury and Chapman, 1968), or in the presence of water, $0\cdot4$ mT (Veksli et al., 1969) in the α-crystal-line gel phase. It was apparent, however, that the headgroups had considerable mobility, especially near the thermal phase transition (Veksli et al., 1969).

In the liquid-crystalline phase, linewidths of 10 μT were obtained from the chain protons and choline headgroup, and lines of $0\cdot1$ mT from the more rigid glycerol backbone. In the liquid crystalline phase, it was found that the $-N^+Me_3$ headgroup in lecithins was highly mobile.

As well as these frequency domain studies, time domain studies have been performed on the gel and liquid-crystalline phases of lecithins and membrane lipids (Oldfield et al., 1971). Spin–spin relation data have been interpreted in terms of a purely dipolar origin of the observed linewidths, at least in fields of up to ~ 2T (Chan et al., 1971).

Spin–lattice relaxation (T_1) in the laboratory frame (Daycock et al., 1971) and rotating frame ($T_{1\rho}$) (Salsbury et al., 1970) have also been applied to studies of lipid mobility, and moderately good correlations with CW data have been obtained. This aspect of spin-dynamics is, however, complicated by the process of spin-diffusion, which occurs in these non-sonicated systems.

2. Proton High-resolution NMR

High-resolution NMR studies of molecular mobility were initially hampered by the low S/N obtained on ^{1}H at 60 MHz. Increased use of higher fields (220 MHz) have effectively solved this problem, and it has been demonstrated that well-resolved high-resolution proton NMR spectra from smectic liquid-

* See note 2, p. 144.
† 1 T=1 Tesla≡10 K gauss.

crystalline lipids, can be obtained. Moderate resolution of an egg-yolk lecithin spectrum has been obtained (Chan *et al.*, 1971), though some lipids give considerable better resolved spectra. For example, (Oldfield and Chapman, 1972) sphingomyelin ($T_m = 40°C$) at 60° shows $-N^+Me_3$, $(CH_2)_n$ and CH_3 signals (Fig. 9),.

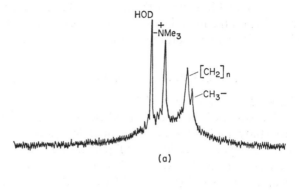

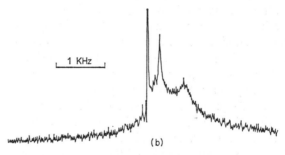

FIG. 9. 220 MHz spectra of 20 wt % hand dispersions in D_2O of (a) sphingomyelin 60°C, and (b) sphingomyelin–cholesterol (1 : 1) 60°C (Oldfield and Chapman, 1972).

A sample of dipalmitoyl lecithin in which the lipid chains were deuterated so as to remove the signal arising from the $[CH_2]_n$ chain protons on heating through the gel–liquid-crystal phase is interesting in that it clearly demonstrates the development of the signal associated with the $N^+(CH_3)$ group (Oldfield *et al.*, 1971). This result is consistent with the wide-line studies which we have previously discussed.

This demonstrates the freedom of the motion of $N^+(CH_3)_3$ group. This contradicts a previous suggestion that this group is fixed in position in lecithin bilayer systems. This is also shown in the transition of dimyristoyl lecithin at different temperatures. As the main transition temperature of 23°C is approached, the $N^+(CH_3)_3$ signal first appears then the signals from the terminal CH_3 and the $[CH_2]_n$ groups (see Fig. 10).

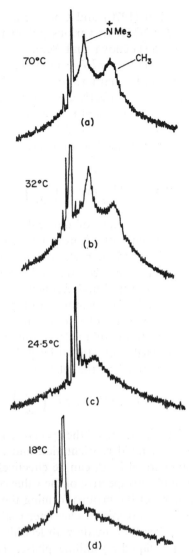

FIG. 10. 220 MHz ^{1}H NMR spectra of dimyristoyl dispersions in D_2O at different
temperatures above and below the transition temperature.

3. *Spinning Phospholipids at the "Magic Angle"*

It has been known for some time that the dipolar broadening of NMR
signals in solids can be largely eliminated by macroscopic rotation of the
sample about an axis inclined at the "magic angle" $\theta = \cos^{-1}(\frac{1}{3})^{1/2} = 54 \cdot 7°$
with respect to the applied magnetic field.

Doskočilová and Schneider (1970) and Cohn *et al.* (1967) have extended these experiments to *proton* NMR in various polymers, e.g. polybenzylglutamate and polystyrene. Doskočilová and Schneider (1970) have recently used Gutowsky and Pake's (1950) *ad hoc* modifield Bloembergen, Purcell and Pound's (1948) treatment to characterize the motional requirements for line narrowing experiments to be effective, and Andrew and Jasinski (1971) have treated the problem in greater depth using the approach of Kubo and Tomita (1954). The simple treatment is as follows.

Line width $\Delta\nu$ (arbitrarily defined with respect of line shape) can be expressed as

$$(\Delta\nu)^2 = A^2 \frac{2}{\pi} \text{ arc } tg \left(\frac{\alpha\Delta\nu}{\nu_c} \right) \tag{1}$$

where ν_c is the correlation frequency of molecular motion, A is the rigid lattice line width and α is a constant of order unity. Partial narrowing of a resonance line may occur by random motion, *isotropic* in space, such that $\nu_c > \Delta\nu$. In such cases, macroscopic rotation of the sample could further reduce the line width only if rotational frequencies $\nu_{rot} > \nu_c$ are reached. This is generally impracticable for protons. But partial narrowing of a resonance line may also occur by rapid internal motion which is *anisotropic* in space (e.g. oscillation of groups of nuclei about fixed axes, random re-orientation of ionic groups fixed at crystal lattice points, etc.). In these cases, line width can usually be described by the relation

$$(\Delta\nu)^2 = B^2 + C^2 \frac{2}{\pi} \text{ arc } tg \left(\frac{\alpha\Delta\nu}{\nu_c} \right) \tag{2}$$

where ν_c is the correlation frequency of the specialized motion, B is the line width in the extreme of very rapid reorientation and $B^2 + C^2 = A^2$. Residual dipolar interactions, summarized in B, can be effectively reduced by macroscopic rotation of the sample, irrespective of the value of ν_c.

Now the motion of many of the groups forming the lipid structure in the liquid-crystalline phase is anisotropic in character (Salsbury and Chapman, 1968). Experiments were, therefore, made with lecithin–water systems in the gel phase and also in the liquid-crystalline phase. The experiments were carried out with a 60-MHz spectrometer.

Dipalmitoyl lecithin at 25°C in excess water is in the α-crystalline gel state (Chapman *et al.*, 1967), and is well below its T_c transition temperature. Only a broad HOD band was detectable at 60 MHz on the sweep widths employed. On rotation at the "magic angle" some narrowing of the adsorption of the DPL gel was observed at 3·5 KHz, yielding a band of 250 Hz width at 6·8 τ assigned to the $N^+(CH_3)_3$ protons (Chapman and Morrison, 1966), in addition to the very sharp HOD line at 5·2 τ (Fig. 11a).

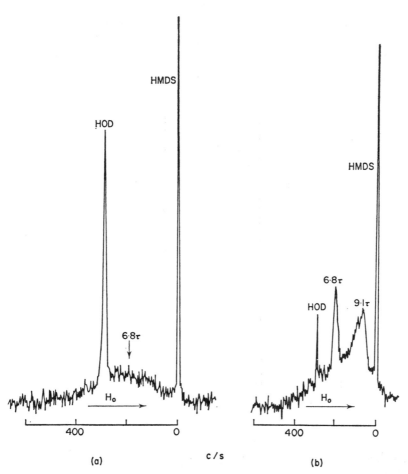

FIG. 11. 60 MHz ^{1}H NMR spectra of (a) dipalmitoyl lecithin (50 wt %)–D$_2$O 25°C (gel state) $\nu_{rot} = 3.5$ KHz $\beta = sec^{-1} \sqrt{3}$; (b) egg-yolk lecithin (33 wt %)–D$_2$O 25°C (liquid-crystalline state) $\nu_{rot} = 3.5$ KHz $\beta = sec^{-1} \sqrt{3}$ (Chapman *et al.*, 1972).

With egg-yolk lecithin, a broad ($\sim$500 Hz) absorption was observed, similar to that reported previously (Penkett *et al.*, 1968). On rotation at the "magic angle", some structure was apparent at frequencies as low as 300 Hz. At 1 KHz the N$^+$(CH$_3$)$_3$ resonance at 6·8 τ reached a linewidth of 17 Hz, which was rotationally invariant at higher spinning speeds. The HOD signal was only 2·5 Hz. A prominent signal was also apparent at 9·1 τ, and is assigned to terminal methyl groups on the lipid acyl chains. The width of this resonance is difficult to estimate because of overlap with resonances from near-terminal methylene groups of the lipid chain. The resolution, however, continues to improve up to 3·5 KHz (Fig. 11b). Any purely mechanical effect upon linewidth is excluded by comparison of spectra measured with rotation about an axis (a) perpendicular and (b) inclined at the "magic angle" with respect of H_0, at equal spinning frequencies, in two consecutive runs spaced by a few seconds (Fig. 12).

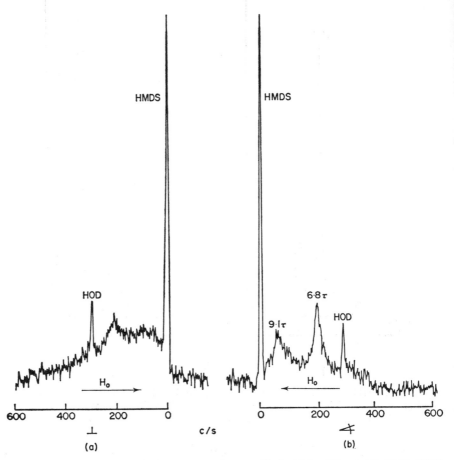

FIG. 12. 60 MHz ^{1}H NMR spectra of egg-yolk lecithin (33 wt %)–D$_2$O 25°C (liquid-crystalline state) $\nu_{rot} = 0\cdot5$ KHz (a) $\beta = 90°$; (b) $\beta = \sec^{-1} \sqrt{3}$ (Chapman et al., 1972).

In the gel phase, only the HOD and $N^+(CH_3)_3$ protons are mobile enough to have static linewidths not exceeding attainable spinning speeds. The residual linewidth of the $N^+(CH_3)_3$ protons, which is a measure of the frequency and form of the rapid reorientations of this group (Doskočilová and Schneider, 1972; Andrew and Jasinski, 1971; Woessner, 1962), is of the order observed in amorphous solid polymers (Doskočilová and Schneider, 1970; Schneider *et al.*, 1972).

In the liquid-crystalline phase of egg lecithin, the $N^+(CH_3)_3$ groups have the narrowest linewidth (~ 50 Hz at 60 MHz), and this is further narrowed by "magic angle" rotation at low frequencies. This indicates (Doskočilová and Schneider, 1972; Haeberlen and Waugh, 1969) that the spectrum of microscopic motions in this liquid-crystalline lipid contains modes with correlation times $> 10^{-4}$ sec. The limiting linewidth of the $N^+(CH_3)_3$ protons is only 17 Hz, more than an order of magnitude less than in the gel phase. This decrease of limiting linewidth may be caused both by increased frequency and by decreased anisotropy of $N^+(CH_3)_3$ group reorientations in the liquid-crystalline phase (Veksli *et al.*, 1969). Of the other proton groups, only the CH_3 and perhaps 1 or 2 terminal CH_2 groups of the aliphatic chains are expected, and observed, to narrow at the rotational frequencies used (max. 3·5 KHz). Since considerable line narrowing occurs at (or near) the "magic angle" at these relatively low rotational frequencies, it is possible that tumbling of *sonicated* lecithin vesicles could be significant in the production of the well-resolved spectra obtained with these systems as suggested by Finer *et al.* (1972) and Penkett *et al.* (1968).

These preliminary results indicate that internal motions of the lipid system can begin to be characterized and that further attempts at line narrowing experiments of these and other lipid membrane systems are well worthwhile. Similar studies on ^{13}C nuclei in higher fields, with lipids and membranes, where resolution is enhanced by an increased chemical shift range, may be particularly useful. Thus the resolution of the different carbon nuclei along the lipid chain in non-sonicated lipid systems and membrane systems should be further improved using the spinning experiment. In the case of cell membranes, useful information about the frequency and type of motion of particular groups in the intact membranes may become available.*

4. Sonicated Systems

Sonicated dispersions of phosphatidylcholine and phosphatidylserine give much narrower lines than unsonicated lipids, and the spectra can be easily observed and studied with a 60 MHz spectrometer. Similar spectra are occasionally obtained from lecithin without sonication, but only after lengthy homogenization. Increasing the time of sonication of a coarse dispersion of egg-yolk phosphatidylcholine causes the high resolution signals to grow at a steady rate (Penkett *et al.*, 1968).

Now the fact that a high resolution NMR spectrum is obtained with sonicated lipids at 60 MHz, whereas normally only broad resonances are observed, could imply that the process of sonication has destroyed *all* the

* See note 3, p. 144.

lamellar or bilayer structure and that a completely random or disordered system has been formed, somewhat akin to a liquid. Sheard (1969) had suggested that this was the case, whilst Chapman *et al.* (1968) considered that sonication broke up the coarse lipid into small fragments but a lamellar arrangement was retained. X-ray scattering studies by Wilkins *et al.* (1971), of sonicated lipids have recently shown quite definitely the existence of lamellar order in these sonicated lipid dispersions (see Chapter 1, p. 14).

It is also of interest to examine a sonicated dispersion of a pure lipid, e.g. dimyristoyl lecithin in D_2O at different temperatures. The high resolution PMR spectra of dimyristoyl lecithin in D_2O at various temperatures as the lipid approaches the *bulk* transition temperature at 23°C are shown in Fig. 13.

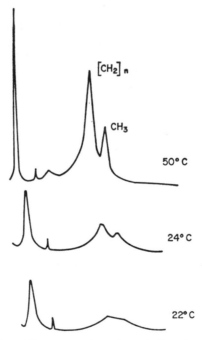

FIG. 13. High resolution ${}^1H$ spectra of sonicated dimyristoyl lecithin in D_2O at temperatures above and below the transition temperature (Chapman, 1972).

At the bulk transition temperature the chain $[CH_2]_n$ signal broadens out corresponding to the crystallization of the lipid chains. This spectral behaviour also appears to be reversible (Chapman, 1972).

The sonication process therefore breaks up the large coarse aggregates into small spherical particles of size about 250 Å (Chapman *et al.*, 1968) in which there is a single bilayer forming the outer shell. Recently these sonicated lipid dispersions have become increasingly popular as model membrane systems. This means that we have a model membrane system providing high resolution signals for all the proton groups in the lipid molecule. It means that in principle one can study the interaction of drug molecules, polypeptides and other molecules with particular groups of the lipid by studying the differential line broadening of the spectrum.

Russian workers (Bergelson et al., 1970) have recently examined these sonicated lipid systems and used the line broadening effects of paramagnetic ions to provide information about this system. They show that adding Mn^{2+} ions to the sonicated lipid–water system causes a decrease in the intensity of the signal associated with the protons of the $N^+(CH_3)_3$ signal—in fact to half its original intensity. This is because the effect of the Mn^{2+} ions is felt by the $N^+(CH_3)_3$ groups on the outside of the lipid particle.

On the other hand, sonication of the lipid in the presence of the Mn^{2+} ions allows these ions to occur inside the spheres and then the whole of the $N^+(CH_3)_3$ signal disappears.

These studies show: (a) that the particles are closed spheres; (b) that the outer shell is indeed in the form of a lipid bilayer; and (c) how one can distinguish between outer and inner $N^+(CH_3)_3$ groups with these model systems.

The use of higher frequency spectrometers, 220 MHz, with these sonicated lipid dispersions has also enabled both inner and outer $N^+(CH_3)_3$ components to be observed, particularly when ions such as $K_3Fe(CN)_6$ are added.

When we consider the reason why these particles give rise to narrow high resolution signals, it has been suggested.

(i) that although the sonicated lipid system is still a bilayer system that there has been some modification of molecular motion in the system. It is interesting from this point of view to note that spin label studies of coarse and sonicated phosphatidylcholine show little difference in hydrocarbon chain reorientation rates (Hubbell and McConnell, 1968; Barratt et al., 1969);

(ii) or/and that the smaller and more symmetrical particles, whilst maintaining the bilayer system, have sufficient isotropic and rapid particle reorientations to average out the residual dipolar interactions. The spinning experiments at the "magic angle" demonstrate experimentally that this can, in principle, occur.

Finer et al. (1972) have presented theoretical arguments in favour of particle rotation being an important effect. Discussion has, however, taken place concerning differences in lipid organization in the sonicated vesicles compared with unsonicated lipid systems. It is clear from the geometry of the small vesicles that some differences in packing may occur and that greater mobility and molecular freedom at the centre of the bilayer system in the sonicated vesicle may occur. Thus a combination of this effect and spinning particle effect may be responsible for the increased resolution observed.

A feature of these sonicated systems which introduces some limitations upon their use for studies of model membrane systems using NMR spectroscopy needs to be considered. This is that if addition of a molecule, such as a polypeptide or a drug molecule, to the sonicated system caused a *marked* slowing of the tumbling rate of the sonicated particle, then this alone could cause a differential line broadening effect to occur. This can be seen by consideration of the results of spinning phospholipids at the "magic angle". This effect would be in addition to any differential effect associated with structural inhibitions of particular molecular groupings due to the drug lipid interaction.

5. *Phosphorus NMR*

Preliminary studies of phosphorus magnetic resonance have also been made with lecithin dispersions (Horwitz and Klein, 1972). Long values of T_1 and shorter values of T_2 were observed. The linewidth was observed to change upon sonication. No detailed interpretation of the results was given. (The results of the spinning experiments, p. 107, need to be considered in interpreting these results.)

6. *Carbon—13 NMR*

Because of the relatively broad and overlapping signals obtained using ^{1}H-NMR, there is at present much interest in the use of ^{13}C-NMR (carbon magnetic resonance, CMR). ^{13}C has a smaller gyromagnetic ratio than ^{1}H, and is less susceptible to dipolar broadening. In addition, the chemical shift range is large (~ 200 ppm typically), and spin diffusion in natural abundance does not occur so that the interpretation of T_1 measurements is simplified.

CMR of smectic liquid-crystalline lecithin in H_2O indicate that the choline-N^+Me_3 group is relatively mobile, though unfortunately a broad envelope of resonances for the methylene carbons is still obtained with the unsonicated lipid. However, unsaturated CH=CH residues in the hydrocarbon acyl groups are well resolved from the main envelope.

Studies of sonicated dispersions have been made by Levine *et al.* (1972). As is observed with proton NMR, these show narrower and better resolved signals than occurs with the unsonicated lipid. The observed T_1 values increase by more than an order of magnitude from the glycerol group towards the terminal methyls of the fatty acid chains and also increase towards the $N^+(CH_3)_3$ group. These workers suggest that chemical shift anisotropy rather than diple–dipole interactions is responsible for the broad envelope of resonances observed with the unsonicated lipid. They also suggest that only the outer lamellae of the unsonicated lipid vesicles can give rise to narrow signals.

7. Deuteron Magnetic Resonance

Nuclei with spin $I \geqslant I$ have an associated electric quadupole moment. Under some circumstances, this gives rise to a well-defined splitting of the nuclear Zeeman levels. For deuterium, $I=1$ and the observed splitting of the NMR absorption line is

$$\Delta\nu_{max} = \frac{3}{4} \frac{e^2qQ}{h} < 3 \cos \theta - 1 >$$

where e^2qQ is the quadrupole coupling constant (170 KHz for CD_2 groups) and θ is the angle between the laboratory field and the electric field gradient tensor at the nucleus.

Chain deuterated dimyristoyl lecithin

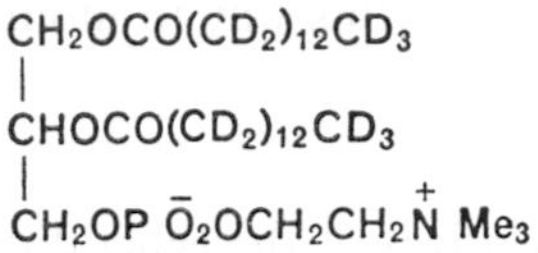

$$CH_2OCO(CD_2)_{12}CD_3$$
$$|$$
$$CHOCO(CD_2)_{12}CD_3$$
$$|$$
$$CH_2OP\ \bar{O}_2OCH_2CH_2\overset{+}{N}\ Me_3$$

(in its gel state) at 10°C gives a broad spectrum, composed of overlapping CD_2 group doublets (Oldfield et al., 1971), implying that different groups are subject to different rates or types of motion, down the chain. Just above the transition at 23·5°C, a maximal quadrupole splitting $\Delta\nu_{max} = 29·8 \pm 1$ KHz is apparent, and at 30°C the splitting is 27 ± 1 KHz. From the shape of the spectrum it is apparent that a relatively wide correlation time distribution is present down the alkyl chains. CD_2 groups near the polar/apolar interface are relatively restricted in their molecular motion, whilst those at the methyl terminal end have greater mobility. This is consistent with earlier studies based on ^1H NMR studies (Chapman and Salsbury, 1966; Salsbury and Chapman, 1968).

C. ELECTRON SPIN RESONANCE STUDIES

The technique of spin-labelling with nitroxide substituent bearing molecules introduced by McConnell and co-workers (Griffith et al., 1965; Stone et al., 1965), has found widespread application in model-membrane structure studies. Three basic types of experiments have been performed—1, solubility studies; 2, studies on the correlation time distribution down spin-labelled molecules, and 3 diffusion of spin-label studies.

1. The Nitroxide TEMPO (2, 2, 6, 6-tetramethyl piperidine-1-oxyl)

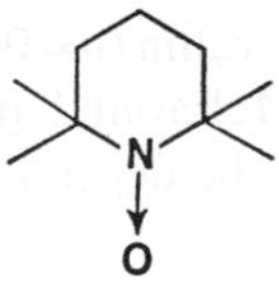

is water soluble. It has been used to study the gel→liquid-crystal phase transition of dipalmitoyl lecithin in H_2O, since it has a high solubility in the fluid liquid crystal, and a low solubility in the gel (Hubbell and McConnell, 1971). The label

(the 2,4,-dinitrophenyl hycrazone derivative of 2,2,6,6-tetramethyl piperid-4-one-1-oxyl) has also been used in this way to monitor the gel→liquid-crystal transition of dipalmitoyl lecithin.

2. *Mobility in Bilayers*
Spin-labelled fatty acid derivatives of the general type

$$CH_3-(CH_2)_m-C-(CH_2)_n-CO_2R$$

(R = H, Me or a lipid residue)

have been used to study the rates and types of motion at different regions along the hydrocarbon chains in bilayers in both liposome dispersions (Hubbell and McConnell, 1971; Oldfield and Chapman, 1971; McFarland and McConnell, 1971) and in oriented systems (Jost *et al.*, 1971a).

When spin labelled fatty acids or lecithins are incorporated into liquid-crystalline egg lecithin bilayers, it is found that the decrease in order parameter S (defined as $=0$ for an isotropic liquid and $=1$ for a perfect crystalline solid) is greater than logarithmic with increasing n, the number of methylene groups separating the label from the polar headgroup (Hubbell and McConnell, 1971). This increased motion towards the centre of the bilayer is difficult to reconcile with a model in which the hydrocarbon chains are in a parallel array, and it has been suggested, using a spin-labelled phospholipid label, that a net tilt of $\sim 30°$ is present in the headgroup region of hydrated egg lecithin multilayers (McFarland *et al.*, 1971), producing a carbon atom density $\sim 12\%$ higher than the carbon atom density near the terminal methyl groups —which is the order of magnitude density change observed between liquid hexadecane ($\rho = 0.77$) and solid paraffin ($\rho = 0.88$). Calculations have suggested this tilted region has $\tau > 10^{-8}$ s (McFarland and McConnell, 1971). The extent of this tilted region is expected to be dependent on many structural factors of the lipid, and the temperature.

3. Diffusion Experiments

(a) *Flip-flop across a bilayer*. The exchange of a spin-labelled phospholipid across a lipid bilayer (single bilayered vesicles produced by prolonged sonic irradiation) has been measured by selectively destroying label on one side of the vesicle by ascorbate (Kornberg and McConnell, 1971). The exchange rate was measured and was found to be slower than 2×10^{-5} times/s.

(b) *Lateral diffusion*. Using a headgroup spin-labelled lecithin in sonicated single bilayer vesicles of di(dihydrosterculoyl)-lecithin (sonication in the presence of unsaturated lipids causes rapid loss of label paramagnetism), it has been shown that by analysis of the line-broadening of the $-N^+Me_3$ group caused by the presence of the label, that the frequency of molecular jumps leading to lateral diffusion must be greater than 3×10^3 jumps/s (Kornberg and McConnell, 1971).

D. INFRARED AND LASER–RAMAN STUDIES

1. Infrared Spectroscopy

Early work indicated that infrared spectroscopy was an excellent technique for studying hydrocarbon chain mobility in lipids, and for the study of their thermotropic mesomorphism. The importance of *cis/trans* isomerism in determining transition temperatures was demonstrated with the dielaidoyl (*trans*-octadec-9-ene-oyl) and dioleyl (*cis*-octadec-9-ene-oyl) phosphatidyl-ethanolamines (Chapman *et al.*, 1966). The former is crystalline at room temperature and the latter is liquid crystalline, and this results in broadened absorption bands characteristic of liquid films. The band at 720 cm^{-1}, assigned to a rock mode of 4 or more all *trans* CH$_2$ groups, decreases in intensity on going from crystal → liquid crystal, and continues to decrease in intensity on further heating, due to the larger number of conformers occurring at higher temperatures.*

Infrared studies in water or D$_2$O are difficult, due to overlapping solvent bands, however, preliminary results have been reported with phosphatidyl-ethanolamines at low water concentrations (Bulkin and Krishnamachari, 1970).

2. Raman Spectroscopy

A particular advantage of (laser) Roman spectroscopy is that the effect depends on the change in polarizability of the molecule during a vibration. Symmetric vibrations that are infrared inactive can thus be seen in the Raman spectrum. In addition the absorption due to H$_2$O around 1600 cm^{-1} is of low intensity.

Laser–Raman studies of the thermal phase transitions in lecithin and

* See note 4, p. 144.

F

lecithin–cholesterol (Lippert and Peticolas, 1971) have shed light on the cooperativity of the transition. Anhydrous dipalmitoyl lecithin has been suggested to undergo a non-cooperative thermal phase transition (a *broad* thermal phase transition is observed for anhydrous distearoyl lecithin, Fig. 96), as does dipalmitoyl lecithin–cholesterol (1 : 1), in excess water. The low cooperativity of the gel → liquid-crystal phase transition at intermediate cholesterol concentrations has been suggested recently by Träuble (1971).

Hopefully, by use of deuterium substitution, it should be possible to shift *selected* resonances to much lower energies so that it might be possible to detect the motions of different parts of a molecule in similar, or even more complex systems.

III. Studies of Lipid–Cholesterol Interactions

Leathes orginally observed (1925) that monolayers of erythrocyte phospholipids are condensed by the addition of cholesterol, and this effect has been confirmed in a variety of model membrane systems (De Bernard, 1958; Chapman and Penkett, 1966; Demel *et al.*, 1967; Cadenhead and Phillips, 1968; Rand and Luzzati, 1968; Chapman *et al.*, 1969; Lecuyer and Dervichian, 1969). Molecular complexes have been suggested by some workers (Finean, 1953; Vandenheuvel, 1963) whilst Shah and Schulman (1967) conclude that there is no interaction between lecithin and cholesterol.

A. THERMAL TECHNIQUES

The effect of cholesterol on the gel → liquid-crystal transition of several lipids has been studied using calorimetric methods. The addition of cholesterol to dipalmitoyl lecithin in water (Fig. 14) causes a lowering of the transition temperature over and above 20 mol % cholesterol and a decrease in the heat of transition (Ladbrooke *et al.*, 1968). At high (50 mol %) cholesterol concentrations the DSC endotherm is completely removed. A condition of "intermediate fluidity" is produced. Below T_c, in the presence of cholesterol, the chains are more mobile than in the absence of cholesterol and above T_c they are less mobile. Above T_c the steroid nucleus effectively prevents flexing of the lipid hydrocarbon chains and below T_c it prevents them from crystallizing into the rigid α-crystalline gel condition. Although the transition from gel → liquid-crystal is not detectable by DSC, laser-Raman evidence suggests that a transition *still* takes place, though over a very wide temperature range, and it is now a *non-cooperative* phenomenon (Lippert and Peticolas, 1971).

The removal of gel → liquid-crystal transitions by cholesterol is not restricted to lecithin, and has been demonstrated for other phospholipids (e.g. sphingomyelin) and glycolipids (e.g. cerebroside (Oldfield and Chapman,

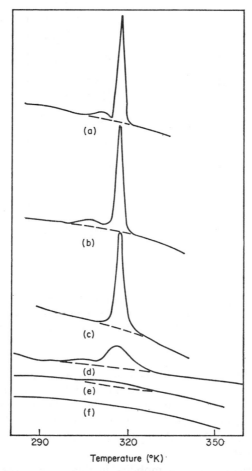

FIG. 14. Differential scanning calorimetry curves of 50% wt dispersions of dipalmitoyl lecithin–cholesterol mixtures containing (a) 0 mole %; (b) 5 mol %; (c) 12·5 mol %; (d) 20 mol %; (e) 32 mol %; (f) 50 mol % cholesterol (Ladbrooke *et al.*, 1968).

1972)). The effect of cholesterol on the "bound water" has also been discussed (Ladbrooke *et al.*, 1968).

B. NUCLEAR MAGNETIC RESONANCE STUDIES

The first application of NMR spectroscopy to the study of lipid–cholesterol interactions was made by Chapman and Penkett (1966) using a sonicated lipid system. It can be seen from Fig. 15 that cholesterol causes differential broadening in the spectrum of sonicated dipalmitoyl lecithin, in that the

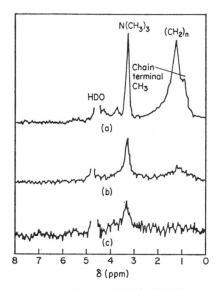

Fig. 15. 60 MHz ^{1}H spectra of dipalmitoyl lecithin (5 % in D_2O sonicated dispersion). (a) at 45°; (b) +cholesterol 1 : 1 at 33°C; (c) at 33°C. The intensities in spectrum (c) have been expanded compared with (a) and (b) (Darke *et al.*, 1972).

$N(CH_3)_3$ signal is broadened far less than the alkyl chain signal. The same effect is observed with egg-yolk lecithin (cf. Chapman and Penkett, 1966) which contains hydrocarbon chains of different lengths and degree of unsaturation. This was interpreted to show that there is a considerable reduction in motion of the majority of the hydrocarbon chain protons compared with egg-yolk lecithin alone.

Since linewidths depend on mobility in this system, the overall state of mobility of dipalmitoyl lecithin molecules in a codispersion with cholesterol, over a wide range of temperature, can be seen from Fig. 15 to be intermediate between the states of lecithin alone above and below its chain-melting temperature (Darke *et al.*, 1972). The choline headgroup still has some mobility in dipalmitoyl lecithin alone just below the chain-melting temperature, from the weak $N(CH_3)_3$ signal remaining (Fig. 15c). This signal disappears on standing, or cooling to 20°C, and reappears on warming back to 33°C; it may be correlated with the small pre-transition peak observed 10°C below the hydrocarbon chain-melting point by Ladbrooke and Chapman (1969), using differential scanning calorimetry. These authors also postulated rotation of the polar part of the lecithin molecule just below the chain-melting temperature.

The residual hump at $\delta 1$ ppm in the lecithin–cholesterol codispersion visible on the normal high resolution scale (Fig. 15b) arises solely from lecithin chain protons. This was shown by replacing the dipalmitoyl lecithin by a lecithin with deuterated chains, when the hump disappeared. Thus some portions of the lecithin chains have considerably more mobility than the cholesterol molecules themselves (cholesterol protons give a signal too broad to be observed on the high resolution scale). The non-polar ends of the chains become more mobile as the temperature is raised above the chain-melting temperature. This was demonstrated by a study of an equimolar egg-yolk lecithin–cholesterol mixture from 10 to 70°C, which showed a steady decrease in linewidth, and increase in area, of the residual hump as the temperature increased.

220 MHz spectra of unsonicated lipids have also been useful in studying lipid–cholesterol interactions (Oldfield and Chapman, 1972). This is shown in Fig. 16.

At 60°C the sphingomyelin is liquid crystalline, and a very well-resolved spectrum is obtained, from a hand-dispersed sample. Both the choline

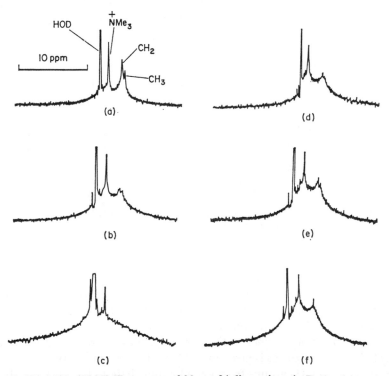

FIG. 16. 220 MHz ^{1}H NMR spectra of 20 wt % dispersions in D_2O of (a) sphingo-myelin 60°C; (b) 40°C; (c) 20°C; (d) sphingomyelin–cholesterol (1 : 1) 60°C; (e) 40°C; (f) 20°C (Oldfield and Chapman, 1972).

— N^+Me_3 and Chain CH_2 and CH_3 groups are apparent (Fig. 16a). The choline — N^+Me_3 group is highly mobile, as has been previously noted for the choline — N^+Me_3 group in lecithin in its smectic liquid-crystalline phase in H_2O, using pulsed NMR and ^{13}C Fourier transform NMR.

On the phase transition, at 40°C (Fig. 16b), there is a decrease in the intensity and a broadening of the chain proton signals. On such a broad thermal phase transition both gel and liquid crystalline regions are present (due to the wide variety of N-acyl groups present), and on the sweep width employed (5000 Hz), only the more mobile and liquid-crystalline regions will be resolved.

At 20°C (Fig. 16c) it is clear that the lipid chains are relatively immobile since no chain signal is observed.

On addition of cholesterol to sphingomyelin above the temperature range of the phase transition, there is a dramatic broadening (Fig. 16d) of the CH_2 signal of the acyl chains, indicating a great restriction in their motional freedom. At 40°C (Fig. 16e) the spectrum is rather similar to the 40°C sphingomyelin spectrum (Fig. 16b), and corresponds to only a limited number of mobile methylene groups, probably near the methyl terminal ends of the N-acyl chains, many of the methylene groups being greatly immobilized by the rigid cholesterol ring structure. Below the range of the sphingomyelin transition, at 20°C (Fig. 16f), the spectrum of sphingomyelin–cholesterol (1 : 1) is rather similar to that obtained at 40°C and 60°C,

Wide line proton spectra of *unsonicated* dipalmitoyl lecithin and equimolar cholesterol–dipalmitoyl lecithin dispersions in deuterium oxide (44 wt %) have also been studied by Darke *et al.*, 1972. The dotted lines in Fig. 17 show the temperature dependence of the linewidths of the hydrocarbon chain proton resonance and polar group proton resonances, respectively, in the absence of cholesterol. The gel-to-liquid crystal transition at 41°C results in a dramatic narrowing of these lines (Veksli *et al.*, 1969). In the presence of equimolar amounts of cholesterol, four lines plus a narrow central component (water plus two "semi-high resolution" lines chemically shifted to high field) are observed. The temperature variation of the linewidths in this spectrum is shown in Fig. 17, and it is apparent that the widths are remarkably temperature independent over the range 10 to 60°C.

When the equivalent spectra were obtained with dimyristoyl lecithin with perdeuterated chains, the same four lines were observed in the proton spectrum, but with different relative intensities.

An assignment of the lines to the various groups of protons in the lecithin and cholesterol molecules is given in Table II. The best intensity fit is obtained by assigning ten CH_2 groups from each lecithin chain to lines (a) and (b) and the others to line (c). The cholesterol backbone and short chain (attached to C_{17}) are considered to give one resonance, as in crystalline cholesterol, where

Assignment of four wide lines observed in equimolar lecithin–cholesterol dispersions

Line	Linewidth at 30°C for 1 : 1 mixture of dipalmitoyl lecithin and cholesterol (gauss)	Assignment	Linewidth in unmixed state (gauss) 30°C	Above lecithin chain-melting temperature	Intensities[a] Dipalmitoyl lecithin and cholesterol 50°C calc.	obs.	Deuterated dimyristoyl lecithin and cholesterol 25°C calc.	obs.
(a)	5·2 ⎱	Cholesterol backbone and chain.	11·5	—	30	—	30	—
		CH₃'s on cholesterol ring system.	4·5	—	6	—	6	—
(b)	3·4 ⎰	10 CH₂'s on each lecithin chain.	4·0	~0·05	40	—	0	—
					‾76	77	‾36	33
(c)	1·2	Other lecithin chain CH₂'s.	4·0	~0·05	16	—	0	—
		Cholesterol chain CH₃'s.	4·5	—	9	—	9	—
					‾25	25	‾9	11
(d)	0·2	Lecithin N(CH₃)₃.	0·4	~0·05	9	—	9	—
		Lecithin chain CH₃'s.	4·0	~0·05	6	—	0	—
		Lecithin polar backbone.	4·0	1·4	9	—	9	—
					‾24	23	‾18	19
Sharp component		Total spectrum from small particles tumbling rapidly.				9%		19 10%

[a]Normalized and expressed in protons/molecule calculated from the known composition of each mixture.
Estimated errors in intensity measurements ±10%.

at 25°C one line of width 11·5 gauss is observed, together with another of 4·5 gauss from the CH3 protons. The measured intensities are best fitted if it is assumed that the methyl groups on the cholesterol chain have more motional freedom than those on the steroid nucleus.

The nuclear magnetic resonance results have been analysed to give a detailed analysis of the location of cholesterol molecules within lecithin bilayers. The variations of mobility with position in the chain in the presence

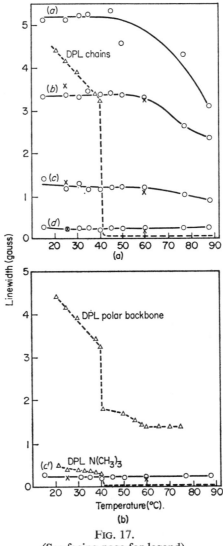

Fɪɢ. 17.
(See facing page for legend).

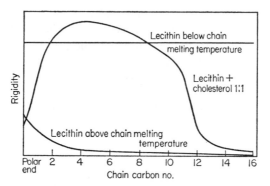

FIG. 18. Diagrammatic representation of variation of fluidity along lecithin hydrocarbon chains in lecithin and lecithin–cholesterol dispersions (Darke *et al.*, 1972).

and absence of cholesterol in equimolar lecithin–cholesterol mixtures are depicted schematically in Fig. 18. The terminal methyl groups, which are the most mobile parts of the chains, are hardly affected by the presence of cholesterol. The ends of the hydrocarbon chains are less rigid than parts lower down towards the polar groups, but all the methylene groups have less motional freedom than in the pure lecithin–water system above its liquid-crystalline transition temperature. Many methylene groups have more motional freedom than when packed in the gel phase of the pure lecithin–water system. Ten methylene groups per hydrocarbon chain have similar mobility to that of the cholesterol steroid nucleus and chain while the rest of the lecithin chain is much freer. A reasonable interpretation is that those methylene groups in direct contract with the steroid nucleus are more

FIG. 17. (a) Temperature dependence of widths of lines containing contributions from lecithin chain protons in wide line spectra of dipalmitoyl lecithin dispersions. The dotted line represents results for a dispersion of dipalmitoyl lecithin alone, while the solid lines represent results for the four lines (a), (b), (c) and (d) observed with a dispersion of equimolar lecithin–cholesterol in deuterium oxide (44 wt %). Lines (a) and (b) together contain contribituions from the ten most rigid methylene groups on each chain, line (c) a contribution from the other methylene protons, and line (d) a contribution from the terminal CH_3 groups. 60 MHz (O) and (Δ); 8 MHz (×).

FIG. 17. (b) Temperature dependence of widths of lines containing contributions from lecithin polar group protons in wide line spectra of dipalmitoyl lecithin dispersions. The dotted lines represent results for the glyceride backbone and short polar chain, and $N(CH_3)_3$ group of dipalmitoyl lecithin when dispersed in deuterium ozide, while the solid line represents results for line (d) observed with a dispersion of equimolar lecithin–cholesterol in deuterium oxide (44 wt %). Line (d) contains contributions from the glyceride backbone and short polar chain protons, together with a contribution from the lecithin $N(CH_3)_3$ group. 60 MHz (O) and (Δ); 8 MHz (×) (Darke *et al.*, 1972).

immobilized than those at the ends of the chains. Molecular models indicate that if the cholesterol hydroxyl group is located adjacent to the phosphate group of lecithin molecule, the remainder of the steroid molecule reaches to about the tenth carbon atom along the lecithin hydrocarbon chain. The presence of cholesterol leads to the glycerol backbones of the lecithin molecules having more freedom than they possess when in either the gel or liquid-crystalline phases of the pure lecithin–water system (Table I). This can be understood in terms of a change in packing of the glycerol backbones and associated water by insertion of cholesterol molecules into the lecithin lattice. Despite this, the presence of cholesterol seems to lessen slightly the freedom of isotropic motion of the $N(CH_3)_3$ protons in the liquid-crystalline phase.

C^{13} magnetic resonance has also been applied to lipid–cholesterol mixtures (Keough *et al.*, 1972). In this case, all the resonances except those associated with the choline headgroup are broadened by the presence of cholesterol. The effect on the methylene chain signal is not dramatic, indicating that there must still be substantial motion in the parts of the chains distant from the steroid nucleus. The signals from the penultimate —CH_2— and terminal —CH_3 groups are still well resolved although they show increased linewidth.

The introduction of the steroid could be expected to cause a substantial broadening of those ^{13}C resonances immediately oppositie it, due to less efficient time-averaging of the ^{13}C—^1H dipole–dipole interactions. Carbons $1 \rightarrow (n-2)$ where n is the total carbon length of the chain in question and is generally 18, are appreciably broadened, though carbons n and $(n-1)$ are still quite free in their motion—and thus give rise to well resolved narrow lines. Carbons 9 and 10 (of the predominant olefinic *oley* residues) are appreciable affected, their half height full-linewidths increasing from 71 ± 15 Hz to 161 ± 15 Hz on addition of equimolar quantities of cholesterol. The carbonyl carbon also appears immobilized on addition of cholesterol.

C. ELECTRON SPIN RESONANCE

The spin label method has been used by many authors (see Jost *et al.*, 1971b) to study lipid–cholesterol interactions. The results are generally consistent with the conclusions based on the NMR technique.

Hubbell and McConnell (1971) showed that in the presence of cholesterol egg lecithin : cholesterol, 2 : 1) the *first 8 carbon atoms* from the bilayer surface can be thought of as a *rigid rod*, with the remaining carbons greatly increasing their motion towards the centre of the bilayer. It has also been demonstrated that cholesterol can have a "dual" role in formation of an intermediate-fluid state with several types of lipid class.

Using the label methyl 4-(2'-(*N*-oxyl-4',4'-dimethyl oxazolidine))-stearate (4NS), the unpaired electron in which residues predominantly at the polar/

apolar interface, values of twice the maximal hyperfine splitting ($2T_m$) of 6·1 mT in the gel state of dipalmitoyl lecithin at 20°C, and 4·1 mT in the fluid liquid-crystalline state of egg lecithin (Fig. 19a, b) have been reported

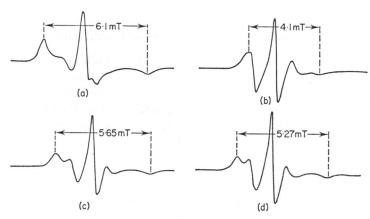

FIG. 19. Electron spin resonance spectra (X-band) of the spin label methyl 4-(2'-*N*-oxyl-4',4'-dimethyl oxazolidine) stearate (4NS) in 7 wt % aqueous dispersions of (a) dipalmitoyl lecithin 20°C; (b) egg-yolk lecithin 20°C; (c) dipalmitoyl lecithin–cholesterol (1 : 1) 20°C; (d) egg-yolk lecithin–cholesterol (1 : 1) 20°C (Oldfield and Chapman, 1971).

(Oldfield and Chapman, 1971). Addition of equimolar quantities of cholesterol results in very similar spectral line-shapes (Fig. 19c, d) with the dipalmitoyl lecithin–cholesterol system having $2T_m = 5\cdot65$ mT and the egg-yolk lecithin–cholesterol system $2T_m = 5\cdot27$ mT. This is consistent with an intermediate fluid liquid-crystalline state being formed from the rigid gel of dipalmitoyl lecithin and the fluid liquid-crystal of egg-yolk lecithin. Noticeably, the egg-yolk lecithin–cholesterol system is *more fluid* than the dipalmitoyl lecithin–cholesterol system at the same temperature (5·27 mT egg-yolk lecithin: cholesterol, 5·6 mT dipalmitoyl lecithin–cholesterol). These interesting results are borne out when the label 12NS is used. In palmitoyl lecithin at 20°C, $2T_m = 5\cdot7$ mT, and in egg-yolk lecithin $2a_N$ (twice the isotropic hyperfine splitting) = 2·8 mT. Addition of cholesterol causes fluidization of the dipalmitoyl lecithin ($2T_m = 4\cdot85$ mT), and immobilization of the egg-yolk lecithin ($2T_m = 4\cdot15$ mT). The egg-yolk lecithin : cholesterol system is more fluid than the dipalmitoyl lecithin : cholesterol system, at the same temperature.

Near the polar groups the mobility of the chains in gel and intermediate fluid liquid crystal is similar (dipalmitoyl lecithin/H_2O/4NS, $2T_m = 6\cdot1$ mT, dipalmitoyl lecithin : cholesterol (1 : 1)/H_2O/4NS, $2T_m = 5\cdot65$ mT, $\Delta T = 0\cdot45$ mT) whereas near the methyl group end of the chains the mobility is quite dissimilar in gel and intermediate fluid liquid-crystal, (dipalmitoyl lecithin/ H_2O/12NS, $2T_m=5·7$ mT, dipalmitoyl lecithin: cholesterol $(1:1)$/H_2O/ 12NS, $2T_m=4·85$ mT, $\Delta T=0·85$ mT).

IV. Studies of Lipid–Polypeptide Interactions

Only a few studies have been made of lipid–polypeptide systems. Although they might be considered as useful model systems to study mainly electrostatic or mainly hydrophobic interaction, the studies that have been carried out are usually associated with ion transport properties. Thus a great deal of work has been carried out on polypeptides and related molecules such as valinomycin, non-actin, gramicidin A and alamethicin.

A. THERMAL STUDIES

In some cases, e.g. gramicidin A and alamethicin, it has been postulated that these molecules actually form channels bridging the width of the bilayer (Urry, 1972; Mueller and Rudin, 1968). Consistent with this idea, freezing the lipid below its transition temperature does not affect the gramicidin A ion-transporting property (Krasne et al., 1971). Thermal calorimetric studies have been made with alamethicin–lipid systems (Ladbrooke and Chapman, 1969) and gramicidin A–lipid systems (Chapman and Saville, 1972). The heating curves obtained in these investigations are shown in Figs. 20 and 21. In both cases, the enthalpy of the lipid transition from gel to liquid crystalline is considerably affected and reduced by increasing the proportion of antibiotic to lipid. Thus the behaviour is somewhat akin to that observed with cholesterol– lipid systems rather than lipid–disiprimine. (In the latter the transition temperature shifts some 20°C, and implies penetration of the lipid chain region by the antibiotic.)

There is good evidence that alamethicin molecules readily penetrate mono- layers and bilayers organization and can modify the lipid. Thus alamethicin molecules are thought to be situated at a lipid–water interface with the polar side chains in the aqueous phase and the non-polar groups among the hydrocarbon chains. The evidence from this orientation comes from the high surface activity of alamethicin observed in monolayer studies and from microelectrophoresis measurements which show that a net negative charge is given to the lecithin particles (Hauser et al., 1970).

The interaction of the molecule gramicidin S with lecithin is an interesting one in that it shows that the polypeptide solubilizes the lipid even when it is at the temperature corresponding to the gel phase (Pache et al., 1972).

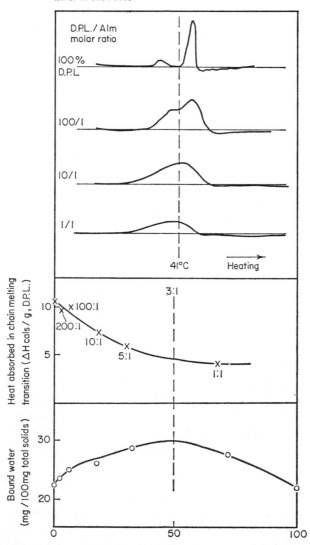

FIG. 20. Differential scanning calorimetry curves of alamethicin–dipalmitoyl lecithin mixtures with associated parameters of heat content (Ladbrooke and Chapman, 1972).

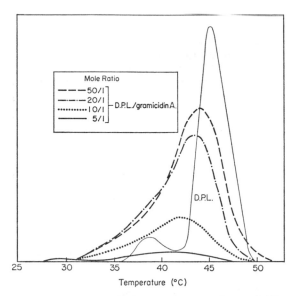

FIG. 21. Differential scanning calorimetry curves of gramicidin A–dipalmitoyl lecithin mixtures (Chapman and Saville, 1972).

V. Studies of Lipid–Protein Interactions

A variety of studies have now been made of model systems incorporating protein into lipid–water systems using a range of physical techniques such as X-ray diffraction, NMR and ESR spectroscopy monolayer and bilayer methods. Here we shall point to only a few of these studies.

Shipley *et al.* (1969) used X-ray diffraction to study the interaction of phospholipids with cytochrome *c* in the aqueous phase. The lipid was a mixture of lecithin and phosphatidylserine with the ratio of phospholipid to cytochrome *c* at about 300 : 1. The presence of cytochrome *c* modifies the swelling characteristics of the lipid dispersion and two lamellar I and II lipoprotein systems were produced which in excess water gave dimensions of 87 Å and 116 Å. These dimensions were consistent with the incorporation of one or two cytochrome *c* molecules (diameter 31 Å) between the lipid bilayers. The presence of water was necessary for the structural integrity of these phases. When the overall ratio of protein to lipid was increased yet another lamellar phase III was observed. The dimensions of these model complexes were observed to be somewhat similar to those of myeline (a prominent second order reflexion at 86 Å). Similarly erythrocyte and mitochondrial membranes on centrifugation to a pellet give diffraction spacings of 115–120 similar to the repeat pattern of lamellar system III (see Chapter 1).

Gulik-Krzywicki *et al.* (1969) also investigated, using X-ray techniques, complexes of ferricytochrome *c* with phosphatidylinositol, cardiolipins, and mitochondria lipids. They also studied lysozome with cardiolipin and also with phosphatidylinositol. A variety of phases were observed. The ferricytochrome *c* complexes were lamellar and their structure were interpreted as indicating protein–water layers between the lipid lamellae, some containing one protein layer and others containing two protein layers. However, other systems such as those containing lysozyme, gave a variety of phases, including hexagonal, dependent upon the nature of the lipid. This work showed that the polymorphism of protein–lipid–water systems may be quite complex.

Rand (1971) has recently studied lecithin–cardiolipin mixtures with bovine serum albumin. He concludes that electrostatic interaction brings the components together so that shorter range polar and hydrophobic interactions can occur causing subsequent gross conformational changes in the protein.

Bilayer methods have also been used to study protein–lipid complexes. Thus Steinemann and Läuger (1971) have studied the interaction of cytochrome *c* with phosphatidylinositol in bilayer systems. At low ionic strength about 10^{13} cytochrome *c* molecules per cm^2 are bound to the lipid surface. It was concluded that the interaction was mainly electrostatic. The fast desorption of the protein after a rise of ionic strength showed that penetration of the protein into the lipid layer does not occur to an appreciable extent.

Clowes *et al.* (1972) have studied the interaction of tetanus toxin with lecithin–ganglioside bilayers. In this case the toxin binds to the bilayer. A measurement of the thickness of the lipid–protein complex showed an increase from 79 ± 2 Å to ~ 119 Å consistent with the toxin maintaining its structure and not forming an extended layer over the surface, as was observed with erythrocyte protein (Cherry *et al.*, 1971).

Cherry *et al.* (1971) studied the interaction of sialic acid free erythrocyte apoprotein with erythrocyte lipids. This showed that the conductance increases up to three orders of magnitude and that the bilayer became-electrically and physically unstable leading to a rupture of the bilayer.

Studies with lipid vesicles have shown that soluble proteins such as lysozyme and cytochrome *c* cause a marked increase in permeability to ions. This has been correlated with the penetration of protein into the hydrocarbon region of the lipid membranes (Kimelberg and Papahadjopolous, 1971a, b). Recently a membrane protein spectrin isolated from erythrocyte ghosts has been studied in these vesicles. When these are added to phosphatidylserine vesicles the ^{23}Na diffusion rate increases. Ca^{2+} and protein exhibit a synergistic effect on phosphatidylserine and also on phosphatidylserine–lecithin vesicles. In contrast to protein alone the effect of calcium plus protein is seen at both neutral and acid pH. The nature of this interplay between protein and calcium is, however, unclear (Juliano *et al.*, 1971).

Thus in a variety of model systems the importance of both ionic and hydrophobic associations has been pointed out. This has also been discussed by Green and Fleischer (1963) for mitochondrial membrane components, by Zwaal and van Deenen (1970) for the recombination of red blood cell lipids and apoproteins, and by Sweet and Zull (1969) for the interaction of albumin and lipid vesicles.

VI. Studies of Membranes

We shall briefly examine chemical evidence and also physical evidence.

A. CHEMICAL EVIDENCE

The ratio of protein to lipid in cell membranes varies from one membrane system to another. This is particularly notable when one compares the composition of myelin—20% protein and 80% lipid—to the composition of the erythrocyte membrane—49% protein and 43·6% total lipid.

With the erythrocyte membrane the total lipid (weight per cent) is very similar in nearly all mammalian erythrocytes. The cholesterol to phospholipid ratio is near 0·9 and the ratio of noncholine to choline containing phospholipid is similar in human, bovine and porcine species. However, the bovine cell contains little lecithin but does have a high content of sphingomyelin whilst in the procine cell there is a particularly high level of glycolipid (12–14%) (Hanahan, 1969). Thus there are differences in the classes of lipids found with various erythrocytes, i.e. with the same type of membrane system. There are also differences within individual classes, e.g. with porcine erythrocytes the level of phosphatidylserine is higher than in other erythrocytes.

There is a marked heterogeneity of the proteins of the erythrocyte membranes. The fraction of nonpolar amino acids present in fractions IV, V and VI is 49 mol % and is greater than in the membrane protein as a whole.

Despite this heterogeneity attempts have been made to determine how protein molecules are arranged in the erythrocyte membrane using chemical methods. Bretscher (1971) has shown, using radioactive labels that attach to proteins with intact cells and cell ghosts, that a major protein of the human erythrocyte membrane of about 105,000 daltons is in part on the external surface of the cell, that different parts of the polypeptide are exposed on the two membrane surfaces. A further part of this polypeptide extends from one side of the membrane to the other. These conclusions are based on the assumption that the detailed molecular structure of the membrane is not substantially different in intact cells and ghosts. It is suggested that this protein has a structural role.

Steck *et al.* (1971) studied vesicles from erythrocyte membranes having normal and inside-out orientation. By treating each species with proteolytic enzymes the two faces of the membrane could be selectively digested. The data suggested a highly asymmetrical arrangement of oriented proteins some of which span the membrane.

An important point which these authors observed was a striking difference in the resistance to proteolysis of the surface of intact erythrocytes and of the isolated membranes. The lipids of intact membranes are much more resistant to phospholipase degradation than those of the isolated membrane (Ibrahim and Thompson, 1965). Zwaal *et al.* (1971) also observed that the phospholipids of intact erythrocytes are resistant to cleavage by highly purified phospholipase-C, whereas the isolated ghost showed cleavage of the lipids. Thus isolation of the erythrocyte membrane increases its reactivity to a variety of agents and shows that caution is required in interpreting studies on isolated membranes.

B. PHYSICAL STUDIES

A whole range of physical techniques have been applied to cell membranes.

1. *Biomembranes with High Cholesterol Levels*

(a) *Thermal studies.* Differential scanning calorimetry studies of both myelin and erythrocyte membrane lipids have been reported (Ladbrooke and Chapman, 1969). With myelin, it was shown that no thermal transition was obtained from "intact" membranes in H_2O, or from total lipids in excess water. Removal of cholesterol from the total lipids resulted in a broad DSC transition encompassing the physiological 37°C, to be observed. This indicates that gel and liquid-crystalline regions are present in the cholesterol depleted lipids, and that an effect of cholesterol is to remove or fluidize the gel lipid areas (and presumably to make the liquid-crystalline regions present, less fluid). Dehydration of the membranes likewise caused a DSC endotherm to be observed, since crystallization of the cholesterol occurred, indicating the importance of water in preserving membrane structure.

Erythrocyte ghosts do not show a phase transition by DSC, although cholesterol depleted lipids do. The observed transition encompasses 37°C. However, an additional small endotherm at low temperatures has been detected (Williams and Chapman, 1970) and attributed to the presence of highly unsaturated species, i.e. the lipids form a kind of monotectic system (Phillips *et al.*, 1970; Bowden, 1954). In the presence of cholesterol, no endotherms are detected from the lipids. Cholesterol may thus have a *dual role* of preventing formation of crystalline gel areas in some membranes whilst also inhibiting the motion of hydrocarbon chains in more fluid, liquid-crystalline, regions.

(b) *Nuclear magnetic resonance studies*. (i) Proton wide-line NMR. Wide-line studies of myelin and erythrocytes have been reported (Veksli *et al.*, 1969; Jenkinson *et al.*, 1969) and show the presence of very wide lines of *ca* 0·2 mT and 0·5 mT (myelin) and 0·2, 0·3, 0·34 and 0·58 mT (erythrocytes), similar to those seen in the lecithin–cholesterol system (Darke *et al.*, 1971), and they may have similar origins.

(ii) Proton high-resolution NMR. It has been reported (McConnell, 1970a) using 200 MHz proton NMR that the choline—N^+Me_3 group in myelin is resolvable, and hence mobile, similar results have been reported for erythrocyte ghosts at 31°C (Kamat and Chapman, 1968). At 18°C (Glaser *et al.*, 1970) no spectrum is observed, though at 75° reversible dissociation of polypeptides from the membrane occurs (Sheetz and Chan, 1970). Care must thus be taken in the treatment of membranes, since it is also known that many ghost proteins are water soluble (Mazia and Ruby, 1968).

Studies on ultrasonically dispersed membranes have also been reported (Jenkinson *et al.*, 1969; Chapman *et al.*, 1968a) and well-resolved spectra have been obtained.

(iii) Carbon-13 NMR. Carbon-13 studies of erythrocyte membranes have been reported (Metcalfe *et al.*, 1971). The resolution is better than that obtained from proton NMR, and appears to indicate relatively mobile choline—N^+Me_3 groups present in the membrane.

(c) *Electron spin resonance*. Spin labelling using labels of the fatty acid or steroid type, e.g. the isoxazolidine derivative of androstan-3-one-17-β-ol

have been used to investigate mobility in erythrocyte ghosts (Hubbell and McConnell, 1969) as has the label TEMPO (Hubbell and McConnell, 1968). TEMPO has a low solubility in erythrocyte membranes, consistent with either strong lipid–protein or lipid–cholesterol interactions. In shear-oriented erythrocyte ghosts (Hubbell and McConnell, 1969) the *N*-oxyl-4′,4′-dimethyl-oxazolidine derivatives of 5α-androstan-3-one-17β-ol and sodium 12-keto-stearate showed relatively high spectral anisotropy. Both spectra were more immobilized than in model sonicated phosphatide dispersion, again indicating possible lipid–cholesterol and/or lipid–protein interactions.

The effect of cholesterol on the mobility of cholestance spin-labelled cholesterol-depleted brain lipids has been reported recently (Butler *et al.*, 1970). It was suggested that the cholesterol caused the hydrocarbon chains to become more "ordered". It may be that there is some preference of the spin label to probe the more fluid liquid-crystalline regions in this type of system (Oldfield and Chapman, 1972; Oldfield *et al.*, 1972b), i.e. spin-label in *liquid-crystalline* regions becomes more ordered.

(d) *Infrared studies*. Infrared studies of myelin and of erythrocyte ghosts, and of their extracted lipids have been reported (Jenskinson *et al.*, 1969; Chapman, 1965; Chapman *et al.*, 1968b).

Erythrocyte lipids show a prominent 720 cm^{-1} band due to a $(CH_3)_n \geqslant 4$ rock mode, consistent with relatively ordered segments of hydrocarbon chain; this is in the presence of cholesterol. Intact membranes lack this distinctive feature, indicating possible lipid–protein interaction.

(e) *Electron microscope studies*. Electron microscope studies using the freeze-etch method have been applied to a number of biological systems.

The fracture faces of most biological membranes (not myelin) appear as smooth sheets interrupted by numerous particles (in erythrocyte membranes, about 85 Å in diameter, uncorrected for shadow thickness). When it became clear that the fracture face represents a plane through the interior of the membrane, it could be concluded that the particles represent localized regions where the bilayer membrane continuum is interrupted. In this context, it seemed legitimate to consider the membrane particles as the morphological representation of entities which are intercalated (but not necessarily seques-tered) in the hydrophobic matrix of the bilayer membrane (Branton, 1971). It is important to note that the particles observed in the erythrocyte membrane account for only about 30% of the total protein of the membrane (see com-ments by Singer, 1971).*

Freeze-etching has been used to follow the course of the proteolytic digestion in red blood cells (Engstrom, 1970). If the particles represent proteins within the membrane matrix, they should disappear when protein is removed from the ghosts before freeze-etching. Red blood cells were incubated in the proteolytic enzyme pronase for varying time periods. The course of protein hydrolysis was monitored by simultaneously measuring protein loss and particle loss. (Preliminary experiments indicated that pronase attacked all of the cells in the preparation with equal vigour.) Loss of membrane protein initially caused extensive particle aggregation with little decrease in particle number. The initial resistance of the particles to pronase digestion is consistent with the view that the particles are buried within the hydrophobic

* See note 5, p. 144.

region of the membrane matrix into which pronase diffusion would be greatly retarded. After more prolonged pronase treatment most of the particles were lost. This eventual removal of particles from extensively digested ghosts is consistent with the hypothesis that the particles are due to protein. These particles are considered to traverse the hydrophobic space of the bilayer and apparently can move when subjected to media of different pH. It is suggested that the plasma membrane is a planar fluid domain formed by a bilayer which is interrupted by localized yet mobile intercalations.

(f) *Circular dichroism.* The circular dichroism spectra of erythrocyte ghosts in the region of 190–230 nm (as well as some other membrane systems) show some of the features characteristic of proteins in a partially helical conformation with certain anomalous features such as low values of (θ) near 190 nm and a red shift about 220 nm. The red shift has been attributed to a number of different structural features of the membrane including protein helix-helix interactions and an apolar environment for the protein helices. There is now convincing evidence that these anomalies are due to optical artefacts arising from the particulate, turbid nature of the samples (Urry, 1972). Urry points out methods of correcting these distortions and suggests that about 50% or more of the red blood cell membrane protein is in an α-helical conformation. This has been interpreted to indicate that the proteins of the membrane are predominently globular rather than spread out over the membrane surface. Models consistent with globular proteins intercalated in the membrane alternating with lipid bilayer structure have been suggested (Lenard and Singer, 1966a, b; Wallach and Zahler, 1966).

2. *Biomembranes with Low Cholesterol Levels*

(a) *Thermal studies.* It has been shown that the "intact" plasma membranes and isolated lipids of *Acholeplasma laidlawii* B exhibit thermal phase transitions from gel → liquid crystal (Steim *et al.*, 1969; Melchior *et al.*, 1970; Chapman and Urbina, 1971). These transitions are all broad, encompassing *ca* 30°C, and thus are of low cooperativity. The thermal transitions of the isolated lipids (Fig. 22) closely resemble the transitions of the membranes themselves* (Steim *et al.*, 1969). Grown in unsupplemented media, the transition range encompasses the growth temperature (Steim *et al.*, 1969; Melchior *et al.*, 1970). Because of the heterogeneity of the membrane lipids (both polar headgroups and hydrocarbon chains) this means that *at the growth temperature, both rigid crystalline gel and fluid liquid-crystalline regions are present in the membrane lipids.*† A similar phenomenon has been observed with the membranes of *Escherichia coli* (Steim, 1970), where for organisms grown at 37°C the thermal transition extends from ∼15°C → 45°C. Again,

* See note 6, p. 144.
† See note 7, p. 144.

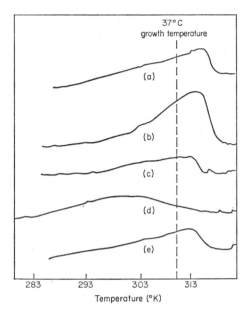

FIG. 22. Differential scanning calorimetry curves of 50 wt % dispersions of lipids extracted from *Acholeplasma laidlawii* B membranes. (a) total lipids; (b) glycolipids; (c) phospholipids; (d) neutral lipids; (e) reconstituted total lipids (Chapman and Urbina, 1971).

gel and liquid-crystalline regions are present in the membrane lipids at their growth temperature. This interpretation is supported by recent X-ray results (Esfahani *et al.*, 1971).

(b) *Nuclear magnetic resonance studies.* Support for the existence of predominantly *rigid* hydrocarbon chains in the membranes of *Acholeplasma laidlawii* B membranes comes from deuteron NMR (Oldfield *et al.*, 1972a). Cells supplemented with deuterated palmitic acid or lauric acid (which is elongated to myristic and palmitic acids) show DMR spectra more characteristic of the gel state of the model system di(perdeutero)myristoyl lecithin, than the liquid-crystalline state of this lipid, or of the model smectic liquid-crystalline soap, potassium perdeuterolaurate—H_2O. ESR spin-labelling results, however, give spectra characteristic of a mobile nitroxide—indicating some preference of the label for the more fluid, liquid-crystalline regions of this heterogeneous membrane system (Oldfield and Urbina, unpublished results).

That rapid growth can occur with almost all the membrane lipids in a rigid gel state is surprising, though the results of Steim *et al.* (1969) indicate that with stearate supplemented *Acholeplasma laidlawii* B, that is the case.

(c) *Electron spin resonance studies.* Spectra characteristic of highly mobile nitroxides have been obtained from a wide variety of membranes (McConnell,

1970b). A central question in interpreting these results is to what extent a spin-label can detect "rigid" regions in the presence of "fluid" ones, i.e. what is the extent of the fluid environments, and also, can the spin-probe artificially create them? These questions are as yet not fully answered. However, indications that spin-labels may probe liquid-crystalline regions in preference to gel regions have recently been obtained, in model systems (Oldfield *et al.*, 1972b), and that the label can act as a perturbant has been inferred by other workers (Chan *et al.*, 1972; Tiddy, 1972; Cadenhead and Katti, 1971; Raison *et al.*, 1971).

(d) *X-ray.* X-ray diffraction has been shown to be a powerful technique with which to obtain information on membrane heterogeneity.

Engelman (1971) has shown that in palmitate supplemented *Acholeplasma laidlawii* B, that the growth temperature both 4·15 Å (gel) and 4·6 Å (liquid-crystalline) short spacings are obtained from the membranes. The gel region only disappears at 45°C, i.e. *well above* the growth temperature of 37°C.

Esfahani *et al.* (1971) have obtained similar results on *Escherichia coli* K12. With elaidate supplemented membranes, the X-ray detected thermal phase transition extends from 30°C → 40°C in intact membranes (grown at 37°C). With linolenate and myristoleate membranes the transitions extend from 36°C → 46°C—here the lipids (phosphatidylethanolamines) would appear to be predominantly in a *gel* state (as evidenced by the sharp 4·2 Å band). It is interesting to note that fatty acids which might be expected to produce highly fluid lipids (i.e. linolenate, myristoleate, oleate) are taken up to a smaller extent than those expected to produce more rigid lipids, e.g. elaidate (Chapman *et al.*, 1966). There would thus appear to be a mechanism controlling fluidity which operates by increasing palmitate concentrations in the presence of linolenate, etc., with this organism.

In some instances, microorganisms lacking cholesterol may regulate "permeability" and mechanical stability by having heterogeneous gel and liquid-crystalline regions in their membranes.*

The possibility that "rigid" as well as "fluid" (or "liquid-like") domains in biomembranes may be a common occurrence, and would appear to be an important factor in constructing models for some biomembranes. That biologically relevant transport processes can occur in rigid systems has recently been shown by Krasne *et al.* (1972), where it was shown that the ion translocating antibiotic, gramicidin, was able to mediate potassium ion transport in both "liquid" and "solid" black lipid membranes.

3. *Sarcoplasmic Reticulum Membranes*

Sarcoplasmic membranes isolated from muscle tissues in the form of vesicular fragments which sustain Ca^{2+} uptake, have been studied by NMR

* See note 8, p 144.

spectroscopy (Davis and Inesi, 1971). A reversible temperature dependent structural transition was observed involving the choline methyl groups of lecithin. The fraction of mobile methyl groups as a function of temperature correlates with the temperature dependent Ca^{2+} efflux. Other agents which alter protein structure and simultaneously increase Ca^{2+} efflux also increase the fraction of mobile choline methyl groups. Irreversible heat denaturation of the membrane protein also alters the temperature dependence of the transition. Thus it was concluded that the observed thermal transition involves protein as well as lipid.

Mitochondrial membranes

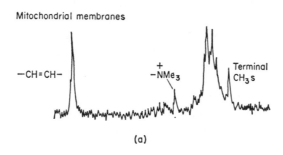

(a)

Erythrocyte ghosts

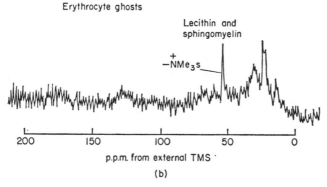

(b)

FIG. 23. 25·2 MHz ^{13}C NMR spectra of (a) mitochondrial membranes 37°C and (b) erythrocyte ghosts (Keough et al., 1972).

4. Mitochondrial Membranes

Mitochondrial membranes give better resolved PMR and DMR spectra than do erythrocyte membranes (Keough et al., 1972). The greatly increased resolution obtained with CMR should thus make this an excellent technique for studying these membranes.* The CMR spectrum of rat liver mitochondrial

* CMR = C^{13} magnetic resonance.

membranes at 37°C (see Fig. 23a) shows excellent comparison to that of the erythrocyte membrane spectrum (Fig. 23b), with sharp signals from $-CH=$, $-N^+(CH_3)_3$ and $-CH_3$ groups being apparent. Resolution in the methylene chain region is good, with the penultimate methylene signal (23·0 ppm) being partially resolved. The carbonyl resonance is very broad in the mitochondrial membrane spectrum indicating relative immobility.

The major contribution to the signals in the high field region of the spectrum is attributable to the lipid superimposed on a large number of overlapping protein side chain resonances. Contributions from aromatic side chains to the well defined resonance at 128·5 ppm are expected to be small. We have assigned the resonance centred at 54·7 ppm to the choline trimethylammonium groups of the lipid. The calculated spectrum shows that a substantial number of protein resonances, mainly from the α-carbons of the amino acids, are however also expected in this region.

5. *Chloroplast Membranes*

The chloroplast membrane spectrum is extremely complicated in the hydrocarbon region, possibly because of resonance contributions from the chlorophyll molecules in this region, but no signal appears at 55 ppm. The chemical shifts of the sphingomyelin $-N^+Me_3$ groups (55 ppm and 54·4 ppm) are close to the 54·7 ppm shift of the $(-N^+Me_3)$ resonance in the mitochondria.

References

Andrew, E. R. and Jasinski, A. (1971). *J. Phys. C.* **4**, 391.
Barratt, M. D., Green, D. K. and Chapman, D. (1969). *Chem. Phys. Lipids* **3**, 140.
Bergelson, L. D., Barsukov, L. I., Dubrovina, N. I. and Bystrov, V. F. (1970). *Dokl. Akad. Nauk SSSR* **194**, 708.
Bethe, H. A. (1935). *Proc. R. Soc. A* **150**, 550.
Bloembergen, N., Purcell, E. M. and Pound, R. V. (1948). *Phys. Rev.* **73**, 679.
Bowden, S. T. (1954). *In* "Phase Rule and Phase Reactions", Chapter 7. MacMillan, London.
Bragg, W. L. and Williams, E. J. (1934). *Proc. R. Soc. A* **145**, 699.
Branton, D. (1971). *Phil. Trans. R. Soc.* **261**, 333.
Bretscher, M. S. (1971). *J. molec. Biol.* **59**, 351.
Brocklehurst, J. R., Freedman, R. B., Hancock, D. J. and Radda, G. K. (1970). *Biochem. J.* **116**, 721.
Bulkin, B. J. and Krishnamachari, N. (1970). *Biochim. biophys. Acta* **211**, 592.
Butler, K. W., Smith, I. C. P. and Schneider, H. (1970). *Biochim. biophys. Acta* **219**, 514.
Cadenhead, D. A. and Katti, S. S. (1971). *Biochim. biophys. Acta* **241**, 709.
Cadenhead, D. A. and Phillips, M. C. (1968). *Adv. Chem. Ser.* **84**, 131.
Chan, S. I., Feigenson, G. W. and Seiter, C. H. A. (1971). *Nature, Lond.* **231**, 110.
Chan, S. I., Seiter, C. H. A. and Feigenson, G. W. (1972). *Biochim. biophys. Res. Comm.* **46** (4), 1488.

Chapman, D. (1968). *In* "Biological Membranes: Physical Fact and Function". (D. Chapman, ed.) Vol. 1, p. 125. Academic Press, London, New York.

Chapman, D. (1965). "The Structure of Lipids." Methuen and Co., London.

Chapman, D. (1966). *Ann N.Y. Acad. Sci.* **137**, 745.

Chapman, D. (1972). *Ann. N.Y. Acad. Sci.* **195**, 179.

Chapman, D., Byrne, P. and Shipley, G. G. (1966). *Proc. R. Soc. A.* **290**, 115.

Chapman, D. Fluck, D. J., Penkett, S. A. and Shipley, G. G. (1968). *Biochim. biophys. Acta* **163**, 255.

Chapman, D., Kamat, V. B., de Gier, J. and Penkett, S. A. (1968a). *J. molec. Biol.* **31**, 101.

Chapman, D., Kamat, V. B. and Levene, R. J. (1968b). *Science, N.Y.* **160**, 314.

Chapman, D. and Morrison, A. (1966). *J. biol. Chem.* **241**, 5044.

Chapman, D., Owens, N. F., Phillips, M. C. and Walker, D. A. (1969). *Biochim. biophys. Acta* **183**, 458.

Chapman, D. and Penkett, S. A. (1966). *Nature, Lond.* **211**, 1304.

Chapman, D. and Salsbury, N. J. (1966). *Trans. Farad. Soc.* **62**, 2607.

Chapman, D. and Saville, J. (1972). To be published.

Chapman, D. and Whittington, S. G. (1964). *Trans. Farad. Soc.* **60**, 1369.

Chapman, D., Williams, R. M. and Ladbrooke, B. D. (1967). *Chem. Phys. Lipids* **1**, 445.

Chapman, D. and Urbina, J. (1971). *FEBS Lett.* **12**, 169.

Cherry, R. J., Berger, K. U. and Chapman, D. (1971). *Biochem. biophys. Res. Comm.* **44**, 644.

Clowes, A. W., Cherry, R. J. and Chapman, D. (1971). *Biochim. biophys. Acta* **249**, 301.

Clowes, A. W., Cherry, R. J. and Chapman, D. (1972). *J. molec. Biol.* **67**, 49.

Cohn, M., Kowalsky, A., Leigh, J. S. Jr. and Maričić, S. (1967). *In* "Magnetic Resonance in Biological Systems". pp. 45–8. Pergamon Press, Oxford.

Darke, A., Finer, E. G., Flook, A. G. and Phillips, M. C. (1972). *J. molec. Biol.* **63**, 265.

Davis, D. G. and Inesi, G. (1971). *Biochim. biophys. Acta* **241**, 1.

Daycock, J. T., Darke, A. and Chapman, D. (1971). *Chem. Phys. Lipids* **6**, 205.

De Bernard, L. (1958). *Bull. Soc. Chim. biol.* **40**, 161.

Demel, R. A., van Deenen, L. L. M. and Pethica, B. A. (1967). *Biochim. biophys. Acta* **135**, 11.

Doskočilová, D. and Schneider, B. (1970). *Chem. Phys. Lett.* **6** (4), 381.

Doskočilová, D. and Schneider, B. (1970). *Chem. Phys. Lett.* **6**, 381.

Doskočilová, D. and Schneider, B. (1972). *Macromolecules*, to be published.

Engelman, D. M. (1971). *J. molec. Biol.* **58**, 153.

Engstrom, L. H. (1970). Ph.D. dissertation, University of California, Berkeley.

Esfahani, M., Limbrick, A. R., Knutton, S., Oka, T. and Wakil, S. J. (1971). *Proc. natn. Acad. Sci. U.S.A.* **68**, 3180.

Finean, J. B. (1953). *Experientia* **9**, 17.

Finer, E. G., Flook, A. G. and Hauser, H. (1972). *Biochim. biophys. Acta* **260**, 59.

Glaser, M., Simpkins, H., Singer, S. J., Sheetz, M. and Chan, S. I. (1970). *Proc. natn. Acad. Sci. U.S.A.* **65**, 721.

Green, D. E. and Fleischer, S. (1963). *Biochim. biophys. Acta* **70**, 554.

Griffith, O., Cornell, D. W. and McConnell, H. M. (1965). *J. chem. Phys.* **43**, 2909.

Gulik-Kryzwicki, T., Shechter, E., Iwatsubo, M., Ranck, J. L. and Luzzati, V. (1970). *Biochim. biophys. Acta* **219**, 1.

Gulik-Kryzwicki, T., Shechter, E., Luzzati, V. and Faure, M. (1969). *Nature, Lond.* **223**, 116.

Gutowsky, H. S. and Pake, G. E. (1950). *J. chem. Phys.* **18**, 162.

Haeberlen, U. and Waugh, J. S. (1969). *Phys. Rev.* **185**, 420.

Hanahan, D. J. (1969). *In* "The Red Blood Cell Membrane". (G. A. Jamieson and T. J. Greenwalt, eds.) Lippincott Co.

Hauser, H., Chapman, D. and Dawson, R. M. C. (1969). *Biochim. biophys. Acta* **183**, 320.

Hauser, H., Finer, E. G. and Chapman, D. (1970). *J. molec. Biol.* **53**, 413.

Hoffman, R. E. (1952). *J. Chem. Phys.* **20**, 541.

Horwitz, A. and Klein, M. P. (1972). *J. Supramolecular Structure* **1**, 19.

Hubbell, W. L. and McConnell, H. M. (1968). *Proc. natn. Acad. Sci. U.S.A.* **61**, 12.

Hubbell, W. L. and McConnell, H. M. (1969). *Proc. natn. Acad. Sci. U.S.A.* **64**, 20.

Hubbell, W. L. and McConnell, H. M. (1971). *J. Am. chem. Soc.* **93**, 314.

Ibrahim, S. A. and Thompson, R. H. S. (1965). *Biochim. biophys. Acta* **99**, 331.

Jenkinson, T. J., Kamat, V. B. and Chapman, D. (1969). *Biochim. biophys. Acta* **183**, 427.

Jost, P., Libertini, L. J., Herbert, V. C. and Griffith, O. H. (1971a). *J. molec. Biol.* **59**, 77.

Jost, P., Waggoner, A. S. and Griffith, O. H. (1971b). *In* "Structure and Function of Biological Membranes". (L. I. Rothfield, ed.) Academic Press, New York, London.

Juliano, R., Kimelberg, H. K. and Papahadjopolous, D. (1971). *Biochim. biophys. Acta* **241**, 894.

Kamat, V. B. and Chapman, D. (1968). *Biochim. biophys. Acta* **163**, 411.

Keough, K. M. and Chapman, D. (1972). To be published.

Keough, K. M., Oldfield, E., Chapman, D. and Beynon, P. (1973). *Chem. Phys. Lipids*, **10**, 37.

Kimelberg, H. K. and Papahadjopolous, D. (1971a). *J. biol. Chem.* **246**, 1142.

Kimelberg, H. K. and Paphadjopolous, D. (1971b). *Biochim. biophys. Acta* **233**, 805.

Kornberg, R. D. and McConnell, H. M. (1971). *Biochemistry* **10**, 1111.

Krasne, S., Eisenman, G. and Szabo, G. (1972). *Science, N.Y.* **174**, 412.

Kubo, R. and Tomita, K. (1954). *J. Phys. Soc. Japan* **9**, 888.

Ladbrooke, B. D. and Chapman, D. (1969). *Chem. Phys. Lipids* **3**, 304.

Ladbrooke, B. D., Williams, R. M. and Chapman, D. (1968). *Biochim. biophys. Acta* **150**, 333.

Larsson, K. (1967). *Z. Phys. Chem.* **56**, 173.

Leathes, J. B. (1925). *Lancet* **208**, 853.

Lecuyer, H. and Dervichian, D. G. (1969). *J. molec. Biol.* **45**, 39.

Lenard, J. and Singer, S. J. (1966a). *Proc. natn. Acad. Sci. U.S.A.* **56**, 1828.

Lenard, J. and Singer, S. J. (1966b). *Science, N.Y.* **159**, 738.

Lesslauer, W., Cain, J. and Blasu, J. K. (1971). *Biochim. biophys. Acta* **241**, 547.

Levine, Y. K., Birdsall, N. J. M., Lee, A. G. and Metcalfe, J. (1972). *Biochemistry* **11**, 1416.

Lippert, J. L. and Peticolas, W. L. (1971). *Proc. natn. Acad. Sci. U.S.A.* **68**, 1572.

Luzzati, V. (1968). *In* "Biological Membranes: Physical Fact and Function". (D. Chapman, ed.) Vol. 1, p. 71. Academic Press, London, New York.

Luzzati, V., Gulik-Krzywicki, T. and Tardieu, T. (1968). *Nature, Lond.* **218**, 1031.

McConnell, H. M. (1970a). *In* "Molecular Properties of Drug Receptors". (R. Porter and M. O'Connor, eds.) J. and A. Churchill, London.

McConnell, H. M. (1970b). *In* "The Neurosciences: Second Study Program". (F. O. Schmidt, ed. in chief.) The Rockafeller University Press, New York.

McFarland, B. G. and McConnell, H. M. (1971). *Proc. natn. Acad. Sci. U.S.A.* **68**, 1274.

Mazia, D. and Ruby, A. (1968). *Proc. natn. Acad. Sci. U.S.A.* **61**, 1005.

Melchior, D. L., Morowitz, H. J., Sturtevant, J. M. and Tsong, T. Y. (1970). *Biochim. biophys. Acta* **219**, 114.

Metcalfe, J. C., Birdsall, N. J. M., Feeney, J., Lee, A. G., Levine, Y. K. and Partington, P. (1971). *Nature, Lond.* **233**, 199.

Mueller, P. and Rudin, D. O. (1968). *Nature, Lond.* **217**, 713.

Oldfield, E. and Chapman, D. (1971). *Biochem. biophys. Res. Comm.* **43**, 610.

Oldfield, E. and Chapman, D. (1972). *FEBS Lett.* **21**, 303.

Oldfield, E., Chapman, D. and Derbyshire, W. (1971). *FEBS Lett* **16**, 102.

Oldfield, E., Chapman, D. and Derbyshire, W. (1972a). *Chem. Phys. Lipids*, **9**, 69.

Oldfield, E., Keough, K. M. and Chapman, D. (1972b). *FEBS Lett.* **20**, 344.

Oldfield, E., Marsden, J. and Chapman, D. (1971). *Chem. Phys. Lipids* **7**, 1.

Pache, W. and Chapman, D. (1972). *Biochim. biophys. Acta* **255** (1), 348.

Pache, W., Chapman, D. and Hillaby, R. (1972). *Biochim. biophys. Acta* **255** (1), 358.

Penkett, S. A., Flook, A. G. and Chapman, D. (1968). *Chem. Phys. Lipids* **2**, 273.

Phillips, M. C. and Chapman, D. (1968). *Biochim. biophys. Acta* **163**, 301.

Phillips, M. C., Ladbrooke, B. D. and Chapman, D. (1970). *Biochem. biophys. Acta* **196**, 35.

Phillips, M. C., Williams, R. M. and Chapman, D. (1969). *Chem. Phys. Lipids* **3**, 234.

Radda, G. K. (1971). *Biochem. J.* **122**, 385.

Raison, J. K., Lyons, J. M., Mehlhorn, R. J. and Keith, A. D. (1971). *J. biol. Chem.* **246**, 4036.

Rand, R. P. (1971). *Biochim. biophys. Acta* **241**, 823.

Rand, R. P. and Luzzati, V. (1968). *Biophys. J.* **8**, 125.

Salsbury, N. J. and Chapman, D. (1968). *Biochim. biophys. Acta* **163**, 314.

Salsbury, N. J., Chapman, D. and Jones, G. P. (1970). *Trans. Farad. Soc.* **66**, 1554.

Schneider, B., Pivcová, H. and Doskočilová, D. (1972). *Macromolecules* **5**, 120.

Shah, D. O. and Schulman, J. H. (1967). *J. Lipid Res.* **8**, 215.

Sheard, B. (1969). *Nature, Lond.* **223**, 1057.

Sheetz, M. P. and Chan, S. I. (1970). *Biophys. Soc. Abs.* **46a**.

Shipley, G. G., Leslie, R. B. and Chapman, D. (1969). *Nature, Lond.* **222**, 561.

Singer, S. J. (1971). *In* "Structure and Function of Biological Membranes". (L. I. Rothfield, ed.) Academic Press, London, New York.

Steck, T. L., Fairbanks, G. and Wallach, D. F. H. (1971). *Biochemistry* **10**, 2617.

Steim, J. M., Tourtellotte, M. E., Reinert, J. C., McElhaney, R. N. and Rader, R. L. (1969). *Proc. natn. Acad. Sci. U.S.A.* **63**, 104.

Steim, J. M. (1970). *In* "Liquid Crystals and Ordered Fluids". (R. S. Porter and J. F. Johnson, eds.) Plenum Press, New York.

Steinemann, A. and Läuger, P. (1971). *J. Mem. Biol.* **4**, 74.

Stone, T. J. Buckman, T., Nordio, P. L. and McConnell, H. M. (1965). *Proc. natn. Acad. Sci. U.S.A.* **54**, 1010.

Stryer, L. (1965). *J. molec. Biol.* **13**, 482.

Sweet, C. and Zull, J. E. (1969). *Biochim. biophys. Acta* **173**, 94.

Tiddy, G. (1972). *J. C. S. Faraday* 1 **68**, 670.

Träuble, H. (1971). *Naturwissenshaften* **58**, 277.

Träuble, H. (1972). Presented at Symposium on "Passive Permeability of Cell Membranes", Rotterdam, 1971. In press.

Turner, D. G. and Brand, L. (1968). *Biochemistry* 7, 3381.

Urbina, J. and Chapman, D. (1971). Unpublished studies.

Urry, D. W. (1972). *Reviews on Biomembranes* 265, 115.

Van Deenen, L. L. M., Houtsmuller, U. M. T., de Haas, G. H. and Mulder, E. (1962). *J. Pharm. Pharmac.* 14, 429.

Vandenheuvel, F. A. (1963). *J. Am. Oil Chem. Soc.* 40, 455.

Vanderkooi, J. and Martonosi, A. (1969). *Archs Biochem. Biophys.* 133, 153.

Veksli, A., Salisbury, N. J. and Chapman, D. (1969). *Biochim. biophys. Acta* 183, 434.

Wallach, D. F. H. and Zahler, P. H. (1966). *Proc. natn. Acad. Sci. U.S.A.* 56, 1552.

Whittington, S. G. and Chapman, D. (1965). *Trans. Farad. Soc.* 61, 2656.

Whittington, S. G. and Chapman, D. (1966). *Trans. Farad. Soc.* 62, 3319.

Wilkins, M. H. F., Blaurock, A. E. and Engelman, D. M. (1971). *Nature, Lond.* 230, 72.

Williams, R. M. and Chapman, D. (1970). *Progress in the Chemistry of Fats and Other Lipids*, Vol. XI (1), 1.

Woessner, D. E. (1962). *J. Chem. Phys.* 36, 1.

Zwaal, R. F. A., Roelofsen, B., Comfurus, P. and van Deenen, L. L. M. (1971). *Biochim. biophys. Acta* 233, 474.

Zwaal, R. F. A. and van Deenen, L. L. M. (1970). *Chem. Phys. Lipids* 4, 311.

NOTES ADDED IN PROOF

1. This effect appears to be an example of how small amounts of a foreign molecule in a membrane surface can influence the polar group packing of a large number of lipid molecules.

2. This study was the first to show the gradation of motion which occurs along the lipid hydrocarbon chains in the liquid-crystalline condition with greater molecular motion occurring at the methyl end of the chain. This occurs with the anhydrous and hydrated liquid crystal phase.

3. A combination of the spinning technique and relaxation measurements of C^{13} nuclei may lead to the production of detailed molecular maps of membrane systems.

4. This technique was used to demonstrate for the first time the fluid nature of the hydrocarbon chains in the liquid crystalline phase of long chain amphipathic molecules. (D. Chapman (1958). *J. Chem. Soc.* 152, 784.)

5. A point of some interest is whether freezing of membranes without cholesterol can cause some of the protein to be excluded from the lipid chain region when chain crystallization occurs.

6. This has only recently been appreciated and discussed. (Oldfield and Chapman (1972). *FEBS Lett.* 23, 285.)

7. The assumptions required in the estimation of the *extent* of bilayer present in this membrane and whether electrostatic interaction of protein and lipid is a predominant factor has been discussed by Chapman and Urbina (1971).

8. Certain transport proteins may be in the fluid lipid regions whilst some structural proteins may be in the more rigid region. When *E. coli* cells are grown in the presence of monosaturated fatty acid then shifted to a polyunsaturated acid, two transition temperatures for enzymatic activity are observed suggesting that the enzyme is located in two separate lipid domains.

Chapter 3

Virus Membranes

HANS-DIETER KLENK

Institut für Virologie, Justus Liebig-Universität,
Giessen, Germany

I. Introduction

Although the key role of viruses as model systems in central fields of biology and biochemistry has been known for a long time and exploited with great profit, it is only recently that they have also been recognized as valuable tools in membrane research. Therefore most of the literature presented in this article has been published in the last three or four years, and with one exception (Franklin, 1962) previous reviews in this field do not exist to the author's knowledge.

The internal component of many viruses consisting of the nucleic acid genome and of protein is surrounded by an envelope which is an integral component of the virion. Upon degradation of the envelope, these viruses lose infectivity and other biological functions. Morphologically the envelopes

145

resemble cellular membranes; on electron micrographs they appear as unit membranes.

The envelopes of most viruses are assembled in continuity with cellular membranes (Table I). Viral nucleocapsids make contact with these membrane areas, and mature virions are then released in a budding process. With this mode of assembly the envelopes may be formed at the plasma membrane, at the membranes of intracellular vesicles or at the inner membrane of the nucleus. Each virus has its typical maturation site. The envelopes of other viruses are formed free in the cytoplasm, i.e. without connexion to preformed cellular membranes, as in the case of the pox viruses or the bacteriophage PM 2.

A. GROUPS OF ENVELOPED VIRUSES

The major groups of enveloped viruses are listed in Table I. Some enveloped viruses contain RNA, others contain DNA. The groups differ over a wide range in particle size and in the size of their genome.

1. *Arboviruses* (alphaviruses) are spherical RNA-viruses with an envelope surrounding an icosahedral nucleocapsid. They are divided into many groups, of which groups A and B have been studied in most detail. Arboviruses multiply in the cytoplasm. Some mature by budding from cytoplasmic membranes, others by budding from intracellular vesicles.

2. *Arenoviruses* belong to a newly constituted virus group. The particles are round, oval, or pleomorphic and have an envelope with rather pronounced surface projections. They mature by budding from plasma membranes.

3. *Coronaviruses*. These are medium-sized somewhat pleomorphic viruses, with widely spaced, club-shaped spikes on their surface.

4. *Myxoviruses* comprise the ortho- and the paramyxoviruses. They occur as spherical or filamentous virions with a helical nucleocapsid within an envelope acquired as they bud from the plasma membrane. The envelope is studded with spikes which are of two kinds, one being the hemagglutinin, the other the neuraminidase.

5. *Rhabdoviruses* are large, bullet-shaped, enveloped RNA-viruses. Rhabdoviruses mature at a variety of cellular membranes. They have a wide host range comprising vertebrates, insects and even plants.

6. *Leukoviruses* are oncogenic RNA-viruses. They mature by budding from the plasma membrane.

TABLE I

Characteristics of the major groups of enveloped viruses

Group	Size of virion (μ)	Nucleic acid		Mode and site of envelope maturation
		Type	MW $\times 10^6$	
Animal viruses				
Arboviruses[a-b]	50	RNA	2–4	Budding from the plasma membrane and from cytoplasmic vesicles
Arenoviruses[c]	50–300	RNA		Budding from the plasma membrane
Coronaviruses[d-f]	70–120	RNA		Budding from membranes of the endoplasmic reticulum and cytoplasmic vesicles
Myxoviruses				
Orthomyxoviruses[g-j]	80–120	RNA	3–5	Budding from the plasma membrane
Paramyxoviruses[k]	100–450	RNA	6–7	Budding from the plasma membrane
Rhabdoviruses[l-n]	70 × 200	RNA	4–6	Budding from the plasma membrane and from cytoplasmic vesicles, free in the cytoplasm
Leukoviruses[o, p]	100–120	RNA	10–12	Budding from the plasma membrane
Herpesviruses[q, r]	120–180	DNA	70–100	Budding from the inner nuclear membrane
Poxviruses[s]	100 × 200 × 300	DNA	160–200	Free in the cytoplasm
Plant viruses				
Rhabdoviruses[t-w]	100 × 300	RNA		Budding from the inner nuclear membrane
Bacterial viruses				
Marine bacteriophage PM2[x]	60	DNA	15–21	Free below the cell membrane

References

[a] Acheson and Tamm, 1967; [b] Matsumara *et al.*, 1971; [c] Murphy *et al.*, 1970; [d] Hamre *et al.*, 1967; [e] Becker *et al.*, 1967; [f] Oshiro *et al.*, 1971; [g] Murphy and Bang, 1952; [h] Morgan *et al.*, 1961; [i] Bächi *et al.*, 1969; [j] Compans and Dimmock, 1969; [k] Compans *et al.*, 1966; [l] Mussgay and Weibel, 1963; [m] Hummeler *et al.*, 1967; [n] Zee *et al.*, 1970; [o] Beard *et al.*, 1963; [p] Gelderblom *et al.*, 1970; [q] Siegert and Falke, 1966; [r] Darlington and Moss, 1969; [s] Dales and Mosbach, 1968; [t] Kitajiama and Costa, 1966; [u] McLeod *et al.*, 1966; [v] Simpson and Hauser, 1966a; [w] Sylvester *et al.*, 1968; [x] Dahlberg and Franklin, 1970.

7. *Herpesviruses* are icosahedral DNA-viruses which multiply in the nuclei of infected cells. They mature by budding through the nuclear membrane, acquiring an envelope in the process.

8. *Poxviruses* are the largest and most complex viruses of vertebrates and contain the largest DNA molecule. The virion consists of a brick-shaped DNA-containing core surrounded by a complex series of membranes of viral origin. Poxviruses mature in the cytoplasm. Envelope assembly takes place without connexion to cellular membranes.

Most of the studies on viral envelopes have been carried out with animal viruses, and this review will therefore deal mainly with these viruses. However, as can be seen in Table I, some *plant viruses* and *bacterial viruses* also contain envelopes. In fact, among the most detailed studies on viral envelopes are those concerning the bacteriophage PM 2.

B. ISOLATION AND PURIFICATION

Viruses are commonly grown in cell cultures from which they can often be recovered in large amounts. Viruses which mature intracellularly are released into the medium by cell lysis, whereas viruses which bud from the plasma membrane are released without significant cell damage and can be isolated from the medium in rather pure form. The purification procedures usually involve a series of differential centrifugations and sedimentations in density gradients. Such preparations of enveloped viruses usually show very little contamination by cellular membranes and can therefore be considered as quite pure membrane preparations.

II. Envelope Constituents

The envelopes of viruses meet all the criteria for cellular membranes. As already mentioned, morphologically they possess a trilaminar or unit membrane structure. Chemically they are composed of proteins, lipids, and carbohydrates, the carbohydrates being covalently linked to proteins or to lipids.

The chemical composition of four groups of RNA-viruses is summarized in Table II. The entire nucleic acid of the virion belongs to the nucleocapsid, whereas the lipids and the non-nucleic acid carbohydrates are located exclusively in the envelope (Compans and Choppin, 1967). 20–30% of the protein of arbo- and myxoviruses is nucleocapsid protein and 70–80% is envelope protein (Strauss *et al.*, 1968; Compans *et al.*, 1970; Schulze, 1970; Klenk *et al.*, 1972a).

TABLE II

Chemical composition of some virus groups

	RNA	Protein (Per cent of dry weight)	Lipid	Carbohydrate
Arboviruses[a]	5·9	66	28	Not determined
Rhabdoviruses[b]	3	65	20	13
Orthomyxoviruses[c]	0·7–1	70–75	20–24	5
Paramyxoviruses[d]	0·9	73	20	6
Leukoviruses[e]	1·9	64	31	6

References

[a] Pfefferkorn and Hunter (1963a); [b] McSharry and Wagner (1971a); [c] Frisch-Niggemeyer and Hoyle (1956); [d] Klenk and Choppin (1969a); [e] Quigley *et al.* (1971).

A. PROTEINS

It has been known for a long time that viral envelopes contain virus-specific proteins. The hemagglutinin and the neuraminidase of myxoviruses, for instance, are virus specific proteins which are located in the envelope. On the other hand, it was a legitimate assumption that proteins of the host cell were incorporated into the viral envelope, because many of these viruses make extensive use of host cell membranes during their assembly. This point seemed to be further strengthened by the frequent finding of host antigens in enveloped viruses (Knight, 1947; Schäfer, 1956; Haukenes *et al.*, 1965; Laver and Webster, 1966; Cartwright and Pearce, 1968; Lee *et al.*, 1969). However, it has never been shown that these host antigens were proteins.

New light on the protein constituents of viral envelopes and on their origin has been shed with the advent of polyacrylamide gel electrophoresis in the presence of strong detergents, a method particularly suitable for the separation of membrane proteins on the basis of their size.

Proteins of the envelope have been investigated and characterized in detail with arboviruses, rhabdoviruses, and myxoviruses. These virus groups are particularly suitable for such investigations because they can be purified to a high degree due to their maturation by budding from the cell surface, and because of the low number of structural proteins due to the limited information content of their genome (Table I).

There is now overwhelming evidence that all envelope polypeptides of arboviruses, rhabdoviruses, and myxoviruses are coded by the viral genome. For instance, the only envelope protein of Sindbis virus, an arbovirus, is synthesized *de novo* after infection in large amounts (Pfefferkorn and Clifford, 1964; Strauss *et al.*, 1969), and temperature-sensitive mutants have been

G

isolated which make viral RNA and kill cells, but do not synthesize a functional membrane protein at non-permissive temperatures (Burge and Pfefferkorn, 1967; Strauss et al., 1968).

When the same strain of vesicular stomatitis virus, which belongs to the rhabdovirus group, is grown in different host cells, the electrophoretic patterns of the virion polypeptides are identical which makes a host protein as structural component unlikely (Wagner et al., 1969). On the same line, host cell proteins radioactively labeled before infection could not be detected in vesicular stomatitis virions.

Similar conclusions have been reached with myxoviruses. The polypeptide pattern of the parainfluenza virus SV5 does not depend on the host cell line in which the virus has been grown (Choppin et al., 1971).

Such results have also been obtained with influenza virus, when it was grown in three different cell types (Compans et al., 1970). Furthermore, it has been calculated on the basis of experiments with prelabeled host cells that influenza virions contain, if at all, less than 1 % host cell protein (Holland and Kiehn, 1970).

The experiments mentioned above strongly indicate that the envelope polypeptides of these viruses are encoded in the viral genome. It is now possible to attribute each peak of the polypeptide profile of influenza virus, which is rather complex, to a specific component of the virion and to show that each of these components is virus-specific (Schulze, 1972).

Therefore, the conclusion seems to be justified that with arboviruses, rhabdoviruses, and myxoviruses the polypeptide chains of the envelope are specified by the viral genome. Although the virus makes use of the plasma membrane of the host cell during assembly, no host protein is incorporated into the virion. The bacteriophage PM 2 is also composed entirely of virus-specified proteins (Datta et al., 1971).

In the case of leuko-, herpes- and poxviruses, however, the situation is not so clear. The genome of these viruses is quite large, it is able to code for many proteins (Table I). Thus, it is not surprising that these viruses have a very complex polypeptide profile (Holowczak and Joklik, 1967; Olshevsky and Becker, 1970; Spear and Roizman, 1972; Abodeely et al., 1971; Bolognesi and Bauer, 1970). On the other hand, it is difficult to ascertain whether host cell proteins are present in these profiles and if so, whether the host proteins are integral components of the virion or just contaminants. However, it is now quite likely that these viruses also contain exclusively virus specific proteins (Spear and Roizman, 1972).

1. Carbohydrate-free Proteins

Certain striking similarities in protein composition among the RNA viruses are becoming apparent.

Besides the nucleocapsid protein and some other proteins associated with the internal component of the virion, arboviruses of group B, rhabdoviruses, and myxoviruses contain additional carbohydrate-free polypeptides which belong with high probability to the viral envelope. The evidence that these proteins are constituents of the virus membrane will be discussed in the chapter on the envelope structure.

There is one major carbohydrate-free polypeptide in the envelopes of rhabdoviruses, orthomyxoviruses, and paramyxoviruses (Table III). With each of these virus groups it is the smallest polypeptide of all structural proteins of the virion. It has a molecular weight of 20,000–30,000 with the rhabdoviruses and orthomyxoviruses and of about 40,000 with the paramyxoviruses.

It is the main protein in influenza virus and comprises 34–40% of the protein of the entire virion and *ca* 50% of the envelope protein.

Rhabdoviruses (McSharry *et al.*, 1971; Wagner *et al.*, 1970; Mudd and Summers, 1970; Sokol *et al.*, 1971) and paramyxoviruses (Caliguiri *et al.*, 1969; Mountcastle *et al.*, 1970; Mountcastle *et al.*, 1971) contain additional carbohydrate-free proteins which are minor components of the virion. It is possible that these proteins are also constituents of the envelope, but their exact localization within the virion has not yet been established.

So far, carbohydrate-free proteins have not been reported to be present with certainty in the envelopes of herpes viruses, pox viruses, and leuko viruses.

2. *Glycoproteins*

It is becoming apparent that glycoproteins are universal constituents of the membrane of all classes of enveloped viruses. This finding throws light on another striking similarity between viral envelopes and cellular membranes where glycoproteins are common constituents, too.

Arboviruses of group A contain a single protein in their envelope, and this protein is a glycoprotein (Table III). With Sindbis virus (Simons and Kääriäinen, 1970) as well as with Semliki Forest Virus (Acheson and Tamm, 1970) this protein has been found to contain a relatively high amount of hydrophobic amino acids. This suggests that this protein is associated with the lipids in the envelope.

In addition to the carbohydrate-free polypeptide, rhabdoviruses also contain a glycoprotein in their envelopes.

A comparative study of paramyxoviruses revealed that the members of this group contain two glycoproteins. Influenza virus which belongs to the orthomyxovirus group has as many as four different glycoprotein species. However, since a precursor–product relationship has been found between some of the influenza glycoproteins, not all of them are primary gene products (Lazarowitz *et al.*, 1971; Klenk *et al.*, 1972b).

TABLE III

Envelope proteins of some virus groups

	Carbohydrate-free polypeptides		Glycoproteins	
	No.	Mol wt	No.	Mol wt
Arboviruses Group A[a–d]	—	—	1	51–53,000
Rhabdoviruses	1[e, g]	25,000[g, h]	1[e, g]	67–80,000[g, h]
Paramyxoviruses	1[j–n]	38–41,000	2[o, p]	I. 65–74,000
				II. 53–56,000
Orthomyxoviruses	1[q–v]	21–26,000	3–4[t–v]	I. 74–78,000
				II. 49–58,000
				III. 45–51,000
				IV. 24–32,000

References

[a] Strauss *et al.*, 1969; [b] Hay *et al.*, 1969; [c] Acheson and Tamm, 1970; [d] Simons and Kääriäinen, 1970; [e] Wagner *et al.*, 1970; [f] Cartwright *et al.*, 1970; [g] Burge and Huang, 1970; [h] Mudd and Summers, 1970; [i] Sokol *et al.*, 1971; [j] Caliguiri *et al.*, 1969; [k] Bikel and Duesberg, 1969; [l] Evans and Kingsbury, 1969; [m] Haslam *et al.*, 1969; [n] Mountcastle *et al.*, 1970; [o] Klenk *et al.*, 1970a; [p] Mountcastle *et al.*, 1971; [q] Joss *et al.*, 1969; [r] Haslam *et al.*, 1970a; [s] Compans *et al.*, 1970; [t] Schulze, 1970; [u] Skehel and Schild, 1971; [v] Klenk *et al.*, 1972a.

To the virus groups listed in Table III other enveloped viruses have to be added which have also been found to contain glycoproteins, namely herpes viruses (Ben-Porat and Kaplan, 1970; Olshevsky and Becker, 1970; Spear *et al.*, 1970; Abodeeley *et al.*, 1971), pox viruses (Holowczak, 1970; Garon and Moss, 1971) and leukoviruses (Duesberg *et al.*, 1970; Bolognesi and Bauer, 1970; Rifkin and Compans, 1970; Mountcastle *et al.*, 1972).

B. CARBOHYDRATES

Carbohydrates have long been recognized as a constituent of enveloped virus particles (Knight, 1947; Schäfer and Zillig, 1954) and they represent a significant proportion of the mass of the envelope (Table II). Detailed carbohydrate analyses have been reported for influenza virus (Frommhagen *et al.*, 1959; Ada and Gottschalk, 1956), Sindbis virus (Strauss *et al.*, 1970; Burge and Strauss, 1970), the parainfluenza virus SV5 (Klenk *et al.*, 1970a), and vesicular stomatitis virus (McSharry and Wagner, 1971b). Less detailed analyses indicate the presence of carbohydrates in Newcastle disease and Sendai virus (Mountcastle *et al.*, 1971), leukoviruses (Bolognesi and Bauer, 1970), herpesviruses (Olshevsky and Becker, 1970; Spear *et al.*, 1970; Ben-Porat and Kaplan, 1970), and vaccinia virus.

The carbohydrate analyses on influenza virus have been carried out by chemical assays (Frommhagen *et al.*, 1959; Ada and Gottschalk, 1956), those on Sindbis virus and on the parainfluenza virus SV5 by a combination of chemical assays and labeling with radioactive isotopes, whereas gas chromatography has been employed for the investigation of vesicular stomatitis virus. Galactose, mannose, glucose, fucose and the hexosamines glucosamine and galactosamine have been found in these viruses. Neuraminic acid has only been found with arbovirus and the rhabdovirus, it has not been detected in the paramyxovirus.

The total carbohydrate content and the proportions of hexoses and hexosamines are remarkably similar in the studies on influenza virus, Sindbis virus, and SV5. They differ from the data on vesicular stomatitis virus where a very high glucose content has been reported.

1. *Carbohydrate Moiety of Glycoproteins*

Ada and Gottschalk (1956) and Frommhagen *et al.* (1959) called attention to the striking similarities in carbohydrate composition of influenza virus and membrane fragments from uninfected cells. At the time, this similarity was taken as evidence for the incorporation into virus particles of some preformed host polysaccharide.

This concept had to be revised, when labeling of the virus with radioactive precursors and separation of the virus proteins on polyacrylamide gels has been employed (Strauss *et al.*, 1970; Klenk *et al.*, 1970a). Since the virus-specified glycoproteins were labeled by monosaccharide precursors added during infection, the viral carbohydrate cannot be polymerized cell material synthesized before infection and passively incorporated into viral membrane. Viral carbohydrates appear instead to be an integral part of virus-specified membrane proteins.

However, carbohydrates in enveloped viruses are not only linked to proteins, but it could be shown that they are also present in the virions as glycolipids (Klenk and Choppin, 1970b). In vesicular stomatitis virus 13% of the neuraminic acid of the virion is bound to lipid and 87% to protein (Klenk and Choppin, 1971). One-third of the total carbohydrate in SV5 grown in MDBK cells is bound to lipid. The remaining carbohydrate of SV5 (6·5 mg/100 mg protein) is covalently linked to the two viral glycoproteins. Together these glycoproteins represent approximately 37% of the total virus proteins, thus they contain about 18 mg carbohydrate per 100 mg protein. If one assumes that these glycoproteins have similar carbohydrate contents, as the ratio of carbohydrate to protein label in the virus would suggest, then each would contain this proportion of carbohydrate. Similar amounts of carbohydrates have been found in the glycoproteins of influenza virus (Laver, 1971).

The monosaccharides found in the glycoproteins of SV5 are galactose, mannose, glucosamine and fucose. In addition, smaller amounts of glucose and galactosamine have been detected. These are constituents of the glyco-lipids in the virion (Klenk *et al.*, 1970a; Klenk and Choppin, 1970b).

Besides the molecular weight and the carbohydrate composition little is known about the structure of the envelope glycoproteins. The glycoproteins of Sindbis and of vesicular stomatitis virus have been digested by proteolytic enzymes (Burge and Strauss, 1970; Burge and Huang, 1970). If neuraminic acid has been removed, the resulting glycopeptides are rather uniform in size and have a molecular weight of about 2500–2700. This would suggest that the carbohydrate is linked in these glycoproteins in the form of uniform oligosaccharide side-chains to a polypeptide backbone.

On the other hand there are indications of heterogeneity in the molecular size of the viral glycoprotein which are most probably based on the carbo-hydrate moiety (Sokol *et al.*, 1971). It is a common feature of glycoproteins of enveloped viruses that they migrate as much broader bands on poly-acrylamide gels than carbohydrate-free proteins. It was observed in the glycoprotein band of vesicular stomatitis virus that the peak of glucosamine label did not coincide with that of the amino acid label (Burge and Huang, 1970; Kang and Prevec, 1970). This finding was interpreted as suggesting the presence in the band of a larger polypeptide and a smaller glycoprotein (Burge and Huang, 1970). Another interpretation would be a variability in size of the oligosaccharide side chains within one glycoprotein molecule, a phenomenon which is known as microheterogeneity.

(a) *Carbohydrates specified by the host cell.* Recent interest has focused on the question whether the structure of the viral carbohydrate is specified by the viral or by the host cell genome. The most pertinent studies on this problem have been carried out again on arboviruses and myxoviruses. A considerable number of sugar transferase enzymes would be required for the synthesis of the glycoproteins of these viruses. It seems doubtful that the relatively small viral genome would specify so many transferases; it therefore appears that at least part of the carbohydrate structure is specified by host transferases.

Different approaches have been used in order to throw some light on this problem: 1. The glycoproteins of the same virus strain grown in different host cells have been analyzed in comparative studies. 2. Glycopeptides derived from glycoproteins of one virus grown in different host cells and of different viruses grown in the same host have been compared. 3. Com-parative studies of the monosaccharides in the glycoproteins of various viruses and in cellular glycoproteins have been carried out. 4. The enzyme activity of glycosyl transferases has been analyzed in infected and in un-infected cells.

The WSN line of influenza A virus contains four different glycoproteins (Compans et al., 1970; Schulze, 1970). On polyacrylamide gels these glycoproteins have a slightly higher mobility when the virus has been grown in baby hamster kidney cells than when it has been grown in bovine kidney cells. On the implicit assumption—the evidence for which has been discussed above—that all polypeptides in the virion are virus coded, these results suggest that the carbohydrate moiety is at least partly specified by the host. This of course does not exclude the possibility that in other virus-cell systems such host-specified modifications are not visible on polyacrylamide gels, as has been observed with the paramyxovirus SV5 when grown in monkey kidney and bovine kidney cells (Choppin et al., 1971).

The other piece of evidence indicating that the carbohydrate moiety of these glycoproteins shows host specificities stems from analyzes of glycopeptides (Burge and Huang, 1970). These glycopeptides were derived from the glycoproteins of an arbovirus and a rhabdovirus which had been grown in chick and in hamster cells. The study suggests that the glycopeptides of both viruses, from either host cell, have the same basic structure in terms of the identities and location of individual sugar residues. The authors favor the concept that the carbohydrate moiety covalently linked to different virus-specified membrane proteins may be specified principally by the host. However, some of their data also indicate virus-directed carbohydrate specification.

Equally ambiguous regarding host or virus specificity are comparative studies of the relative amounts of monosaccharides in viral and cellular glycoproteins (Strauss et al., 1970; Klenk et al., 1970a). These studies emphasize, however, the basic structural similarity of viral and cellular glycoproteins.

More conclusive evidence in favor of a host-controlled carbohydrate moiety comes from experiments, where enzyme activities which catalyze transfer of monosaccharides to glycoprotein (neuraminosyl and fucosyl transferases) have been assayed (Grimes and Burge, 1971). Comparison of particulate enzyme preparations from infected and uninfected cells showed no difference in either the specific activity or acceptor specificity of these enzymes. This is impressive in view of the fact that the Sindbis membrane glycoprotein is the only glycoprotein synthesized in the infected cell. It was also determined that neuraminosyl transferase from uninfected cells is capable of transferring ^{3}H-neuraminic acid to an acceptor prepared from Sindbis membrane glycoprotein. These results imply that glycosylation of viral glycoproteins is catalyzed by cellular enzymes and that at least some of the carbohydrate of the virus glycoprotein can arise by host modification.

Taken together all these data indicate that the carbohydrate moiety of the viral glycoproteins is assembled not on a template but by a family of sugar transferases which act in consort to produce the complete carbohydrate structure. Thus, depending on the disposition of these enzymes in the host cell and the peculiarities of synthesis and maturation of the membrane glycoprotein of each virus, structures identical in terms of sequence and linkage of sugars may reach a degree of completion which is characteristic of both virus and cell.

Although it is not absolutely proven it seems reasonable to assume that in these glycoproteins the virus-determined polypeptide bears carbohydrate side chains in covalent linkage, which are specified by the host and are thus the site of possible host modifications. To proceed with the assumption of host modification, one must infer that the virus-specified protein has evolved such an amino acid sequence to serve as substrate for the first of the host-glycosylating enzymes, or, alternatively, that the virus specifies one sugar transferase to catalyze the formation of the initial sugar–amino acid bond, and the host enzymes then complete the structure (Burge and Huang, 1970).

These hypotheses are in agreement with previous reports by Haukenes *et al.* (1965), Laver and Webster (1966), and Lee *et al.* (1969), who isolated host specific glycopeptides from influenza virus. It is thus possible that viral glycoproteins are hybrid antigens, part host-specified, part virus-specified.

The presence of glycoproteins seems to be essential for the structure of viral envelopes but their function is, as yet, unknown. However, it should be mentioned here that besides host-dependent variations the glycoproteins particularly also show strain-specific differences with influenza viruses (Klenk *et al.*, 1972a). This suggests that these constituents might be the molecular basis for the strain specific antigenicity of these viruses.

(b) *Carbohydrates specified by the virus.* The evidence for a host specificity of the carbohydrate moiety of viral glycoproteins has been obtained exclusively with viruses which mature by budding from cellular membranes. In contrast, data obtained from studies on vaccinia virus which matures without connexion to cellular membranes suggest that here the carbohydrates of the viral glycoproteins might be virus-specific (Garon and Moss, 1971). Analysis of a pronase digest of glycoproteins labeled in uninfected and vaccinia-infected cells provides evidence that the virion glycopeptide is made only in the latter. Furthermore, the vaccinia glycoprotein contains only glucosamine; in contrast to arboviruses, rhabdoviruses, and myxoviruses it does not contain any other sugars. These data suggest that vaccinia virus might have its own glycosylating enzymes. Since cellular glycosyltransferases exist as multienzyme complexes associated with smooth membranes, these host enzymes may not be available for glycosylation of proteins in the so-called virus factories where assembly of the vaccinia envelope takes place.

C. LIPIDS

1. *Origin of Lipids in Viruses*

For a variety of reasons it is generally accepted that enveloped viruses acquire their lipids by utilization of host cell lipids:

(1) Enveloped viruses contain most or all of the lipid classes present in membranes of the host cell. Only in a few instances lipids have been found in viruses which could not be detected in the uninfected host cell.

(2) Viruses show differences in their lipid pattern, if they are grown in different host cells.

(3) Radioactively labeled cellular lipids—and in many cases most significantly those which had been labeled before infection—are incorporated into virions.

(4) The lipid pattern of viruses is as complex as that of the host cells. The information capacity of the genome of many viruses, however, appears to be too small to code for virus-specific lipid-synthesizing enzymes.

Pfefferkorn and Hunter (1963b) searched for the source of the phospholipids of Sindbis virus. When cells were labeled with P^{32} and then incubated for 30 hours in unlabeled medium before infection, each individual phosphatide in the infected cell and in purified virus had a nearly identical specific activity. It was concluded that the viral phospholipid is largely derived from preexisting cellular phosphatides. The main lipid classes in the virion were sphingomyelin, phosphatidylcholine, phosphatidylethanolamine, and cholesterol, and these were also all present in the host cell (Pfefferkorn and Hunter, 1963a). Similar results have been obtained with Semliki Forest virus, another arbovirus (Friedman and Pastan, 1969).

The lipids of vesicular stomatitis virus (VSV), which belongs to the rhabdovirus group, have been investigated in a detailed analysis (McSharry and Wagner, 1971a). Thin-layer chromatography revealed no unusual neutral lipids or phospholipids and gas chromatography revealed no unusual fatty acids incorporated into VSV.

The lipid composition of myxoviruses was found to be similar to that of the host cell (Armbruster and Beiss, 1958; Frommhagen *et al.*, 1958, 1959). Host cell lipids, labeled radioactively before infection, are incorporated into virus particles (Wecker, 1957). Furthermore, Kates and co-workers (1962) found in influenza virus grown in different cell lines host-specific modifications of the lipid composition.

The lipid pattern of leukoviruses was found to be similar to that of other RNA-containing viruses budding from the cell surface, and it was assumed that the viral phospholipids originate from the host cell (Rao et al., 1966; Quigley et al., 1971).

Herpes virus-infected cells, prelabeled with radioactive choline or phosphate, yielded virus particles which contained radioactively labeled cellular lipids. From this it was concluded that most of the phospholipids which become part of the virions are present in the cell at the time of infection (Asher et al., 1969; Ben-Porat and Kaplan, 1971).

On the basis of these findings it is generally accepted that the lipids of viruses budding from preformed cellular membranes are derived from the host cell. Evidence for the host cell origin of lipids is not as obvious in viruses which are assembled without continuity to cellular membranes. For instance, it has been shown that the envelopes of vaccinia virus derive their lecithin not from a cellular pool, already present before infection, but primarily from synthesis prevailing at the time of membrane biogenesis (Dales and Mosbach, 1968).

However, since all lipids of vaccinia are also found in uninfected cells, it may be concluded that lipid biosynthesis is not basically altered in infected cells and that the virus structure does not require novel lipids.

Such unusual lipids not present in the host cell prior to infection have been found in two viruses in significant amounts.

Fowlpox virus contains squalene (White et al., 1968), and the bacteriophage PM2 phosphatidic acid (Braunstein and Franklin, 1971). Both lipids are metabolic precursors of major lipids in the uninfected host cells of these viruses: squalene of cholesterol and phosphatidic acid of phosphatidyl ethanolamine and phosphatidylglycerol. This suggests that squalene and phosphatidic acid are the products of an impaired cellular lipid biosynthesis rather than of a true virus-specific lipid biosynthesis.

2. Lipid Classes in Viral Envelopes

Detailed lipid analyses exist on quite a variety of viruses which acquire their envelopes by budding from plasma membranes. It is therefore possible to draw some general conclusions on the lipid profile of these types of enveloped viruses. Although the lipid pattern of the individual host cell has a significant influence on the lipid composition of the virus, as will be shown later, a series of typical features is common to all of these viruses.

(a) *Phospholipids*. The phospholipids of these viruses have invariably been found to be composed of sphingomyelin, phosphatidylcholine, phosphatidyl ethanolamine, phosphatidyl serine, and phosphatidylinositol. Sphingomyelin, phosphatidylcholine, and phosphatidylethanolamine are the predominant phospholipids, whereas phosphatidylserine and phosphatidylinositol are usually present in smaller amounts. Of particular interest is the constant finding of a high sphingomyelin content and frequently comparatively large proportions of phosphatidylethanolamine have been reported.

(b) *Cholesterol.* Viral envelopes also contain a significantly higher cholesterol content than does the host cell. Therefore the molar ratio of cholesterol to phospholipid is usually close to 1 in viral envelopes, whereas it is about 0·2 in host cells. Cholesterol has been found mostly in the free form in these viruses; very little cholesterolesters are present. Lipid patterns showing these features have been found in the envelopes of arboviruses (Pfefferkorn and Hunter, 1963a; Heydrick *et al.*, 1971; Renkonen *et al.*, 1971), rhabdoviruses (McSharry and Wagner, 1971a), orthomyxoviruses (Frommhagen *et al.*, 1958; Kates *et al.*, 1962; Blough *et al.*, 1967; Klenk *et al.*, 1972a), paramyxoviruses (Soule *et al.*, 1959; Blough and Lawson, 1968; Klenk and Choppin, 1969b, 1970a), and leukoviruses (Quigley *et al.*, 1971).

(c) *Fatty acids.* The fatty acid pattern of the phospholipids consists predominantly of saturated and unsaturated acids with a chain length of 16, 18, and 20 carbon atoms. Compared with whole cells the phospholipids of the viruses contain larger amounts of saturated fatty acids (Klenk and Choppin, 1969b, 1970a; McSharry and Wagner, 1971a; Renkonen *et al.*, 1971) presumably due to the high content of sphingomyelin which possesses mainly saturated fatty acids (Blough, 1971).

(d) *Glycolipids.* Glycosphingolipids consisting of sphingosine, fatty acids, and carbohydrates have been detected in viral envelopes. It was found that glucosylceramide and N-acetyl-galactosaminyl-galactosyl-galactosyl-glucosyl-ceramide were constituents of the parainfluenza virus SV5 grown in bovine and in monkey kidney cells (Klenk and Choppin, 1970b). Arboviruses and rhabdoviruses have been found to possess in addition to neutral glycolipids gangliosides, i.e. neuraminic acid containing glycolipids (Klenk and Choppin, 1971; Renkonen *et al.*, 1971). Glycolipids may represent a substantial portion of the total polar lipids of viral envelopes. They are clearly host specific. In addition to their significance as structural components of the viral envelope, these substances may be antigenic, e.g. they may possess blood group or Forssman activity (Koscielak *et al.*, 1968; Makita *et al.*, 1966; Martensson, 1969). Such activities have been detected by immunological means in preparations of myxoviruses and other lipid-containing viruses (Springer and Tritel, 1962; Isacson and Koch, 1965; Rott *et al.*, 1966). These results, and the finding of glycopeptide host antigens in influenza virions, suggest that if host cell antigens are present in enveloped viruses, carbohydrate moieties will be the antigenic determinants.

From the chemical data on the lipid composition some conclusions can be drawn as to the physical properties of viral envelopes. The lipid pattern with relatively high proportions of sphingolipids and of saturated and mono-unsaturated fatty acids favors the formation of closely packed membranes.

The presence of large amounts of cholesterol in the envelopes would be an additional factor which facilitates the formation of close packing of lipid molecules, because cholesterol has a well-known condensing influence on the packing of phospholipids (for ref. see Malhotra, 1970). Viral membranes would thus show a high degree of order and rigidity. These features are not only found in envelopes which bud from the plasma membrane, but they are also shown by the vaccinia virus envelope which is assembled without connexion to cellular membranes and contains relatively large amounts of saturated fatty acids (Dales and Mosbach, 1968) and cholesterol (for ref. see Holowczak, 1970).

2. Factors Determining the Lipid Composition

It is clear from the findings reported above that the lipids of viral envelopes originate from cellular sources, where they may be synthesized before or during infection. It could therefore legitimately be assumed that the lipid pattern of the host cell plays an important role in determining the lipid composition of the viral envelope. Furthermore it could be expected that a virus showed host specific modifications in its lipid composition, if it was grown in different host cells. In the envelopes of myxoviruses such host-induced modifications were suggested by the different susceptibility to the action of phospholipase C (Simpson and Hauser, 1966b). Virus was found to be highly sensitive to inactivation by the enzyme when grown in cultures of chick embryo fibroblasts but was resistant if propagated in the allantoic sac. Host-specific differences in the buoyant density of myxoviruses have been reported which were also interpreted as manifestations of variations in the viral lipid composition (Stenbeck and Durand, 1963). Kates and co-workers (1962) analyzed the lipids of influenza virus grown in chick embryo and in calf kidney cells and found that the lipids of the virus particles resembled those of the particular cell in which the virus was grown.

On the other hand, it seems now to be an established fact that all proteins in the envelopes of many viruses are virus-specific. Therefore it has been postulated that these proteins might exert a selectivity in utilizing the available cellular lipids and thus might have a directing influence on the viral lipid composition. In the following a critical survey will be given on the arguments in favor and against each of these modes of lipid determination.

(a) *Lipid composition of host membranes as determinant of envelope lipids.* It had long been thought that enveloped viruses contain cellular lipids, and the assembly of most of these viruses by budding from cellular membranes has suggested these membranes as the logical source of the viral lipids. However, isolation of plasma membranes and comparison of their lipids with those of purified virions was necessary to prove this hypothesis.

The first in a series of such studies has been carried out on the paramyxovirus SV5 which had been grown in three different host cells (Klenk and Choppin, 1969b). Plasma membranes have been isolated from these viruses by the fluorescein mercuric acetate method of Warren and co-workers (1966). Detailed lipid analyses of the virus have been carried out and compared with those of the respective host cells and their plasma membranes. It was found that the plasma membranes contained more cholesterol, sphingomyelin, and saturated fatty acids than whole cells and that in each of these aspects the lipids of the viral envelope resembled those of the plasma membrane of the particular cell type in which the virus was grown.

In a comparable study on Semliki Forest virus, Renkonen and co-workers (1971) obtained similar results. Plasma membranes were prepared according to the method of Wallach and Kamat (1966) by gas cavitation. The virus lipids were found to consist of 31% neutral lipid which was predominantly cholesterol, 61% phospholipid with a high proportion of sphingomyelin, and 8% glycolipid.

Whereas significant differences were detected between the lipid pattern of the virus and that of the whole cell and of endoplasmic reticulum, almost all the lipids of the virion resembled closely those of the plasma membrane.

McSharry and Wagner (1971a) in a comparative study analysed the lipids of two different strains of vesicular stomatitis virus grown in two different host cells with the respective plasma membranes which were prepared by two different methods. The phospholipids in both virus strains and their fatty acid pattern were quite similar. Furthermore, the data indicated that the lipid composition of vesicular stomatitis viruses primarily reflected that of the plasma membranes as the site of their maturation.

Quigley and co-workers (1971) analyzed the phospholipids of leukovirus, two paramyxoviruses, and an arbovirus, and compared them with the lipids of the plasma membrane of chick embryo fibroblast in which all of these viruses were grown. They found that the phospholipid pattern of these viruses was remarkably similar and resembled that of the plasma membrane of their common host cell.

Comparative data on the lipids of viruses with maturation sites other than the cell surface and on the membranes from which they bud are scarce. Ben-Porat and Kaplan (1971) determined the phospholipids of a herpes virus and compared it to the inner nuclear membrane, the membrane from which this virus buds. Their experiments showed that the phospholipid composition of the inner nuclear membrane is different from the bulk of cellular membranes mainly by a higher sphingomyelin content. The phospholipid composition of the viral membrane closely mimics that of the inner nuclear membrane. This suggests that the mechanisms involved in the assembly of herpes virus on the inner nuclear membrane are similar to those responsible for the assembly of other viruses on the plasma membrane.

Cellular membranes show organelle specificities and species specificities (Rouser *et al.*, 1968). Organelle specificities are cell-independent general features of a certain organelle and in the case of plasma membranes they consist in high proportions of cholesterol and sphingolipids. It has been shown above that in this respect virus envelopes clearly resemble the membranes on which they are assembled. On the other hand, species specificities are also found in plasma membranes. For instance, the individual phospholipids are present in the plasma membranes of a variety of cells in different proportions (Klenk and Choppin, 1969b). Such quantitative species-specific differences were reflected in the paramyxovirus SV5, when it was grown in these cells.

Studies on glycolipids reveal such host specificities in an even more convincing manner, because here the qualitative differences between various cell membranes are also reflected in the virions, e.g. the glycolipid galactosyl-galactosylglucosylceramide is found in plasma membranes of monkey kidney cells and in virions grown in these cells, but not in membranes or virions from bovine kidney cells (Klenk and Choppin, 1970b). Such differences in the lipids of virions which, though grown in different cells, contain the same virus-specific proteins emphasize the importance of the host cell membrane in determining the lipid composition of the virion.

(b) *Viral proteins as determinants of the envelope lipids.* The lipoprotein-complex theory of membrane structure (Benson, 1966) suggests that the hydrocarbon chains of the lipids are bound specifically to a polar region of the membrane proteins. In order to test this hypothesis Blough and co-workers undertook comparative lipid analyses of various strains of myxoviruses which were grown in the same host, the embryonated egg (Blough and Lawson, 1968; Tiffany and Blough, 1969a, b; Blough, 1971). In these studies the authors found primarily differences in the fatty acid patterns of these viruses. From these results they concluded that by hydrocarbon chains the envelope proteins select certain lipid species. Therefore the viral envelope proteins were considered to be the most important determinant of the lipid composition of the envelope.

However, differences in the fatty acid pattern should be cautiously interpreted. Since with one exception (Tiffany and Blough, 1969a) lipid analyzes of host material have not been carried out, variations in the fatty acid spectrum of the host cannot be ruled out. It is a well-known fact that the fatty acid pattern in tissues can be influenced by dietary measures (for instance, in an egg) or by alterations in the composition of the culture medium. Such changes in the cell culture medium have been found to have a direct effect on the fatty acid pattern of cultured cells and virus grown in these cells (Klenk and Choppin, 1970a).

It should also be noted here that McSharry and Wagner (1971a) were not able to detect significant differences in the fatty acid patterns of the phospholipids of two strains of vesicular stomatitis virus. These experiments were carried out in tissue cultures, a host more suitable for such studies than the chicken embryo, since it can be maintained under quite well-defined conditions. On the same line Renkonen and co-workers (1971) could not detect significant differences between the fatty acids of an arbovirus and its host cell membrane.

Compared to the phospholipids even more striking differences were found in the fatty acid patterns of neutral lipids, when different myxoviruses were analyzed (Tiffany and Blough, 1969a, b). However, the significance of neutral lipids for the structure of the virus is dubious: despite many reports suggesting the presence of triglycerides and sterol esters as components of cell membranes, these lipids do not appear to be membrane constituents. Triglycerides occur as aggregates (fat droplets, vacuoles) of various sizes particularly in virus-infected cells and thus may be present in organelle and virus preparations as contaminants (Rouser et al., 1968).

McSharry and Wagner (1971a) had found that two strains of vesicular stomatitis virus grown in the same host cell differed by their neutral lipid content. For the reasons outlined above this finding does not seem to be convincing evidence for determination of the envelope lipids by virus proteins.

There are other indications for a virus directed modification of cellular lipids during envelope assembly. Although in general the phospholipids of viruses which bud from the cell surface resemble those of the host cell membrane, the envelopes frequently contain relatively higher amounts of phosphatidyl ethanolamine. This has been found first with the parainfluenza virus SV5 (Klenk and Choppin, 1970a). This phenomenon has been taken as evidence that the virus might prefer a high phosphatidylethanolamine content. This finding has been confirmed with other viruses (McSharry and Wagner, 1971a) and careful examination of a series of investigations (McSharry and Wagner, 1971a; Renkonen et al., 1971; Quigley et al., 1971) reveals that viruses budding from plasma membranes have in general a slightly higher sphingolipid and phosphatidylethanolamine and a slightly lower phosphatidylcholine content than the plasma membrane of the host cell. It is conceivable that this shift might be a virus-directed alteration of the lipid pattern. However, there might be a more trivial explanation, if one realizes that the same differences, though much more pronounced, are found between plasma membranes and whole cells. This suggests that the phospholipids in plasma membranes and in viral envelopes might be in fact identical and that the apparent differences might just be due to contamination of plasma membrane preparations by other cellular membranes. This hypothesis is supported by the finding that certain plasma membranes which can be obtained in very pure form like erythrocyte ghosts (Malhotra, 1970) and the membranes of milk globules (Huang and Kuksis, 1967) contain almost equal amounts of sphingomyelin, phosphatidylcholine, and phosphatidyl ethanolamine, as do viral envelopes.

In a recent study the lipids of Sindbis virus grown in chick embryo fibroblasts and in BHK cells have been analyzed and compared to those of the plasma membrane of the host cell (David, 1971). Whereas the phospholipid as well as the fatty acid patterns were very similar in virus grown in both host cells and in BHK plasma membranes, significant differences were found between plasma membranes of chick embryo fibroblasts and the respective virus. From these results it was concluded that the lipid composition of viral envelopes reflects the lipid affinities of the viral envelope peptide and not the lipid composition of the host cell membrane. A closer look at these data reveals, that the lipid composition of the plasma membrane preparation of chick embryo fibroblasts is very similar to that of total cells, particularly in its unusually low sphingomyelin content. However, a high sphingomyelin content seems to be a typical feature of plasma membranes in general, and has indeed been found also in plasma membranes of chick embryo fibroblasts by others (Quigley et al., 1971). Thus, it is conceivable that these results, which are not in agreement with those of others (Quigley et al., 1971), do not indicate lipid determination by the viral envelope protein but are rather due to contamination of the plasma membrane preparation of chick embryo fibroblasts by other cellular organelles.

General principles, according to which envelope proteins would selectively utilize and rearrange lipids of cell membranes, do not emerge from the studies where significant differences in the lipids of viral envelopes and of their host cell membranes have been reported. In some studies differences in the fatty acid pattern have been found (Tiffany and Blough, 1969a, b), in some studies differences in the phospholipids (David, 1971), and in others differences in the neutral lipids (McSharry and Wagner, 1971a). There is no consistency in these findings. On the other hand, it has been found now by several groups in a variety of different virus-cell systems that viral envelopes share characteristic features with the plasma membranes from which they bud (Klenk and Choppin, 1969b, 1970a, b; Renkonen et al., 1971; McSharry and Wagner, 1971a; Quigley et al., 1971).

Thus, the lipids of viruses budding from preformed cellular membranes are in general very similar to the lipids of these membranes. On the other hand, viral envelopes which mature without connexion to cellular membranes contain lipid patterns logically quite distinct from those of host membranes. This has been found with vaccinia virus (Dales and Mosbach, 1968) as well as with the bacteriophage PM2 (Braunstein and Franklin, 1971).

III. Envelope Structure

A. ARBOVIRUSES, RHABDOVIRUSES, MYXOVIRUSES

Detailed studies on the arrangement of lipids, proteins, and carbohydrates in viral envelopes employing a variety of physical and chemical techniques have been carried out in the past years. The investigations which have been greatly facilitated by the preparation and analysis of subviral particles, have been carried out, mainly on influenza viruses and on arboviruses of group A, but it appears that many conclusions reached in these studies are of general validity for all viral envelopes.

Electron microscopy revealed in numerous studies that viral envelopes consist of two or three principal layers: (1) an outer fringe of surface projections, (2) a central layer which has the trilaminar appearance of a unit membrane, and (3) an additional inner leaflet which has been found in some studies (Compans and Dimmock, 1969; Bächi et al., 1969) but is not present in all viruses.

1. The Lipid Layer

Since on electron micrographs viral envelopes show generally the structural feature of a unit membrane, the lipids have long been thought to be present in the virion as a bimolecular leaflet. The first direct evidence that a continuous lipid bilayer forms the central leaflet of the viral envelope stems from experiments with virus particles devoid of spikes. If influenza virions are treated with proteolytic enzymes, particles can be recovered which have lost the surface projections and are surrounded by a smooth membrane (Kendal et al., 1969; Compans et al., 1970; Schulze, 1970). Lipid antigens in intact virions not accessible to antibodies are unmasked after removal of the spikes (Klenk et al., 1972a). This indicates that lipids are exposed on the surface of this smooth membrane. On the other hand, all viral proteins except the spike proteins are well protected from the enzymatic degradation. Therefore it seems likely that the lipids are present as a continuous layer which clearly separates the spikes from the other components of the virion. Removal of the outer layer of the viral envelope, i.e. the spikes, leaves the lipid layer completely unchanged, as can be judged from the amount and the composition of the lipid in the stripped particles (Klenk et al., 1972a). This suggests that the lipid layer in the virion is a structural entity, which is able to maintain its integrity at least to some extent without support of the peripheral proteins. The membrane model which provides the best explanation

H

for these observations is that of a bimolecular leaflet. It is interesting that the treatment with proteolytic enzymes removes the spikes completely and that no split products are left in the stripped particles. This suggests that the spikes are attached to the surface of the lipid layer and that they do not extend through it.

The presence of a lipid bilayer has been also substantiated in Sindbis virus by X-ray diffraction (Harrison *et al.*, 1971a). The radial electron density distribution in the Sindbis virus particle has been determined to a resolution of 28 Å from measurements of spherically averaged X-ray diffraction. The most striking feature of the density profile is a deep minimum at a radius of 232 Å, from which the authors infer that the lipids of Sindbis virus are organized in a bilayer at about that radius. The polar groups are localized near 210 and 258 Å, as indicated by the density maxima at these radii. The interpolar-group distance of 48 Å is typical of phospholipid–cholesterol bilayers.

By incorporating spin-label analogs of stearic acid and androstan into the envelope of influenza virus and into the erythrocyte membrane, Landsberger and co-workers (1971) have shown that these probes are ordered in the lipid layer in such a way that the hydrocarbon axis is preferentially oriented perpendicular to the plane of the layer. Striking increases in the mobility of the nitroxide group could be shown, as it is moved away from the carboxyl end of the fatty acid. This relationship suggests that the lipid of the virus is organized into a bilayer.

In a comparable study similar evidence was provided for the presence of a lipid bilayer in the envelope of Rauscher murine leukemia virus (Landsberger *et al.*, 1972). Thus, also in this respect the lipid component of an oncogenic virus cannot be distinguished from that of a non-oncogenic enveloped virus.

Tiffany and Blough (1970a) have presented several models of the influenza viral envelope in which the lipid is arranged in the form of spherical micelles having the radius of about the length of a lipid molecule. These models are based on a calculation that estimates that there is about twice as much lipid as is needed to form a single bilayer covering the spherical virion. However, Landsberger *et al.* (1971) have made similar calculations using more recent data and come to different conclusions. They calculated the area of the lipid layer in the influenza virus envelope to be 147×10^4 Å^2. The lipid layer was found to consist of 3.6×10^4 molecules of cholesterol and 3.6×10^4 molecules of phospholipids. From these data they calculated that the area of bilayer that can be formed by the total lipid in each virion is 148×10^4 Å^2, a value in agreement with that required to coat the virion with a single bilayer.

Comparable data are available on the area covered by the lipid bilayer of Sindbis virus which is smaller than influenza virus. The surface area of all polar groups in the bilayer has been calculated in this case to be $13 \cdot 2 \times 10^5$ Å^2 (Harrison et al., 1971a). The sum of the areas of spherical surfaces with radii of 210 Å and 258 Å, i.e. the inner and the outer surface of the bilayer, is $13 \cdot 9 \times 10^5$ Å^2. The two figures agree within the uncertainty of about 10% in the estimate based on polar-group area and compositional data. It was therefore concluded that also with Sindbis virus the lipid bilayer extends completely around the core.

In the closely related Semliki Forest virus, Renkonen et al. (1971) determined about 10,000 cholesterol–phospholipid pairs, which also can be arranged in a bilayer type of structure in the space available in the envelope.

2. The Spikes

When influenza viruses are exposed to proteolytic enzymes, spike-free particles can be obtained which have lost all the viral glycoproteins (Compans et al., 1970; Schulze, 1970). This striking observation has not only been found with influenza, but with all enveloped viruses studied to date. These include an arbovirus (Compans, 1971), a paramyxovirus (Chen et al., 1971), a rhabdovirus (McSharry et al., 1971), and a leukovirus (Rifkin and Compans, 1971). These observations are consistent with the finding that isolated spikes of influenza virus, which can be split from intact virions by treatment with ether and adsorption to red blood cells (Hoyle, 1952; Schäfer and Zillig, 1954) consist of the viral glycoproteins (Klenk, 1971; Klenk et al., 1972a).

From these results it can be concluded that the components forming the spikes are only the glycoproteins and none of the other major viral proteins. Since the spikes of influenza virus contain the hemagglutinin and the neuraminidase, one has to conclude that these structural entities are composed of glycoproteins. For instance, the hemagglutinin is composed of two glyco-protein molecules of the molecular weight of about 50,000 and two glyco-protein molecules of the molecular weight of about 30,000 (Skehel and Schild, 1971; Laver, 1971; Klenk et al., 1972a). These form a particle with a molecular weight of approximately 150,000, which has been reported to be the size of the functional hemagglutinating unit (Laver and Valentine, 1969; Webster, 1970).

The other functional subunit of the influenza virus envelope also located in the spike layer but clearly distinct from the hemagglutinin is the viral neuraminidase (Rott et al., 1969). More information is required before the structure of the neuraminidase as it exists on the virion can be determined. Its morphology when isolated by detergent treatment (Drzeniek et al., 1968; Laver and Valentine, 1969), its sedimentation coefficient, and current information about its polypeptide composition (Haslam et al., 1970b; Webster, 1970; Skehel and Schild, 1971) suggest that it is composed of several subunits. Enzymes released from the virions by proteolysis and by detergents have similar sedimentation coefficients (7·8–10·8) (Noll *et al.*, 1962; Laver, 1963; Drzeniek *et al.*, 1966; Kendall *et al.*, 1968; Laver and Valentine, 1969; Webster and Darlington, 1969), suggesting that these two methods yield molecules with the same polypeptide composition. Proteolytic enzymes may, therefore, foster release of the active molecule by altering interactions between the glycoprotein subunits and the lipid layer (Schulze, 1972).

Schulze (1972) has calculated that a virion of the WSN strain of influenza virus contains 553 spikes. This is in good agreement with the value of 550 spikes calculated from interspike distances for spherical virions of the PR8 strain (Tiffany and Blough, 1970b).

By the use of an interspike spacing of 75 Å (Nermut and Frank, 1971) and a spike radius of 20 Å (Drzeniek *et al.*, 1968; Laver and Valentine, 1969), Landsberger and co-workers (1971) calculated that about 25% of the particle surface of influenza virus is covered with spikes, if a hexagonal arrangement is assumed (Nermut and Frank, 1971; Almeida and Waterson, 1967). Complete removal of the glycoprotein spikes has no detectable effect on the lipid organization, as monitored by the electron spin resonance studies of Landsberger and co-workers (1971). The fact that such a large change in surface structure as spike removal can occur with no detectable alteration of lipid structure suggests once more that with influenza virus the spikes do not penetrate through the lipid phase. The same conclusion has been reached with Sindbis virus (Harrison *et al.*, 1970a).

It has been shown previously in this article that the carbohydrates in viral envelopes can be either lipid-bound (Klenk and Choppin, 1970b) or protein-bound (Klenk *et al.*, 1970a) and that the constituent sugars of the glycolipids and of the glycoproteins are different. These findings are confirmed by studies on influenza virus with phytagglutinins and blood group specific antisera (Klenk *et al.*, 1972a). Concanavalin A reacts only with the virions, if the glycoprotein containing spikes are present on the particle. The sugar determinant for this type of interaction is most likely mannose which has only been found in viral glycoproteins (Klenk *et al.*, 1970a) but not in viral glycolipids (Klenk and Choppin, 1970b). On the other hand, antisera against blood-group A substance and phytagglutinin of *Dolichos biflorus* with the same specificity interact with the stripped virions. Since the glycoproteins are completely removed from these particles, the reaction can involve only the lipid-bound carbohydrates which are still present in these particles as shown by chemical analysis. The observation that Concanavalin A interacts only with the spikes whereas phytagglutinin from *Dolichos biflorus* interacts only with the stripped particles provides further evidence that the glycoproteins are located in the spikes and the glycolipids

in a separate lipid layer. The interaction of Concanavalin A with myxoviruses is particularly interesting, since agglutinability by this phytagglutinin
has been thought to be rather specific for transformed cells (Burger, 1969;
Inbar and Sachs, 1969). However, cells infected with a large variety of
non-oncogenic enveloped viruses show the same effect (Becht et al., 1971,
1972; Oram et al., 1971). These findings suggest that Concanavalin A induces
agglutination by bridging spikes appearing at the surface of infected cells
and carrying the determinant sugar in their carbohydrate moiety.

3. *Inner Protein Layer*

The envelopes of rhabdoviruses, orthomyxoviruses and paramyxoviruses
contain in addition to the glycoproteins a major polypeptide which is carbohydrate-free and has the lowest molecular weight of all viral proteins.

There is evidence that this protein completely coats the inner surface
of the lipid bilayer. With influenza virus it remains with the internal component and with the lipid, when the spike proteins are removed by proteolytic enzymes (Compans et al., 1970; Schulze, 1970). This suggests that it is
protected from the enzymatic degradation by the lipid layer. It is not removed when these stripped particles are fixed with glutaraldehyde and then
treated with the detergent NP–40 (Schulze, 1970). The latter treatment
removes all the phospholipid from the particle, but does not reveal strands
of ribonucleoprotein. Rather, it removes only the smooth layer, leaving
oblong to spherical particles with a rough surface (Schulze, 1972). Subviral
particles similar in density, chemical composition and appearance have also
been obtained from influenza virus by treatment with desoxycholate without
the use of a fixative (Skehel, 1971).

Other evidence that the internal component of influenza virus is surrounded
first by a layer of carbohydrate-free protein and then by a layer of lipid
comes from electron microscopical studies of negatively stained stripped
particles. The walls of these particles consist of two layers. In thin sections,
an electron dense layer about 60 Å in thickness encloses the nucleocapsid of
stripped particles (Schulze, 1971); this same structure is between the ribonucleoprotein and the spike layer in intact virus particles (Apostolov and
Flewett, 1969; Compans and Dimmock, 1969). The carbohydrate-free
membrane protein is the only protein present in the virion in sufficient
quantity to form such a structure. It has been calculated that 3300–3700
molecules are required to enclose the nucleocapsid in a spherical shell of
protein with an outside diameter of 800 Å and a thickness of 60 Å (Schulze,
1972). Similar studies of a rhabdovirus and an arbovirus of group B revealed
that here the envelope also contains a protein leaflet below the lipid bilayer
(Cartwright et al., 1970; Stollar, 1969).

The envelopes of arboviruses of group A do not contain a carbohydrate-free protein, the glycoprotein of the spikes is the only membrane protein (Table III). The surface of the internal component must therefore make direct contact with the inner polar groups of the lipid bilayer. Here the nucleocapsid protein is in this sense also a "membrane protein".

B. BACTERIOPHAGE PM2

Several studies on the envelope structure of this virus have been carried out and they suggest that it might be different from that of the envelopes of the animal viruses described above in an important point. From electron microscopic studies, the virus appears to be icosahedral in shape, 600 Å in diameter, and has small brush-like "spikes" at the vertices (Silbert et al., 1969). From low angle X-ray diffraction studies, the particle can be divided into three spherically symmetric compartments: (1) an internal sphere containing protein, and most likely nucleic acid as well, with a radius of 200 Å; (2) a lipid layer 40 Å thick centered at a radius of 220 Å; and (3) an external protein shell extending radially from 240 Å to the surface at 300 Å (Harrison et al., 1971b). Without making any specific assignments, at least three proteins are structurally required; one for the external shell, another for the spikes and still a third for the inner sphere. Datta and co-workers (1971) have found four different polypeptides in the total virion. It has not been determined yet, whether any of them contains carbohydrates, nor have they been attributed to specific subviral structures. However, the authors reason with some justification that one of the structural proteins which has the molecular weight 34,000 and comprises about 60% of the total virus protein can be assigned to the external shell. To be more specific, since this protein accounts for close to 50% of the total weight of the PM2 particle and one-third of the particle volume is inside the lipid layer, it would be very unlikely that this protein, together with the nucleic acid would occupy the inner sphere. On the other hand, the protein could easily cover the outside of the lipid layer with cylinders 30 Å in diameter and 60 Å in length (Harrison et al., 1971b). 650–710 molecules of this protein have been calculated to be present in the PM2 particle (Harrison et al., 1971b; Datta et al., 1971).

Harrison and co-workers (1971b) have calculated that about 2×10^4 lipid molecules would be required to form a complete bilayer in the particle. The measured phospholipid content of PM2 is about $1 \cdot 3 \times 10^4$ molecules per virion. It seems from the correlation of chemical and structural results that lipid may occupy only about 65% of the area at $r = 220$ Å. The bilayer may be fenestrated or separated into patches; in particular, these authors expect special structures at the icosahedral vertices and it is conceivable that here the spikes penetrate through the lipid bilayer. Thus, unlike the enveloped viruses described above, PM2 would contain protein structures penetrating the lipid layer, a situation which has been found in erythrocyte membranes (Bretscher, 1971).

IV. Biogenesis of Viral Envelopes

The biogenesis of membranes remains one of the fundamental unresolved problems in biology. Viral envelopes offer certain advantages over other systems in such studies. Because of their virus specificity envelope proteins can be expected to be easily traced from their site of origin to the site of envelope assembly. Moreover, on the basis of morphological criteria, at least in certain stages of their assembly, viral envelopes can be identified and discriminated from cellular membranes. Nevertheless, our knowledge on the biosynthesis of envelope components and on their assembly is far from complete. Therefore, in this chapter interest will be focussed only on certain aspects which seem to be of major importance, and many of the conclusions which will be drawn have to be largely hypothetical.

A. BIOSYNTHESIS OF ENVELOPE COMPONENTS

1. *Envelope Proteins*

With some viruses several distinct stages in the formation of the envelope glycoproteins have been identified. There is evidence that a precursor–product relationship exists between these structural envelope proteins and non-structural proteins which are also formed in infected cells.

In cells infected with fowl plague virus, an influenza A virus, D-glucosamine inhibits the synthesis of the viral hemagglutinin and the formation of mature virions (Kaluza *et al.*, 1972). Analysis by polyacrylamide gel electrophoresis revealed that under these conditions the virus-specific carbohydrate-free polypeptides are still being formed. However, all viral glycoproteins are missing, due to a lack of activated sugars which serve as carbohydrate donors in the glycosylation reaction. Instead of the glycoproteins a single polypeptide (MW 64,000) can be detected. After removal of the glucosamine-block this polypeptide is glycosylated to a large non-structural glycoprotein (MW 76,000) (Klenk *et al.*, 1972b). This glycoprotein is cleaved into two smaller structural glycoproteins which represent the subunits of the viral hemagglutinin (Lazarowitz *et al.*, 1971).

That envelope constituents have to migrate considerable distances from one cell compartment to another in order to get from the site of their biosynthesis to the site of envelope assembly has been known for a long time. Breitenfeld and Schäfer (1957) have shown in their classical study that in cells infected with fowl plague virus the viral hemagglutinin can first be seen throughout the cell and is located in a higher concentration in a juxta-nuclear locus. Later the hemagglutinin accumulates in the peripheral region of the cell and also may be demonstrated in fine filaments which protrude from the cell margin.

With herpes virus electron microscopic data indicate that the nucleocapsid is assembled in the nucleus (Morgan et al., 1954; Siegert and Falke, 1966), and that the nucleocapsid acquires its envelope at the inner layer of the nuclear membrane (Siegert and Falke, 1966; Darlington and Moss, 1969). Sydiskis and Roizman (1967) showed that cytoplasmic protein synthesis accounts for most, if not all, of the proteins synthesized in herpes simplex virus-infected cells. Subsequently, it has been found that proteins synthesized in the cytoplasm were transferred into the nucleus (Fujiwara and Kaplan, 1967; Spear and Roizman, 1969; Ben-Porat et al., 1969). In these studies it has not been analysed whether these were proteins forming the nucleocapsid or proteins of the viral envelopes. However, it can be assumed that at least some of them were envelope proteins.

2. Lipid Metabolism in Virus-infected Cells

Wecker (1957), Kates and co-workers (1962), and Pfefferkorn and Hunter (1963a) showed that the specific activity of the phospholipids of arboviruses and myxoviruses closely resembled that of the phospholipids of the cells in which the viruses were grown. Asher and co-workers (1969) found no change in incorporation of choline after infection of cells with herpes simplex virus. It was concluded from these studies that lipid metabolism was not altered in virus-infected cells.

On the other hand, Ben-Porat and Kaplan (1971) described virus-induced changes in lipid metabolism. After infection of rabbit kidney cells with herpes virus, they observed an increase in the incorporation of ^{32}P, choline-^3H, and myoinositol-^3H into the phospholipids. An increase in the incorporation of choline-^3H into the cytoplasmic fractions of the infected cells occured, but this increase was especially marked in the nuclear membrane, from which this virus buds. The distribution of the radioactivity incorporated into the different phospholipids was also changed by infection; approximately two or three times more appeared in sphingomyelin in infected than in uninfected cells. Thus, infection with herpes virus seems to stimulate lipid metabolism as does infection with vaccinia virus (Dales and Mosbach, 1968). The difference to the work of Asher and co-workers (1969) may reside in the different experimental designs—in one case (Asher et al., 1969) the viral phospholipids were compared with the bulk of the cellular phospholipids, whereas in the other case (Ben-Porat and Kaplan, 1971) the viral phospholipids were compared with those of the membrane from which the virus buds.

Phospho- and glycolipid metabolism was also studied in cells infected with the paramyxovirus SV5 (Scheid and Choppin, 1971). The high virus yield obtained in bovine kidney cells suggests an active turnover of the lipids of the plasma membrane where this virus acquires its envelope. Incorporation studies with radioactive choline and glucosamine, and chemical analysis revealed a decrease of sphingomyelin and an increase of glycolipids in infected cells compared to uninfected cells. The results show alterations in sphingolipid metabolism in SV5 infected cells which is of interest because these lipids are predominantly found in the plasma membrane from which the virus acquires its envelope.

B. ENVELOPE ASSEMBLY

1. *Free Assembly*

Vaccinia belongs to the group of pox viruses which acquire their envelope in an assembly process free in the cytoplasm without connexion to preformed cellular membranes. Its biogenesis has been studied in detail by Dales and Mosbach (1968).

In synchronously infected cells, viral membranes are the first identifiable morphological component of vaccinia to appear in the viroplasmic matrix of factories. Such membranes can be observed for the first time at 3–3·5 hours after infection. Initially they exist as short, 500–1000 Å, segments, each possessing a unit membrane coated externally with a dense layer of 100 Å-long spicules. In cells sampled 3·5 hours after infection the sectioned membranes appear as arcs or circles. Therefore, in three dimensions they are developed by increasing their surface area and become envelopes of spherical particles. The unit membrane of vaccinia is 50–55 Å wide and indistinguishable in appearance or width from those of the host cells.

Sequential appearance of viral envelopes and immature and mature particles could be arrested at defined stages by means of actinomycin D or streptovitacin A. Application of these compounds in advance or following morphogenesis indicated that transcription into the requisite RNA precedes by 60 min, and translation into protein by less than 30 min, the assembly of membranes. Proteins required for maturation are synthesized within 30 min following morphogenesis of immature virus. Experiments with isotopically labeled choline indicated that the nascent phospholipid is preferentially integrated into the progeny, and analyses using thin-layer and gas-chromatography revealed that the fatty acid composition of vaccinia can be distinguished from that of host cell membranes. These combined results imply that unique membranes of vaccinia can condense *de novo* from precursors to become the envelope surrounding immature particles. Further differentiation into mature virus occurs inside this envelope (Dales and Mosbach, 1968).

Grimley and co-workers (1970) studied the envelope formation of vaccinia virus in the presence of rifampicin. This compound interrupted envelope assembly in HeLa cells. The primary action of rifampicin on vaccinia morphogenesis appeared to occur during the stage of envelope formation. When envelopes and immature particles were already present, maturation could continue, even in the presence of rifampicin. It was demonstrated that trilaminar membranes of irregular contours which accumulate in the presence of rifampicin are precursors of virus envelopes. When rifampicin was removed under controlled conditions, synchronous transitions were observed as the precursor membranes rapidly converted into uniformly curved units with a 100–120 Å coat on the convex surface.

It was found that the proteins necessary for envelope formation are made but not assembled in the presence of rifampicin (Moss et al., 1971). Although the drug does not appear to prevent virus envelope formation by inhibiting protein or nucleic acid synthesis, the question on its mode of action on virus assembly remains. Since coating of membranes is the earliest change so far detected after reversal of the antiviral effects of rifampicin, this may be the step that is directly affected by the drug. These findings also suggest that in vaccinia envelope assembly the formation of the unit membrane structure precedes the apposition of the surface coat.

2. *Assembly by Budding*

The basic principles of envelope assembly in a budding process are derived from studies on myxoviruses (Choppin et al., 1971). The first morphologically identifiable step in virus assembly is the appearance of the nucleocapsid in the cytoplasm (Fig. 1a). This structure is synthesized and assembled in the

FIG. 1. Envelope assembly by budding from the plasma membrane. Monkey kidney cells 22 hours after inoculation with paramyxovirus SV5. (a) Strands of nucleocapsid free in the cytoplasmic matrix near the surface of the cell (arrows). Magnification: ×47,500. (b) Nucleocapsid strands, many in cross section, closely aligned under a long region of the cell membrane. A layer of dense material resembling the spikes found in the viral envelope is present on the outer surface of the membrane (arrow). Magnification: ×73,000. (c) Two particles budding at the cell surfaces which contain nucleocapsid and possess spikes on the outer surface of the unit membrane (arrows). Magnification: ×52,500. (d) A row of eight budding particles showing many cross sections of nucleocapsid. Magnification: ×63,000. (e) and (f) Long filaments protruding to the cell surface. Viral surface projections and nucleocapsid strands are present along the entire length of filaments but are absent on the adjacent cell membrane; the layers of the cell membrane are continuous with those of the unit membrane in the viral envelope (arrows). Magnification: (e) ×94,000; (f) ×131,000 (from Compans et al., 1966).

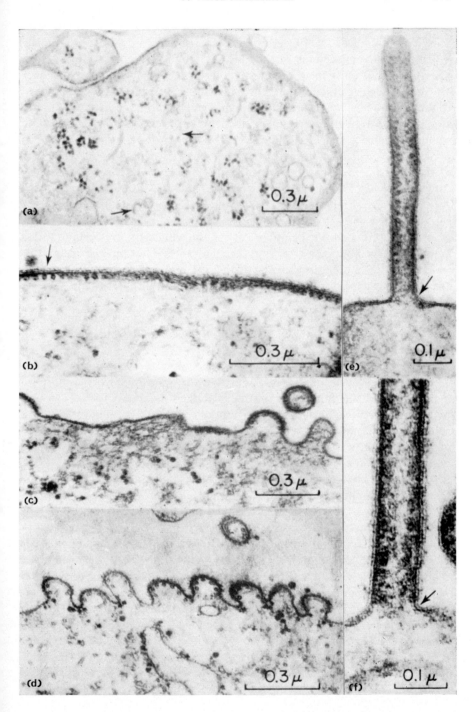

cytoplasmic matrix. The nucleocapsid then becomes aligned in a regular array beneath areas of cell membrane which contain viral envelope proteins. When seen in thin sections in the electron microscope, those areas of cell membrane which have the nucleocapsid aligned beneath in an ordered manner also have on their outer surface the projections or spikes which are typical of the virion (Fig. 1b). Neither the aligned nucleocapsids nor the spikes are seen alone. Although the complete sequence of events is not yet clear, the available evidence suggests that the viral envelope proteins are first incorporated into regions of the cell membrane, and the nucleocapsid recognizes these regions and aligns beneath them (Compans and Dimmock, 1969; Compans et al., 1966). This conclusion is based on several points: (a) the very regular arrangement of nucleocapsid beneath only certain areas of the cell membrane implies that it is recognizing some very specific sites rather than associating with the membrane at random (Fig. 1b); (b) ferritin-labeled antiviral antibody attached to viral antigens at sites on the cell membrane which have no nucleocapsid beneath them (Choppin et al., 1971); and (c) virus-specific hemadsorption occurs on regions of the membrane beneath which no nucleocapsid is seen (Choppin et al., 1971). Thus, viral hemagglutinin and perhaps other viral proteins appear to be incorporated into the membrane, and then the nucleocapsids align under these regions of membrane.

The fact that the viral spikes are detected only above areas of membrane with the nucleocapsid beneath suggests that the appearance of the morphologically identifiable spikes occurs after aligning of the nucleocapsid, perhaps by some cooperative phenomenon (Choppin et al., 1971).

These conclusions are supported by studies on rhabdoviruses (Cohen et al., 1971). The appearance of the various proteins of the virus at the plasma membrane has been analyzed in puls–chase experiments by polyacrylamide gel electrophoresis. At the end of a 5 min pulse period the plasma membranes do not contain nucleocapsid protein, while substantial amounts of the carbohydrate-free envelope protein and of the glycoprotein of the spikes are present. With chase times up to 60 min, the amount of nucleocapsid protein found at the plasma membranes increases, although from 20 min on, the amount of envelope proteins attached to the membrane remains approximately the same. From these and other findings (McSharry et al., 1971) it can be concluded that first the envelope proteins are incorporated into the plasma membrane and that then the nucleocapsids are bound to these areas by specific interaction with the acrbohydrate-free envelope proteins.

The next step in virus assembly is the well-known budding process (Fig. 1c–f). An important feature of the assembly of the virus is that during the budding process the membrane of the viral envelope is continuous with and morphologically similar to the plasma membrane of the host cell. Not only does the membrane of virions resemble that of the cell, but the membranes of virions grown in different cells, whose membranes show distinctive staining properties, can also be distinguished (Choppin *et al.*, 1971).

It has been pointed out previously in this report that the lipids in these envelopes resemble very closely those of the host cell membrane and frequently are synthesized after infection. There are several hypothetical explanations how lipids and proteins could assemble in these envelopes:

(i) Host proteins could be replaced by viral proteins on those areas of the host membranes where budding occurs.

(ii) Viral proteins attach to membrane lipids on areas devoid of proteins. Such areas have been suggested to occur on cellular membranes (Coleman *et al.*, 1970).

These hypotheses both imply that the viral lipids antedate envelope proteins in the plasma membrane (Renkonen *et al.*, 1971). In alternative modes of assembly envelope lipids and proteins would be simultaneously inserted into the membrane.

(iii) The envelopes originate from a nucleation point in the preformed membrane and spread by appositional growth within this membrane as has been suggested by Holland and Kiehn (1970).

(iv) The envelopes are assembled at a different site and then migrate to the membrane from which they bud.

With herpes virus a mode of assembly similar to the two latter ones has been suggested by Ben-Porat and Kaplan (1971). These authors found that most of the phospholipids which become part of the virions are present in the cell at the time of infection. They believe, however, that the viral envelope is not derived from unchanged segments of the nuclear membrane preexisting in the cell at the time of infection. The specific activities of the lipids of the viral envelope and of the inner nuclear membrane—the budding site—of infected cells were compared under various experimental conditions. The results of these experiments have been interpreted to indicate that the virions bud from areas on the membrane where the lipids have been assembled after infection. Furthermore, it could be shown that the phospholipids which become associated with the nuclear membrane during infection are derived in most part from other cellular membranes and migrate to the nuclear membrane (Ben-Porat and Kaplan, 1972). These results suggest that the envelope of this herpes virus is assembled *de novo* in the budding process.

More information will be required before it is clear whether the lines along which assembly of the herpes virus envelope seems to take place are also valid for other viruses.

Obviously interactions between envelope lipids and proteins play an essential role in the assembly and structure of viral envelopes. Little is known about these interactions. However, the data presented in the chapter on envelope structure suggest that at least the envelopes of arboviruses, rhabdoviruses, and myxoviruses are formed by a central lipid bilayer which is coated on the inner surface by a protein layer and on the outer surface by the spike glycoproteins. Such a model would agree with the trilaminar hypothesis of Danielli and Davson (1935) and implies that is is primarily the polar groups of the lipids which interact with the envelope proteins. These proteins have a strong affinity to lipids as suggested by experiments of Cohen *et al.* (1971). Moreover, they seem to require a specific lipid pattern to interact with which they find in certain cellular membranes such as the plasma membrane. It cannot be ruled out that the envelope proteins might exert some limited selectivity in utilizing these lipids. On the other hand, a certain flexibility in the interaction of the proteins and the lipids seems to be possible. Therefore, variations in the lipid composition of the membranes of different host cells are reflected in the virions.

On the other hand, there are indications that enzymatic events may be involved in the formation of myxovirus envelopes. These viruses contain neuraminidase which is the only truly virus specific enzyme found so far to be associated with an envelope. It has been found that the surface of the myxovirus envelopes are devoid of neuraminic acid (Klenk and Choppin, 1970b; Klenk *et al.*, 1970b). In viral envelopes which do not contain the enzyme this carbohydrate is a common constituent (Klenk and Choppin, 1971; McSharry and Wagner, 1971b; Renkonen *et al.*, 1971). Moreover, in electron microscopic studies it has been observed that neuraminic acid is present everywhere over the surface of the infected cell except on the budding myxovirus particles (Klenk *et al.*, 1970b). These results suggest that the absence of neuraminic acid on the surface on myxovirus envelopes is a characteristic feature of these viruses and that the enzyme neuraminidase residing in the envelope is responsible for the appearance of this feature.

V. Conclusions

Viral envelopes share many features in their morphological appearance and in their chemical composition with cellular membranes. Structural analysis of the envelopes of some viruses revealed that they consist of a central lipid bilayer which is coated on its inner surface by a protein layer, whereas its outer surface is covered by a fringe of glycoprotein spikes. There are only very few proteins present which are all virus specific. The simplicity of the protein pattern greatly facilitates studies on envelope assembly and on the interactions of proteins and lipids. Therefore, viral envelopes represent valuable tools for the investigation of membrane structure and bio-genesis in general.

VI. Acknowledgements

I am grateful to Drs. H. J. Eggers and R. Rott for their valuable criticism. Frau M. Ebert has kindly provided efficient secretarial help.

References

Abodeely, R. A., Palmer, E., Lawson, L. A. and Randall, C. C. (1971). *Virology* **44**, 146–152.

Acheson, N. H. and Tamm, I. (1967). *Virology* **32**, 128–143.

Acheson, N. H. and Tamm, I. (1970). *Virology* **41**, 321–329.

Ada, G. L. and Gottschalk, A. (1956). *Biochem. J.* **62**, 686.

Almeida, J. D. and Waterson, A. P. (1967). *J. gen. Microbiol.* **46**, 107.

Apostolov, K. and Flewett, T. H. (1969). *J. gen. Virol.* **4**, 365–370.

Asher, Y., Heller, M. and Becker, Y. (1969). *J. gen. Virol.* **4**, 65–76.

Armbruster, O. and Beiss, U. (1958). *Z. Naturf.* **13b**, 75.

Bächi, T., Gerhard, W., Lindenmann, J. and Mühlethaler, K. (1969). *J. Virol.* **4**, 769–776.

Beard, J. W., Bonar, R. A., Heine, U., de Thé, G. and Beard, D. (1963). *In* "Viruses, Nucleic Acids, and Cancer", pp. 340–373. The Williams and Wilkins Company, Baltimore.

Becker, W. B., McIntosh, K., Dees, J. H. and Chanock, R. M. (1967). *J. Virol.* **1**, 1019–1027.

Becht, H., Rott, R. and Klenk, H.-D. (1971). *Z. med. Mikrobiol. u. Immunol.* **156**, 305–308.

Becht, H., Rott, R. and Klenk, H.-D. (1972). *J. gen. Virol.* **14**, 1–8.

Ben-Porat, T. and Kaplan, A. S. (1970). *Virology* **41**, 265–273.

Ben-Porat, T. and Kaplan, A. S. (1971). *Virology* **45**, 252–264.

Ben-Porat, T. and Kaplan, A. S. (1972). *Nature, Lond.* **235**, 165–166.

Ben-Porat, R., Shimono, H. and Kaplan, A. S. (1969). *Virology* **37**, 56–61.

Benson, A. A. (1966). *J. Am. Oil Chem. Soc.* **43**, 265.

Bikel, I. and Duesberg, P. H. (1969). *J. Virol.* **4**. 388–393.

Blough, H. A. (1971). *J. gen. Virol.* **12**, 317–320.

Blough, H. A. and Lawson, D. E. M. (1968). *Virology* **36**, 286.

Blough, H. A., Weinstein, D. B., Lawson, D. E. M. and Kodicek, E. (1967). *Virology* **33**, 459.

Bolognesi, D. B. and Bauer, H. (1970). *Virology* **42**, 1097–1112.

Braunstein, S. N. and Franklin, R. M. (1971). *Virology* **43**, 685–695.

Breitenfeld, P. M. and Schäfer, W. (1957). *Virology* **4**, 328–345.

Bretscher, M. S. (1971). *Nature New Biology* **231**, 229–232.

Burge, B. W. and Huang, A. S. (1970). *J. Virol.* **6**, 176–182.

Burge, B. W. and Strauss, J. H., Jr. (1970). *J. molec. Biol.* **47**, 449–466.

Burge, B. W. and Pfefferkorn, E. R. (1967). *J. Virol.* **1**, 956.

Burger, M. (1969). *Proc. natn. Acad. Sci. U.S.A.* **62**, 994.

Caliguiri, L. A., Klenk, H.-D. and Choppin, P. W. (1969). *Virology* **39**, 460–466.

Cartwright, B. and Pearce, C. H. (1968). *J. gen. Virol.* **2**, 207–214.

Cartwright, R., Talbot, P. and Brown, F. (1970). *J. gen. Virol.* **7**, 267–272.

Chen, C., Compans, R. W. and Choppin, P. W. (1971). *J. gen. Virol.* **11**, 53–58.

Choppin, P. W., Klenk, H.-D., Compans, R. W. and Caliguiri, L. A. (1971). "Perspectives in Virology VII." (M. Pollard Ed.) pp. 127–156. Academic Press, New York.

Cohen, G. H., Atkinson, P. H. and Summers, D. F. (1971). *Nature, Lond.*, **231**, 121.

Coleman, R., Finean, J. B., Knutton, S. and Limbrick, A. R. (1970). *Biochim. biophys. Acta* **219**, 81–92.

Compans, R. W. (1971). *Nature, Lond.* **229**, 114.

Compans, R. W. and Choppin, P. W. (1967). *Proc. natn. Acad. Sci. U.S.A.* **57**, 949–956.

Compans, R. W. and Dimmock, N. J. (1969). *Virology* **39**, 499–515.

Compans, R. W., Holmes, K. V., Dales, S. and Choppin, P. W. (1966). *Virology* **30**, 411.

Compans, P. W., Klenk, H.-D., Caliguiri, L. A. and Choppin, P. W. (1970). *Virology* **42**, 880–889.

Dahlberg, J. E. and Franklin, R. M. (1970). *Virology* **42**, 1073–1086.

Dales, S. and Mosbach, E. H. (1968). *Virology* **35**, 564.

Danielli, J. F. and Davson, H. (1935). *J. cell. comp. Physiol.* **5**, 495–508.

Darlington, R. W. and Moss, C. H. (1969). *J. Virol.* **2**, 48–55.

Datta, H., Camerini-Otero, R. D., Braunstein, S. N. and Franklin, R. M. (1971). *Virology* **45**, 232–239.

David, A. E. (1971). *Virology* **46**, 711–720.

Drzeniek, R., Seto, J. T. and Rott, R. (1966). *Biochim. biophys. Acta* **128**, 547.

Drzeniek, R., Frank, H. and Rott, R. (1968). *Virology* **36**, 703–707.

Duesberg, P. H., Martin, G. S. and Vogt, P. K. (1970). *Virology* **41**, 631–646.

Evans, M. J. and Kingsbury, D. W. (1969). *Virology* **37**, 597–604.

Franklin, R. M. (1962). *In* "Progress in Medical Virology", Vol. 4. S. Karger, Basel, New York.

Friedman, R. M. and Pastan, I. (1969). *J. molec. Biol.* **40**, 107–115.

Frisch-Niggemeyer, W. and Hoyle, L. (1956). *J. Hyg.* **54**, 201.

Frommhagen, L. H., Freeman, N. K. and Knight, C. A. (1958). *Virology* **5**, 173.

Frommhagen, L. H., Knight, C. A. and Freeman, N. K. (1959). *Virology* **8**, 176.

Fujiwara, S. and Kaplan, A. S. (1967). *Virology* **32**, 60–68.

Garon, C. F. and Moss, B. (1971). *Virology* **46**, 233–246.

Gelderblom, H., Bauer, H. and Frank, H. (1970). *J. gen. Virol.* **7**, 33–45.

Grimes, W. J. and Burge, B. W. (1971). *J. Virol.* **7**, 309–313.

Grimley, P. M., Rosenblum, E. M., Mims, S. J. and Moss, B. (1970). *J. Virol.* **6**, 519–533.

Hamre, D., Kindig, D. A. and Mann, J. (1967). *J. Virol.* **1**, 810–816.

Harrison, S. C., David, A., Jumblatt, J. and Darnell, J. E. (1971a). *J. molec. Biol.* **60**, 523–528.

Harrison, S. C., Caspar, D. L. D., Camerini-Otero, R. D. and Franklin, R. M. (1971b). *Nature New Biology* **229**, 197–201.

Haslam, E. A., Cheyne, L. M. and White, D. O. (1969). *Virology* **39**, 118–129.

Haslam, E. A., Hampson, A. W., Egan, J. A. and White, D. O. (1970a). *Virology* **42**, 555–565.

Haslam, E. A., Hampson, A. W., Radiskevics, I. and White, D. O. (1970b). *Virology* **42**, 566–575.

Haukenes, G., Harboe, A. and Mortensson-Egnund, K. (1965). *Acta path. microbiol. scand.* **64**, 534.

Hay, A. J., Skehel, J. J. and Burke, D. C. (1968). *J. gen. Virol.* **3**, 175–184.

Heydrick, F. P., Comer, J. F. and Wachter, R. F. (1971). *J. Virol.* **7**, 642–645.

Holland, J. J. and Kiehn, E. D. (1970). *Science, N.Y.* **167**, 202–205.

Holowczak, J. A. (1970). *Virology* **42**, 87–99.

Holowczak, J. A. and Joklik, W. (1967). *Virology* **33**, 717–725.

Hoyle, L. (1952). *J. Hyg.* **50**, 229.

Huang, T. C. and Kuksis, A. (1967). *Lipids* **2**, 453.

Hummeler, K., Koprowski, H. and Wiktor, T. J. (1967). *J. Virol.* **1**, 152–170.

Inbar, M. and Sachs, L. (1969). *Proc. natn. Acad. Sci. U.S.A.* **63**, 1418.

Isacson, P. and Koch, A. E. (1965). *Virology* **27**, 129.

Joss, A., Gandhi, S. S., Hay, A. J. and Burke, D. C. (1969). *J. Virol.* **4**, 816–822.

Kaluza, G., Scholtissek, C. and Rott, R. (1972). *J. gen. Virol.*, in press.

Kang, C. Y. and Prevec, L. (1970). *J. Virol.* **6**, 20–27.

Kates, M., Allison, A. C., Tyrrell, D. J. A. and James, A. T. (1962). *Cold Spring Harb. Symp. quant. Biol.* **67**, 293.

Kendal, A. P., Biddle, F. and Belyavin, G. (1968). *Biochem. biophys. Acta* **165**, 419–431.

Kendal, A. P., Apostolov, K. and Belyavin, G. (1969). *J. gen. Virol.* **5**, 141–143.

Kitajiama, E. W. and Costa, A. S. (1966). *Virology* **29**, 523–539.

Klenk, H.-D. (1971). *In* "The Dynamic Structure of Cell Membranes" (D. F. Hölzl Wallach, H. Fischer, eds.) pp. 97–118. Springer Verlag, Berlin, Heidelberg, New York.

Klenk, H.-D. and Choppin, P. W. (1969a). *Virology* **37**, 155–157.

Klenk, H.-D. and Choppin, P. W. (1969b). *Virology* **38**, 255–268.

Klenk, H.-D. and Choppin, P. W. (1970a). *Virology* **40**, 939–947.

Klenk, H.-D. and Choppin, P. W. (1970b). *Proc. natn. Acad. Sci. U.S.A.* **66**, 57–64.

Klenk, H.-D. and Choppin, P. W. (1971). *J. Virol.* **7**, 416.

Klenk, H.-D., Caliguiri, L. A. and Choppin, P. W. (1970a). *Virology* **42**, 473–481.

Klenk, H.-D., Compans, R. W. and Choppin, P. W. (1970b). *Virology* **42**, 1158–1162.

Klenk, H.-D., Rott, R. and Becht, H. (1972a). *Virology* **47**, 579–591.

Klenk, H.-D., Scholtissek, C. and Rott, R. (1972b). *Virology* **49**, 723–734.

Knight, C. A. (1947). *J. exp. Med.* **85**, 99–116.

Koscialek, J., Hakomori, S. and Jeanloz, R. W. (1968). *Immunochemistry* **5**, 411.

Landsberger, F. R., Lenard, J., Paxton, J. and Compans, R. W. (1971). *Proc. natn. Acad. Sci. U.S.A.* **68**, 2579–2583.

Landsberger, F. R., Compans, R. W., Paxton, J. and Lenard, J. (1972). *J. Supramolecular Structure* **1**, 50–54.

Laver, W. G. (1963). *Virology* **20**, 251–262.

Laver, W. G. (1971). *Virology* **45**, 275–288.

Laver, W. G. and Valentine, R. C. (1969). *Virology* **38**, 105–119.

Laver, W. G. and Webster, R. G. (1966). *Virology* **38**, 104–115.

Lazarowitz, S. G., Compans, R. W. and Choppin, P. W. (1971). *Virology* **46**, 830–843.

Lee, L. T., Howe, C., Meyer, K. and Cho, H. U. (1969). *J. Immunol.* **102**, 1144.

Makita, A., Suzuki, O. and Yorizawa, Z. (1966). *J. Biochem.* **60**, 502.

Malhotra, S. K. (1970). *In* "Progress in Biophysics and Molecular Biology" (J. A. V. Butler and D. Noble, eds.) Vol. 20, pp. 67–131. Pergamon Press, Oxford.

Mårtensson, E. (1969). *In* "Progress in the Chemistry of Fats and Other Lipids" (R. T. Holman, ed.) Vol. X, pp. 367–407. Pergamon Press, Oxford.

Matsumara, T., Stollar, V. and Schlesinger, R. W. (1971). *Virology* 46, 344–355.

McLeod, R., Black, L. M. and Moyer, F. H. (1966). *Virology* 29, 540–552.

McSharry, J. J., Compans, R. W. and Choppin, P. W. (1971). *J. Virol.* 8, 722–729.

McSharry, J. J. and Wagner, R. R. (1971a). *J. Virol.* 7, 59–70.

McSharry, J. J. and Wagner, R. R. (1971b). *J. Virol.* 7, 412–415.

Morgan, C., Ellison, S. A., Rose, H. M. and Moore, D. H. (1954). *J. exp. Med.* 100, 195–202.

Morgan, C., Hsu, K. C., Rifkind, R. A., Knox, A. W. and Rose, H. M. (1961). *J. exp. Med.* 114, 825–832.

Moss, B., Rosenblum, E. N. and Grimley, P. M. (1971). *Virology* 45, 123–134.

Mountcastle, W. E., Compans, R. W., Caliguiri, L. A. and Choppin, P. W. (1970). *J. Virol.* 6, 677.

Mountcastle, W. E., Compans, R. W. and Choppin, P. W. (1971). *J. Virol.* 7, 47.

Mountcastle, W. E., Harter, D. E. and Choppin, P. W. (1972). *Virology*, in press.

Mudd, J. A. and Summers, D. F. (1970). *Virology* 42, 328–340.

Murphy, F. A., Webb, P. A., Johnson, K. M., Whitfield, S. G. and Chappell, W. A. (1970). *J. Virol.* 6, 507–518.

Murphy, J. S. and Bang, F. B. (1952). *J. exp. Med.* 95, 259–268.

Mussgay, M. and Weibel, J. (1963). *J. Cell Biol.* 16, 119.

Nermut, M. V. and Frank, H. (1971). *J. gen. Virol.* 10, 37.

Noll, H., Aoyagi, T. and Orlando, J. (1962). *Virology* 18, 154–157.

Olshevsky, U. and Becker, Y. (1970). *Virology* 40, 948–960.

Oshiro, L. S., Schieble, J. H. and Lennette, E. H. (1971). *J. gen. Virol.* 12, 161–168.

Oram, J. D., Ellwood, D. C., Appleyard, G. and Stanley, J. L. (1971). *Nature New Biology* 233, 50–51.

Pfefferkorn, E. R. and Hunter, H. S. (1963a). *Virology* 20, 433.

Pfefferkorn, E. R. and Hunter, H. S. (1963b). *Virology* 20, 446.

Pfefferkorn, E. R. and Clifford, R. L. (1964). *Virology* 23, 217.

Quigley, J. P., Rifkin, D. B. and Reich, E. (1971). *Virology* 46, 106–116.

Rao, P. R., Bonar, R. A. and Beard, J. W. (1966). *Exp. Molec. Pathol.* 5, 374–388.

Renkonen, O., Kääräinen, L., Simons, K. and Gahmberg, C. C. (1971). *Virology* 46, 318–326.

Rifkin, D. B. and Compans, R. W. (1971). *Virology* 46, 485–489.

Roizman, B. and Spear, P. G. (1971). *Science, N.Y.* 171, 298–300.

Rott, R., Drzeniek, R., Saber, S. and Reichert, E. (1966). *Arch. ges. Virusforsch.* 19, 273.

Rott, R., Drzeniek, R. and Frank, H. (1969). *In* "The Biology of Large RNA Viruses" (R. D. Barry, B. W. J. Mahy, eds.) Academic Press, London.

Rouser, G., Nelson, G. J., Fleischer, S. and Simon, G. (1968). *In* "Biological Membranes", Vol. I. (D. Chapman, ed.) Academic Press, London.

Schäfer, W. (1956). "Ciba Foundation Symposium on the Nature of Viruses", pp. 91–103.

Schäfer, W. and Zillig, W. (1954). *Z. Naturf.* 9b, 779–788.

Scheid, A. and Choppin, P. W. (1971). *Bact. Proc.*, p. 202.

Schulze, I. T. (1970). *Virology* 42, 890–904.

Schulze, I. T. (1972). *Virology* 47, 181–196.

Siegert, R. and Falke, D. (1966). *Arch. ges. Virusforsch.* **19**, 230.

Silbert, J. A., Salditt, M. and Franklin, R. M. (1969). *Virology* **39**, 666–681.

Simons, K. and Kääriäinen, L. (1970). *Biochem. biophys. Res. Comm.* **38**, 981–988.

Simpson, R. W. and Hauser, R. E. (1966a). *Virology* **29**, 654–667.

Simpson, R. W. and Hauser, R. E. (1966b). *Virology* **30**, 684.

Skehel, J. J. (1971). *Virology* **44**, 409–417.

Skehel, J. J. and Schild, G. (1971). *Virology* **44**, 396–408.

Sokol, F., Stancek, D. and Koprowski, H. (1971). *J. Virol.* **7**, 241–249.

Soule, D. W., Marinetti, G. V. and Morgan, H. R. (1959). *J. exp. Med.* **110**, 93–102.

Spear, P. G. and Roizman, B. (1969). *Virology* **36**, 545–555.

Spear, P. G. and Roizman, B. (1972). *J. Virol.* **9**, 143–159.

Spear, P. G., Keller, J. M. and Roizman, B. (1970). *J. Virol.* **5**, 123–131.

Springer, G. F. and Tritel, H. (1962). *Science, N.Y.* **138**, 687–688.

Stenbeck, W. A. and Durand, D. B. (1963). *Virology* **20**, 545–551.

Stollar, V. (1969). *Virology* **39**, 426–438.

Strauss, J. H., Jr., Burge, B. W., Pfefferkorn, E. R. and Darnell, J. E. (1968). *Proc. natn. Acad. Sci. U.S.A.* **59**, 533.

Strauss, J. H., Jr., Burge, B. W. and Darnell, E. R. (1969). *Virology* **37**, 367–376.

Strauss, J. H., Jr., Burge, B. W. and Darnell, E. R. (1970). *J. molec. Biol.* **47**, 437.

Sydiskis, R. J. and Roizman, B. (1967). *Virology* **32**, 678–686.

Sylvester, E. R., Richardson, J. and Wood, P. (1968). *Virology* **36**, 693–696.

Tiffany, J. M. and Blough, H. A. (1969a). *Science, N.Y.* **163**, 573–574.

Tiffany, J. M. and Blough, H. A. (1969b). *Virology* **37**, 492–493.

Tiffany, J. M. and Blough, H. A. (1970a). *Proc. natn. Acad. Sci. U.S.A.* **65**, 1105–1112.

Tiffany, J. M. and Blough, H. A. (1970b). *Virology* **41**, 392–394.

Wagner, R. R., Schnaitman, T. A. and Snyder, R. M. (1969). *J. Virol.* **3**, 395–403.

Wagner, R. R., Snyder, R. M. and Yamazaki, S. (1970). *J. Virol.* **5**, 548–558.

Wallach, D. F. H. and Kamat, V. B. (1966). *Meth. Enzym.* **8**, 164.

Warren, L., Glick, M. C. and Nass, M. K. (1966). *J. Cell Physiol.* **68**, 269–288.

Webster, R. G. (1970). *Virology* **40**, 643–654.

Webster, R. G. and Darlington, R. W. (1969). *J. Virol.* **4**, 182–187.

Wecker, E. (1957). *Z. Naturf.* **12b**, 208–210.

White, H. B., Powell, S. S., Gafford, L. G. and Randall, C. C. (1968). *J. biol. Chem.* **243**, 4517–4525.

Zee, Y. C., Hackett, A. J. and Talens, L. (1970). *J. gen. Virol.* **7**, 95–102.

Chapter 4

The Plasma Membrane Initiator Systems for the Actions of Insulin and Glucagon

JOSEPH AVRUCH AND STEPHEN L. POHL

Department of Medicine, Washington University School of Medicine, St. Louis, Missouri, U.S.A.

I. Introduction

The mechanisms of hormone action have been under investigation for over a century. As the study of endocrine physiology progressed from the organism to the cell, it became apparent that many observed hormonal effects could be ascribed to earlier changes in cellular function induced by the hormone. This led to a search for the earliest, or "primary" action of the hormone, i.e. those cellular events which initiated all subsequent effects of the hormone.

The first formulations of this concept arose from physiologic and pharmacologic studies of the actions of catecholamines (Ahlquist, 1948, 1959; Furchgott, 1959). It was appreciated early that relatively minor alterations in the structure of the catecholamine molecule resulted in marked quantitative changes in the observed response to this hormone. To explain this finding, the existence of a cellular constituent which recognized the molecular structure of the catecholamine with great specificity was postulated; however, this cognitive function could be appreciated experimentally only by measurement of a distal effect, such as myocardial contractility. Thus, the concept of an "adrenergic receptor" evolved which included both a sensing system for the molecular structure of the catecholamine and a specific biologic effector system (Ahlquist, 1968; Moran, 1966).

In recent years, the events involved in the initiation of the cellular response to peptide hormones have been investigated in biochemical rather than physiological terms, and two major insights into the molecular basis of these events have emerged. The first resulted from the demonstration by Sutherland and Rall (1957) that the adenylate cyclase system in mammalian liver functions as the effector system for glucagon and epinephrine. Both hormones stimulate the activity of adenylate cyclase (Sutherland and Rall, 1960), and the increased intracellular concentrations of cyclic 3'5'-adenosine monophosphate (cyclic AMP), the product of this enzyme, leads to most of the known physiologic actions of these hormones in liver (Robison et al., 1967). This effector system has subsequently been shown to be operative in the actions of a large number of peptide hormones (Jost and Rickenberg, 1971), and to date is the only such system clearly defined for peptide hormones. Adenylate cyclase appears, however, not to be the primary effector of insulin's action.

The second major insight resulted from the demonstration that the plasma membrane of the target cell is the location of both the recognition system and the effector system for each of the peptide hormones. In the case of hormone sensitive adenylate cyclase systems, subcellular fractionation of a number of tissues demonstrated that this activity copurified with other plasma membrane markers (Davoren and Sutherland, 1963; Rodbell, 1967a; Pohl et al., 1971a; McKeel and Jarett, 1970). With respect to insulin, it was observed that certain actions of this hormone on isolated fat cells could be mimicked by a variety of enzymatic and chemical perturbations of the cell surface (Rodbell et al., 1968). Later, insulin coupled to agarose beads, and presumably restricted to the plasma membrane in its interaction with the cell, was shown to be capable of producing many of the known effects of native insulin on isolated fat cells (Cuatrecasas, 1969). Thus, both the recognition and subsequent effects on cellular metabolism of polypeptide hormones are mediated by molecularly distinct yet tightly coupled components of the same cell organelle, rather than, for example, by a specific transport mechanism in the plasma membrane which facilitates access of the hormone to an intracellular site of action.

As a consequence of the rapid progression of the study of hormone action to the subcellular, or even molecular level, the usefulness of the "receptor" concept as first applied to catecholamines has declined. In contrast to earlier work, both parts of the classical receptor, recognition and effect, may now be observed independently in a single experiment, and are thus distinguishable operationally as well as theoretically.

In light of these considerations, we have chosen to discard the classical definition of "receptor" with respect to polypeptide hormones. Instead, the initial events in the alteration of cellular functions produced by these hormones will be examined in terms of a three component system comprising: (1) *The Receptor* capable of specifically recognizing the hormone, (2) *The Effector*, which, consequent to the hormone receptor interaction, mediates the alterations in cellular function, and (3) *The Transducer*, which couples the receptor to the effector.

The three components will be collectively referred to as the "initiator" system for peptide hormone action, i.e. those plasma membrane constituents involved from the initial contact of the cell with a peptide hormone through the activation of the processes which mediate the effects of the hormone. While this definition provides a useful organizational basis for the present discussion, it must be emphasized that the semantic problems in this field are not yet solved, and no attempt is made to extend this definition to other recognition effector systems, such as those for steroid hormones, thyroid hormones, drugs, etc.

This review will attempt to summarize recent studies of the initiator systems for insulin and glucagon. These represent the most extensively studied of all such systems for polypeptide hormones, and in addition furnish contrasting mechanisms. However, before examining studies involving specific hormones, certain features of plasma membrane methodology should be considered.

II. Plasma Membrane Preparation

The cellular action of peptide hormones may be studied at several levels of cell organization. To obtain an integrated view of the sequential changes in cellular function, hormone action is best analyzed in the intact cell. Rodbell, in 1964, described a method for the preparation of fat cells free of the stromal-vascular components of adipose tissue. These free fat cells are devoid of basement membrane and their plasma membrane is in direct contact with the bathing medium. Furthermore, they have been shown to be morphologically and metabolically intact. Such cells are responsive to a large number of hormones and have been used extensively in the study of hormonal effects on lipid and cyclic AMP metabolism, and more recently in studies of the interaction between hormone and receptor. A source of continuing concern in the use of isolated fat cells arises from the use of crude collagenase in their preparation. Highly purified bacterial collagenase does not disperse adipose tissue; an additional proteolytic activity, such as trypsin, must be supplied (Kono, 1969). Crude bacterial collagenase contains a variety of contaminating proteases of largely undefined specificity (Mandl, 1961; Kono, 1968). Whether the plasma membrane of the fat cell is altered during the incubation with crude collagenase is unknown. Certainly the plasma membrane of the adipocyte, when isolated, is highly sensitive to proteolytic attack. This question has been approached by comparing the properties of the hormone sensitive functions in intact adipose tissue and isolated cells (Gliemann, 1969). The quantitative differences described reflect the favorable effects of removal of the intercellular diffusion barriers. While the qualitative differences thus far reported are minor, this point represents an ever-present hazard.

Nevertheless, preparations of free adipocytes have been extremely useful in establishing the importance of the plasma membrane in the initiation of peptide hormone action. The detailed structural and functional characterization of this process, however, can be pursued only with relatively purified preparations of plasma membrane, wherein secondary effects are eliminated.

Rodbell (1967a) used hypotonic lysis of intact fat cells as a means of releasing intracellular contents. This technique yielded large, closed vesicles of plasma membrane, which he called "ghosts". These particles are essentially devoid of neutral fat, but retain some metabolic activities as well as hormonal responsiveness. While the fat cell ghost preparation has proved useful in the characterization of the hormone sensitive adenylate cyclase system (Rodbell et al., 1970a; Birnbaumer and Rodbell, 1969; Bär and Hechter, 1969; Birnbaumer et al., 1969), it is heavily contaminated by mitochondria, microsomes, nuclei and some cytoplasmic constituents, which are trapped in the intravesicular space (Rodbell, 1967a).

The isolation of purified plasma membrane remains one of the more formidable problems in cell fractionation. This organelle may assume a wide variety of sizes and densities on cellular disruption, depending on the method of homogenization employed, the resistance of various areas of the membrane to the disrupting force, the presence of areas of specialized surface structure, and the tendency to spontaneous vesiculation (Steck and Wallach, 1970). The first useful method for the isolation of a purified mammalian plasma membrane fraction was that described by Neville (1960, 1968) for liver.

Histologically, the plasma membrane of hepatic parenchymal cells exhibit important structural specializations in the region of the bile canaliculus. The walls of this structure are formed by the surface membranes of adjoining hepatocytes, which are firmly interconnected by lateral terminal bars. In addition, the subjacent surface of adjoining hepatocytes are further coupled by desmosomes, junctional complexes and interdigitating projections (Steck and Wallach, 1970). When liver is homogenized, this reinforced area of the hepatocyte surface remains largely intact, and can easily be isolated in large quantities as an intact canalicular structure by a series of differential and density gradient centrifugations. This preparation binds peptide hormones (Rodbell et al., 1971a; Johnson et al., 1972; Freychet et al., 1971a; Cuatrecasas et al., 1971) and contains a hormone sensitive adenylate cyclase (stable in the frozen state) (Pohl et al., 1971a); its utility in the study of peptide hormone action is evident. It should be noted, however, that these canalicular membranes represent only a selected portion of the hepatocyte surface, and in fact with least access to circulating hormones.

The preparation of plasma membranes from fat cells has proved more exacting. This membrane contains no detectable areas of specialized structure, and upon homogenization is disrupted predominantly into closed semipermeable vesicles of varying size (Avruch and Wallach, 1971). Since the smooth endoplasmic reticulum assumes a similar form upon cell breakage, the purification of plasma membrane free of endoplasmic reticulum is difficult. McKeel and Jarett (1970) have described a procedure, starting with isolated cells, which yields a fraction of plasma membrane approximately 80% pure. While the method is rapid and simple, the conditions of homogenization and centrifugation must be followed closely to ensure similar results. These particles can mediate the facilitated uptake of glucose (Carter et al., 1972a) and are enriched in adenylate cyclase (McKeel and Jarett, 1970). However, the hormonal responsiveness of the cyclase system differs from that observed in the intact cell (Jarett et al., 1971) and glucose transport by these vesicles is not stimulated by the direct addition of insulin (Avruch et al., 1972). Despite the reservations cited, preparations of purified plasma membranes have proved of great value in the study of the initial events in peptide hormone action.

III. The Receptor System

Operationally, the receptors for insulin and glucagon have been identified and characterized through the study of hormone binding to intact cells or to isolated plasma membrane preparations. Such binding is easily detected and characterized with respect to its time course, saturability, and response to changes in temperature, pH, ionic strength, etc. However, in order for the observed binding to be considered as an integral step in the hormone's action, certain criteria must be fulfilled:

Firstly, and most importantly, both the hormone and the cellular assembly to which it binds must be functionally intact. This can only be assured by the demonstration that binding of the tracer hormone is accompanied by the activation of a measurable biologic effect. Furthermore, these processes must occur in a congruent manner with respect to both dose-response relationships and time course. In addition, binding (and bioactivity) must occur at hormone concentrations expected under physiologic conditions *in vivo*.

Secondly, in order for hormone binding to be considered specific, structural analogues and fragments of the hormone, and the structurally modified hormone must exhibit parallel potency in binding and bioactivity, either as agonists or antagonists. On the other hand, it is expected that alterations in membrane structure are much more likely to yield discordant effects on hormone binding and bioactivity, since the hormone recognition site is only the first component in the complex membrane bound initiator system.

Once the functional significance of the hormone binding has been validated, the following questions may be posed:

(1) What is the structure of the hormone in its physiologically active form, and which are the portions critical to cellular recognition? This problem is largely beyond the scope of the present review and will be considered only briefly.

(2) What is the molecular structure of the receptor?

(3) What are the forces of interaction between the hormone and its receptor? Some information on this point has been obtained from the study of the kinetics of hormone–membrane interaction. Conclusions based on currently available work should be evaluated cautiously, as hormone degradation may introduce significant artefacts into the data.

IV. The Insulin Receptor System

The binding of insulin to tissues has been studied for over twenty years. In 1949 Stadie and collaborators demonstrated that when rat hemi-diaphragms were briefly exposed to high concentrations of insulin and then washed, the utilization of glucose was enhanced during subsequent incubation, relative to an unexposed diaphragm (Stadie *et al.*, 1949). In subsequent studies using ^{131}I insulin (Stadie *et al.*, 1952, 1953) these workers showed that even after brief exposure to the hormone, insulin remained detectable in the muscle in significant amounts, despite prolonged washing. It was concluded that the initial step in insulin action was a firm binding of the hormone to a tissue site.

Over the next twenty years, numerous studies corroborated in large part the pioneering observations of Stadie. However, until recently, because of a number of formidable methodologic difficulties as well as the fact that the effector system for insulin has remained undefined, rigorous support for his conclusions has been lacking. In the last few years these obstacles have apparently been overcome, as demonstrated in an elegant series of reports by Cuatrecasas. Before reviewing this work it is instructive to examine the problems which so persistently clouded the study of insulin binding.

While the binding of labeled insulin to intact muscle and adipose tissue can be easily measured, the question as to whether this binding is relevant to the hormone's action is more difficult. Specificity of binding may be safely inferred if the quantitative aspects of hormone binding and biologic activity are congruent. This requirement may be too restrictive, since within a given tissue the various biologic responses to a single hormone may exhibit different dose response curves (Crofford, 1968). Nevertheless, binding measurements obtained at hormone concentrations far above levels which are obtainable *in vivo*, under physiologic conditions, must be viewed skeptically. In excised muscle (either rat hemi-diaphragm or soleus), stimulation of glucose uptake by insulin can be demonstrated at concentrations as low as 10–100 microunits/ml (Vallence-Owen and Hurlock, 1954; Wohltmann and Narahara, 1966), and maximal stimulation is achieved at 500–1000 microunits/ml. A number of authors (Wohltmann and Narahara, 1966; Ferrebee *et al.*, 1951; Newerly and Berson, 1957; Garratt *et al.*, 1966) have reported that the binding of ^{131}I insulin to muscle rises progressively as the hormone concentration in the medium is increased, and shows no evidence of a plateau, even at medium concentrations of 5×10^7 microunits/ml (Newerly and Berson, 1957). Such results have been interpreted in two ways. Newerly and Berson (1957), pointing to the highly surface active and adsorbent properties of insulin, conclude that observed binding is entirely non-specific. On the other hand, Wohltmann and Narahara (1966) and Garratt *et al.* (1966) with similar data, postulate the presence of saturable specific sites in addition to a relatively large background of non-specific binding. Similar observations have been made using intact rat epididymal fat pads for the study of insulin binding (Malaisse and Franckson, 1965; Bewsher *et al.*, 1966). A major difficulty in these experiments stems from the use of intact tissues. In addition to the insulin sensitive cells, a significant volume is occupied by connective tissue and vascular elements, which may bind insulin (Stein and Gross 1959; Worthington *et al.*, 1964) without being capable of responding to the hormone metabolically. Furthermore, significant diffusion barriers to macromolecules exist in such intact tissues. The use of the isolated fat cell preparation introduced by Rodbell (1964) has obviated both these problems. For example, with intact fat pads, the response of glucose uptake to insulin is first demonstrable at 10 microunits/ml and is maximal at 320 microunits/ml; with isolated fat cells, an initial response can be shown at 1·25 microunits/ml insulin, and the process is saturated at 20 microunits/ml (Gliemann, 1968). The use of isolated fat cells, while representing a major breakthrough, serves to uncover an even more serious defect of the earlier studies, i.e. the validity of iodinated insulin as a tracer for physiologically significant hormone binding. In order for iodinated insulin to serve as a reliable tracer, the bio-activity of the modified hormone must be identical to that of native insulin. Iodination of insulin has been generally accomplished by chemical (rather than enzymatic) means (Hunter and Greenwood, 1962). While the technique appears straightforward, the nature of the final product is subject to a large number of variables (Yalow and Berson, 1966). In addition to radiation damage (Berson and Yalow, 1957), which may occur during the iodination or only become manifest upon storage, the oxidizing agents (such as chloramine T) usually present in the incubation mixture can inactivate the hormone, even in the absence of iodine. This problem has been especially well demonstrated in the case of gastrin (Stagg et al., 1970). The extent of iodination, which is critical to bio-activity, must be carefully monitored. Of special importance, in addition to the total duration of the reaction, is the use of $Na^{131}I$ preparations of high specific activity (Yalow and Berson, 1966; Izzo et al., 1964a), as well as techniques that ensure thorough mixing (Brunfeldt, 1965). Several groups (DeZoeten and Van Strik, 1961; Izzo et al., 1964b; Garratt, 1964; Rosa et al., 1967) have shown that iodinated insulin preparations with average ratios of 1·0 iodide atom per molecule of insulin or less, exhibit 80–100 % full biologic potency, and this is the degree of iodination usually sought. More highly labeled preparations invariably show significant inactivation (DeZoeten and van Strik, 1961; Izzo et al., 1964b; Rosa et al., 1967). Even at ratios of 1·0 iodide atom per insulin molecule or less, it must be ensured that the iodide atoms are relatively uniformly distributed and that significant pools of unlabeled and highly labeled hormone do not exist. The careful studies of DeZoeten and collaborators (DeZoeten and Van Strik, 1961) indicate that insulin molecules containing one atom of iodide, and probably those containing two atoms, retain full bio-activity. However, Arquilla et al. (1968) have clearly demonstrated that even iodoinsulin preparations containing less than one iodide per insulin molecule may be biologically inactive. Thus, each preparation must be validated by a sensitive bio-assay; studies of insulin binding that lack careful validation of the bio-activity of the labeled hormone must be interpreted cautiously.

In an effort to circumvent the problems arising from the use of iodinated insulin, Crofford (1968) studied the uptake of native (unlabeled) insulin by free fat cells, by measuring removal of hormone from the bathing medium. Using relatively concentrated cell suspensions he found that uptake proceeded in two phases: (1) a rapid, temperature insensitive uptake which could be abolished by pre-treatment of the cells with 10^{-3} M maleimide, and (2) a slower, temperature sensitive uptake, not abolished by maleimide. The physiological significance of this uptake is uncertain. Although the stimulation of glucose uptake by insulin was maximal at 20 microunits/ml, the rapid component of binding showed no evidence of saturation at insulin concentrations up to 10^4 microunits/ml. Furthermore, while the biologic effect of insulin on dilute fat cell suspensions could be reversed by several washes in an insulin free medium, or by the addition of anti-insulin serum, fat cells exposed to insulin did not release biologically or immunologically detectable hormone into insulin-free buffer under conditions where a release of 10% of the bound hormone would have been detectable. Finally, if cells exposed to a given concentration of insulin were transferred, without washing, to a fresh medium containing the same insulin concentration, insulin uptake from the second medium was undiminished. The contribution of hormone degradation to the insulin "uptake", as measured in this study, is difficult to assess. Certainly the slow component of insulin disappearance from the medium is suggestive of enzymatic breakdown of the hormone, a process known to occur in adipose tissue homogenates (Rudman et al., 1966, 1968). The lack of correlation between the biologic activation and rapid component of insulin binding, in terms of reversibility and saturability, is suggestive of nonspecific binding. On the other hand, the concomitant abolition of insulin uptake and biological effect by pretreatment of the fat cells with trypsin (Crofford et al., 1970) does suggest that some portion of the rapid component of uptake is involved in the sequence of hormonal activation.

Recent studies by Cuatrecasas and co-workers (Cuatrecasas, 1969, 1971a, b, c, d, 1972; Cuatrecasas and Illiano, 1971; Cuatrecasas et al., 1971), have significantly advanced the understanding of insulin the receptor, Cuatrecasas (1969) demonstrated that insulin covalently bound to sepharose beads, and presumably in a form inaccessible to the cell interior, retained essentially full biologic potency when incubated with isolated fat cells. This strongly supports the contention that the initiator system for insulin is located in the plasma membrane. In subsequent studies he examined insulin binding to isolated fat cells. The pivotal features of these studies, lacking in earlier reports, are the direct measurement of the binding of biologically intact radioiodinated insulin to physiologically responsive cells, which are in free contact with the bathing medium. The iodinated insulin used gives a dose-response curve for the stimulation of glucose uptake by fat cells identical to that of native hormone (Cuatrecasas, 1971a). The measurement of binding (Cuatrecasas, 1971a) employs rapid millipore filtration and washing of cell or particulate suspensions; radioactivity retained with the cells on the filter is determined. "Specific" binding is ascertained by subtraction of counts retained after filtration of replicate incubation mixtures which contain an excess of unlabeled native insulin. Measured in this way, specific binding of insulin shows excellent quantitative correlation with the dose-response curve for the enhancement of glucose oxidation. This observation, as well as the fact that derivatives of insulin compete for binding in direct proportion to their biologic activity (Cuatrecasas, 1971c), strongly supports the contention that so-called specific binding reflects an important physiologic event in the sequence of hormone action. The kinetics of association conform to a second order equation, consistent with a bimolecular reaction between insulin and its receptor, while disassociation is strictly linear, with $t\frac{1}{2}$ of 16 min. The disassociation constant calculated from the kinetic data is 5×10^{-11} molar, and there are approximately 11,000 receptor sites per fat cell. The insulin binding capacity of fat cells is quantitatively recovered in a particulate fraction after homogenization, and its properties are qualitatively unchanged. No direct evidence of insulin degradation is apparent in these studies. From 68–75% of the bound insulin can be eluted from intact cells (Cuatrecasas, 1971a) or membranes (Cuatrecasas, 1971c). This material, as well as the hormone remaining unbound after the original incubation, is unaltered in its physical, biological and binding properties. The state of the residual iodoinsulin not eluted from the membrane is unknown.

Cuatrecasas examined the effect of a variety of chemical and enzymatic perturbations of intact fat cells and fat cell membranes on their ability to bind and respond to insulin. Trypsin (Cuatrecasas and Illiano, 1971) at low levels, markedly and irreversibly diminished the binding affinity of the insulin receptor, but the maximal binding capacity is unaltered. The metabolic response to the hormone parallels this altered binding. More vigorous proteolysis further diminishes receptor affinity, such that total binding capacity can not be practicably measured. Under these circumstances, the parallelism between binding and metabolic response is lost. Incubation of fat cells with phopholipase C or phospholipase A (Cuatrecasas, 1971b) markedly increases the apparent number of binding sites without altering their affinity; hormonal responsiveness is lost before the unmasking of receptor sites is demonstrable. Solvent extraction of membrane lipids (Cuatrecasas, 1971b) yields effects on insulin binding similar to phospholipase treatment, and both were partially reversible by readdition of phospholipids. Treatment of membranes with substances which interact with phospholipids (Cuatrecasas, 1971b) generally resulted in increased insulin binding, whereas those agents which bound chiefly to sterols were without effect. Of a variety of protein functional reagents (Cuatrecasas, 1971c), those directed primarily at tryptophan, SH, and COOH did not alter insulin binding, despite previous suggestions that the first two species were involved in the insulin–receptor interaction (Rieser, 1967). Treatment of membranes with diazonium 1-H-tetrazole (specificity for histidyl, tyrosyl and tryptophanyl residues) and tetranitromethane (specificity for tyrosyl, SH, tryptophanyl residues) greatly reduced insulin binding. Finally, maximal binding capacity was reversibly increased by exposure of the membranes to high ionic strength and divalent cations, whereas a variety of hormones and nucleotides did not affect insulin binding (Cuatrecasas 1971c). Cuatrecasas has also reported that the characteristics of insulin binding to isolated liver cell membranes are virtually identical to those observed with fat cell membrane (Cuatrecasas et al., 1971).

Using a carefully validated preparation of monoiodoinsulin (Freychet *et al.*, 1971b) to study insulin binding to isolated plasma membrane from liver, Freychet *et al.* (1971a) reported qualitatively similar data. While this group did not report binding parameters, examination of the curves describing displacement of membrane-bound radioinsulin by unlabeled hormone indicates that the insulin binding to liver membranes measured by Freychet *et al.* (1971b) exhibits an affinity similar to that described by Cuatrecasas for both liver and fat cell membranes. Furthermore, a preliminary report indicates that the degradation of insulin by liver membranes is independent and separable from the binding of hormone to its receptor (Freychet, 1971c). The physiological importance of degradation to the onset and/or termination of hormone action is uncertain.

The purification and molecular characterization of the insulin receptor system from liver and adipose plasma membranes appears to be well underway. Using 1% Triton X100, Cuatrecasas has "solubilized" the insulin binding activity from these membranes (Cuatrecasas, 1972). In close agreement with the findings in the intact fat cell, the disassociation constant for insulin binding to the solubilized receptor is 13×10^{-11} molar and the binding is readily abolished by trypsin. The term "solubilized" is operational; the material contains no particles, nor does the binding activity sediment at approximately 1×10^8 g min. However, removal of the detergent leads to progressive precipitation of the binding activity. Furthermore, whereas a preliminary estimate of the molecular weight of the solubilized insulin receptor complex is 300,000 daltons, polyacrylamide gel electrophoresis of adipocyte plasma membranes in 1% sodium dodecyl sulfate fails to reveal significant quantities of membrane protein in this weight range (Avruch, unpublished). Consequently, the receptor complex as isolated may be a multimolecular, and possibly heterogeneous, assembly of proteins. Despite the great scarcity of the insulin receptor, it is anticipated that a clear picture of this highly important biologic structure will shortly emerge.

V. The Glucagon Receptor System

The investigation of the glucagon receptor has not been hampered by many of the methodologic difficulties involved in studies of insulin binding. Prior to the first attempts at measuring binding of glucagon, the liver parenchymal cell plasma membrane was established as a site of action of the hormone (Davoren and Sutherland, 1963; Pohl *et al.*, 1969; Reik *et al.*, 1970). Since methods for preparation of this organelle were available (Neville, 1960, 1968) the problems inherent in working with whole cells or tissues were circumvented. Glucagon labeled with ^{125}I using chloramine-T (Hunter and Greenwood, 1962) has been shown to be fully active biologically and therefore a suitable tracer (Rodbell *et al.*, 1971a). Several laboratories have reported studies of binding of ^{125}I glucagon to subcellular particles (Rodbell *et al.*, 1971a; Johnson *et al.*, 1972; Tomasi *et al.*, 1970; Goldfine *et al.*, 1972; Krug *et al.*, 1971).

Binding of glucagon to liver plasma membranes can be demonstrated by incubating the labeled hormone with membranes in a medium containing buffer and albumin followed by separation of bound and free labeled material either by centrifugation or by filtration (Rodbell *et al.*, 1971a). The binding is diminished by the presence of unlabeled glucagon but not by a variety of other peptides including insulin, secretion, adrenocorticotropin, and the 1–21 and 22–29 peptide fragments of glucagon. The number of binding sites in the liver membrane preparation has been estimated to be 2·6 pmoles per mg of membrane protein. The binding is saturated at approximately 10^{-7} M glucagon and is half-maximal at 4×10^{-9} M. These findings establish that the binding of glucagon to liver membranes is specific and saturable.

Under usual assay conditions, 10 to 30% but never more than 40% of the labeled hormone added is bound (Rodbell *et al.*, 1971a). The binding is temperature dependent with the rate of binding measured at $0\,^{\circ}C$ being about one-third that observed at $30\,^{\circ}C$. Treatment of the membranes with trypsin, urea, detergents, or phospholipase A reduces the ability of membranes to bind glucagon, indicating that binding is dependent on both lipid and protein components of the membrane.

The time course of binding of glucagon under all conditions so far examined is non-linear. In a typical experiment, 30–40% of the total binding occurs in the first 1–2 min of incubation. However, binding continues to occur at a progressively diminishing rate until a maximum is reached at 10–20 min. After this point, the amount of glucagon bound to the membranes remains constant for several hours. Under the minimal assay condition, i.e. membranes and labeled hormone incubated in a medium containing only buffer and albumin, the binding is effectively irreversible. After incubation for a period of time sufficient for maximal binding to occur, addition of a thousand-fold excess of unlabeled glucagon displaces only a small fraction of the labeled glucagon from the membranes, even if incubation is then continued for several hours (Rodbell *et al.*, 1971a). When the binding assay is carried out under adenylate cyclase assay conditions, i.e. in the presence of ATP and Mg^{2+}, two important characteristics of the binding process are changed: the maximal binding of labeled glucagon is reduced by over 50%, and the binding under this condition is completely and rapidly reversible (Rodbell *et al.*, 1971b). Both of these changes are caused by the presence of ATP, and the effect is relatively non-specific since several nucleoside di- and tri-phosphates are effective. Of these, GTP and GDP are by far the most potent; effects of these nucleosides can be observed at concentrations as low as 10^{-8} M.

Hydrolysis of the terminal phosphate is not essential for this effect since the nonhydrolyzable phosphonate analogue of GTP is effective (Rodbell *et al.*, 1971b). These nucleosides also affect the process of activation of adenylate cyclase by glucagon as is described below.

Glucagon bound to purified rat liver plasma membranes can be disassociated with 2 M urea and recovered in biologically active form (Rodbell *et al.*, 1971a). However, under both minimal and adenylate cyclase assay conditions, the unbound glucagon remaining in the medium is rapidly altered by the membranes in such a way that it will no longer bind to, or activate adenylate cyclase in the membranes (Pohl *et al.*, 1972). This inactivation process is specific for glucagon in that it is not blocked by the addition of large excesses of other small peptide hormones. In common with the glucagon stimulated adenylate cyclase in the purified membranes, the inactivation process is stimulated by EDTA and inhibited by urea, trypsin, digitonin, and phopholipase A. The specific activity of the inactivation process co-purifies with the glucagon sensitive adenylate cyclase in the purified membranes compared to a crude liver homogenate. The change in the glucagon molecule has not been determined, but it must be very minor since the inactivated glucagon co-chromatographs with active glucagon on sephadex, DEAE-cellulose, and a thin layer partition chromatography system. The product is not des-His-glucagon (see below), and glucagon inactivation is not due to the presence of cathepsin C, a soluble dipeptidyl aminopeptidase which inactivates glucagon (McDonald *et al.*, 1969a,b).

Thus, the liver plasma membrane, an organelle which contains the glucagon initiator, also contains a specific glucagon inactivation system which shares several properties with the initiator. The inactivation process is clearly distinct from the glucagon binding sites since bound glucagon can be eluted in an active form. Johnson *et al.* (1972) have recently reported that glucagon covalently attached to agarose beads is biologically active but not susceptible to inactivation by liver plasma membranes. If substantiated, this finding would clearly distinguish inactivation of glucagon from the mechanism through which glucagon stimulates adenylate cyclase.

In order to justify further study, or attempt to isolate the glucagon binding components of the liver plasma membrane, it is necessary to demonstrate that this observed binding process is actually involved in the mechanism of the action of the hormone on the liver. Several characteristics of the binding of glucagon are similar or identical to the glucagon stimulated adenylate cyclase activity. These are specificity, dose-response relationship, similar susceptibility to several destructive agents, and alteration by nucleotides. However, the non-linear and relatively slow kinetics of the binding process (Rodbell, 1971a) are difficult to reconcile with the linear kinetics of the glucagon stimulated adenyl cyclase (Pohl *et al.*, 1971a). Furthermore,

Birnbaumer *et al.* have recently shown that des-His-glucagon competitively and completely inhibits glucagon stimulated adenylate cyclase, but displaces only 10–15% of the glucagon bound to membranes (Birnbaumer *et al.*, 1972; Birnbaumer, pers. comm.). Thus, the biological significance of the binding of glucagon to liver membranes must not be considered to be fully established.

Despite this theoretical limitation, at least two laboratories are proceeding with attempts to isolate and purify the glucagon binding material by the technique of affinity chromatography (Johnson *et al.*, 1972; Krug *et al.*, 1971). This method, which involves highly specific binding of biological material to columns of agarose containing covalently attached substrate or ligand has been applied recently with dramatic success to several biochemical problems (Cuatrecasas and Anfinsen, 1971), and it will doubtless prove extremely useful in the membrane receptor field. However, the glucagon receptor appears to be a complex lipoprotein and is rather unstable. Furthermore, using the estimate that there are about 2 pmoles of binding sites per mg of plasma membrane protein (Rodbell *et al.*, 1971a), and guessing that the molecular weight of a receptor is about 50,000, it appears that the receptor constitutes of the order of 0·1% of the plasma membrane mass. Consequently, major technical and theoretical obstacles must be overcome before this approach can be expected to give significant information regarding the structure and mechanism of action of the glucagon receptor.

As an alternative to the isolation-purification approach, the structure of the glucagon molecule has been examined in an effort to gain insight into its interaction with the receptor. Glucagon is a single chain polypeptide with 29 amino acids. The sequences for bovine, porcine and human hormone are known and all three are identical (Bromer *et al.*, 1957; Bromer *et al.*, 1971; Thomsen *et al.*, 1972). Both the amino and carboxy termini are unblocked. All of the naturally occurring amino acids are represented with the notable exceptions of cystine, cysteine, proline and isoleucine, and there are no unusual amino acids.

The amount of primary structure of glucagon required for biological activity has been investigated in several ways. Spiegel and Bitensky have shown that glucagon treated with cyanogen bromide is biologically active (Spiegel and Bitensky, 1969). Cyanogen bromide cleaves two amino acid residues from the carboxy terminus of the peptide and converts the third amino acid, methionine, to homoserine. However, synthetic 1–23 glucagon is biologically inactive (Spiegel and Bitensky, 1969). Therefore, at least two and less than six amino acids can be removed from the carboxy terminal portion of the peptide without loss of biological activity. Removal of the amino terminal amino acid, histidine, from glucagon has been accomplished in several laboratories by a one step Edman degradation (Felts *et al.*, 1970; Sundby, 1970; Lande *et al.*, 1972). The resulting peptide, des-His-glucagon is biologically inactive. Rodbell *et al.* (1971c) have reported that des-His-gluca-gon is a competitive inhibitor of glucagon on the liver plasma membrane adenylate cyclase system. Lande *et al.* (1972) have recently called attention to the fact that the presence of the NE-phenyl isothiocarbamyl derivative of the single lysine residue in glucagon was not rigorously excluded in the studies of Rodbell *et al.* (1971c). Although it is tempting to speculate that the amino terminal histidine of glucagon is critical for the action of glucagon but not for the binding, the state of this lysine residue must first be resolved.

Higher structure of glucagon has been studied by a variety of techniques including UV absorption, fluorescence spectroscopy, circular dichroism, nuclear magnetic resonance, sedimentation velocity, sedimentation equili-brium, and X-ray crystallography (King, 1965; Blanchard and King, 1966; Gratzer and Beaven, 1969; Gratzer *et al.*, 1967, 1968; Edelhoch and Lippold, 1969; Srere and Brooks, 1969; Swann and Hammes, 1969; Schiffer and Edmundson, 1970; Patel, 1970; Bornet and Edelhoch, 1971; Epand, 1972). At concentrations of the order of 10^{-4} M, glucagon probably exists in a compact globular structure. In addition, under certain conditions, self asso-ciation occurs with further changes in tertiary structure (Gratzer and Beaven, 1969; Swann and Hammes, 1969). However, these changes do not appear to occur in the cyanogen bromide treated glucagon, which is biologically active, and therefore are probably not critical for activity of the hormone (Epand, 1972). The higher structure of glucagon is highly dependent upon the com-position of the solvent, and structural changes would be likely to occur coincident with translocation of the peptide from the aqueous extracellular fluid to the more ordered environment at the membrane surface.

Unfortunately, the concentrations of glucagon required for these physical techniques are 5 to 7 orders of magnitude higher than the concentrations of the hormone *in vivo*. Consequently, the information obtained in these studies is of limited usefulness in understanding the interaction of glucagon with its receptor.

Since both glucagon and the glucagon receptor represent distinct chemical entities, it must be possible ultimately to describe their interaction in terms of a set of chemical forces. These forces include formation of covalent, electro-static and hydrogen bonds and hydrophobic interactions. Furthermore, because of the size of the glucagon molecule and the great specificity of the system for glucagon it is reasonable to postulate that these forces of inter-action are multiple in both number and kind. Formation of a covalent bond between glucagon and its receptor can be safely excluded because of the rapid reversibility of the stimulation of adenylate cyclase by glucagon (Birnbaumer *et al.*, 1972) and the susceptibility of the system to relatively low concentrations of urea (Pohl *et al.*, 1971a; Rodbell *et al.*, 1971a).

The amino acids sequence of glucagon can be divided roughly into three parts (Rodbell et al., 1971c). The segment of the peptide representing the first one-third from the amino terminus contains mainly polar but uncharged amino acids, including two serine, two threonine, one histidine and one glutamine. The middle one-third of the sequence contains a large number of residues which are charged at neutral pH including three aspartic acids, two arginines and one lysine. The third of the sequence at the carboxy terminus is intensely hydrophobic, containing phenylalanine, valine, tryptophan, leucine and methionine. Thus, it is clear that the potential exists for formation of many different bonds between glucagon and its receptor. Sorting out these possibilities would be almost unapproachable were it not for the similarity between glucagon and another peptide hormone, secretin (Mutt and Jorpes, 1967). Both glucagon and secretin activate adenylate cyclase in the adipocyte plasma membrane but do so through different receptors (Rodbell et al., 1970a). Proof of the existence of different receptors is based on the demonstration that des-His-glucagon is a competitive inhibitor of glucagon but not secretin-stimulated adenylate cyclase in the fat cells membrane (Rodbell et al., 1971c; Pohl and Rodbell, unpublished), and conversely des-His-secretin is a competitive inhibitor of secretin but not glucagon-stimulated adenylate cyclase (Rodbell, pers. comm.). Secretin does not stimulate adenylate cyclase in liver plasma membranes. Secretin is also a single chain polypeptide but has 27 instead of 29 amino acid residues. Fourteen of these residues in sequence are identical to the corresponding amino acids in glucagon (Mutt and Jorpes, 1967). In addition, six amino acid pairs in the two peptides are very similar, for example aspartic acid in glucagon and glutamic acid in secretin at position nine, and valine in glucagon and leucine in secretin at position 23. Thus, there are only seven major differences in the sequences of the two peptides. These are at positions 3, 10, 13, 14, 17, 21, and 25, mainly in the central portion of the two peptides. Two of the differences, at position 10 and 13, involve tyrosine in glucagon and leucine in secretin, and four of the differences involve net charge differences (Pohl et al., 1971b; Rodbell et al., 1971d). Thus, there is ample reason for suspecting that both hydrogen bonds and electrostatic forces are involved in the interactions between these peptide hormones and their receptors, and that these interactions determine the specificity. However, there is as yet no direct experimental support for this contention.

An attractive hypotheses which has been approached experimentally is that the hydrophobic carboxy end of the glucagon peptide is involved in a hydrophobic interaction with a site on the plasma membrane. Several kinds of evidence support this hypothesis. Firstly, this region of the glucagon peptide binds to aqueous dispersions of detergents and phospholipids (Bornet and Edelhoch, 1971). Secondly, liver plasma membranes can be modified with detergents or lipases in such a way that adenylate cyclase remains intact, as shown by its ability to respond to stimulation by fluoride ion, but is unresponsive to glucagon stimulation (Birnbaumer *et al.*, 1971). Thirdly, part of this loss of glucagon sensitivity can be restored by incubating the treated membranes with dispersions of phospholipids (Pohl *et al.*, 1971c; Levey, 1971; Réthy *et al.*, 1971). Fourthly, under very carefully controlled conditions of digestion of the membranes with phospholipase A, the dose-response relationship between glucagon and adenylate cyclase activity can be shifted to the right, i.e. the apparent affinity of the system for glucagon can be decreased without a major reduction in the maximal stimulation obtainable with glucagon (Pohl, unpublished).

In summary, a hydrophobic interaction between the carboxy terminal portion of glucagon and membrane phospholipids provides part of the energy for the glucagon–receptor interaction but not the specificity. The specificity appears to be determined by the formation of electrostatic and hydrogen bonds between the central portion of the glucagon molecule and the receptor. The amino terminal region, particularly the terminal histidine, may be involved in the action of the hormone on its receptor.

VI. The Effector System

As defined in the introduction, the only clearly established effector system for peptide hormones is adenylate cyclase. Strong evidence has been presented indicating that the stimulation of cyclic AMP production initiates the effects of glucagon, vasopressin, oxytocin, parathyroid hormone, thyrocalcitonin, ACTH, MSH and secretin in a variety of tissues (Jost and Rickenberg, 1971). The cellular effects of cyclic AMP have been extensively discussed elsewhere (Jost and Rickenberg, 1971), and will not be re-examined. In the case of pancreatic glucagon, essentially all actions of the hormone thusfar studied can be explained by this mechanism. The possibility exists, however, that other effector systems, mediating certain of the actions of the above mentioned peptide hormones, remain to be delineated. Certainly the complex relationship of adenylate cyclase to the cellular effects of insulin suggests that other effector systems are operative for this hormone (Steiner and Freinkel, 1972).

In the following section, the properties of the glucagon sensitive adenylate cyclase, especially that found in liver, will be reviewed. The nature of the effector system(s) for insulin will then be examined.

VII. The Glucagon Effector System

Following the initial description of adenylate cyclase (Sutherland *et al.*, 1962), it was established, first with cell fractionation studies (Davoren and Sutherland, 1963) and later with purified plasma membranes (Pohl *et al.*,

1969; 1971; McKeel and Jarett, 1970), that glucagon sensitive adenylate cyclase systems are located in the surface membranes of adipocytes and hepatic parenchymal cells. Glucagon sensitive adenylate cyclase activity has also been identified in heart (Murad and Vaughan, 1969; Levey and Epstein, 1969), but detailed sub-cellular fractionation studies have not been performed with these tissues (Goldfine et al., 1972; Malaisse et al., 1967).

The plasma membrane localization of the glucagon sensitive adenylate cyclase system was an exciting finding because it explained, in part, how a relatively bulky peptide hormone could influence intracellular events despite the barrier to entrance of the hormone posed by the membrane. However, this finding also raised two important questions: (1) Given the ubiquity of hormone sensitive adenylate cyclase systems in nucleated mammalian cells (Sutherland et al., 1962; Robinson et al., 1968), how could a hormone sensitive enzyme located in the surface membrane, and exposed, presumably to all circulating hormones, confer the required specificity on the receptor system? (2) How could a single enzyme serve to transmit information across the membrane, a barrier approximately 75 Å in width (Benedetti and Emmelot, 1968)?

Although several theories were offered to answer these questions, the first relevant experimental observations were offered independently by Birnbaumer and Rodbell (1969) and by Bär and Hechter (1969). Both groups established that in the adipocyte at least three hormones, glucagon, epinephrine, and adrenocorticotropin, stimulate a single adenylate cyclase. Then by examining the response of adenylate cyclase to each hormone in the presence of specific competitive inhibitors of individual hormones as well as non-specific agents such as chelators and trypsin, they established that the three hormones stimulate this enzyme by different mechanisms. Although experimental proof is lacking, it is reasonable to postulate that these mechanisms are mediated by distinct receptor molecules. The list of hormones having discrete receptors in the adipocyte adenylate cyclase system has been extended to include secretin (Rodbell et al., 1970a). In addition, thyrotropin and luteinizing hormone stimulate the enzyme (Birnbaumer and Rodbell, 1969), but the relatively small effects of these hormones have precluded establishing whether or not they also have distinct receptors. This interesting work with the adipocyte is reviewed in Rodbell et al. (1968, 1969, 1970b) and Birnbaumer et al. (1970).

The relationship of the adenylate cyclase system to membrane structure was first approached experimentally by Kono, who showed that limited digestion of adipocytes with trypsin resulted in selective destruction of the sensitivity of the system to glucagon but not to epinephrine or F⁻ (Kono, 1969). Therefore, only the structures mediating glucagon sensitivity and not the catalytic site of adenylate cyclase are modified by trypsin under these conditions. Since the integrity of the membrane was not compromised by such treatment (Fain and Loken, 1969), this finding indicates that the hormone receptor is located on or near the extracellular surface of the plasma membrane. Rodbell *et al.* extended this observation by showing that trypsin treatment of fat cell ghosts, the hypotonic lysate of isolated adipocytes, results in destruction of the catalytic site of adenylate cyclase (Rodbell *et al.*, 1970a). Since the procedure for preparing ghosts changes the barrier properties of the plasma membrane, the difference in the effects of trypsin on ghosts compared to intact fat cells probably reflects access of the proteolytic enzyme to the interior of the cell. These observations suggest that the catalytic site of adenylate cyclase is located on or near the intracellular surface of the plasma membrane. Such studies have led to the following formulation of the organization of hormone sensitive adenylate cyclase systems. The catalytic site, adenylate cyclase as such, is located on or near the intracellular surface of the plasma membrane and is not in itself hormone specific. The activity of the catalytic site is regulated by a hormone specific receptor, a separate molecular structure located on or near the extracellular surface of the membrane. The nature of the hormone–receptor interaction and the mechanism of coupling of the receptor to the catalytic site have not been revealed by these studies.

Enzymologic investigation of adenylate cyclase was hindered for many years by the lack of a simple, reliable assay method for the enzyme. This problem was solved by Krishna *et al.* who devised a method based on the production of ^{32}P-labeled cyclic AMP from ATP-$\alpha^{32}P$ (Krishna *et al.*, 1968). The labeled product is separated from the substrate by a combination of ion exchange chromatography and adsorption of substrate to a nascent $BaSO_4$ precipitate. Recently, two laboratories have described methods based upon the preferential adsorption of polyvalent anions to hydrous aluminum oxide; cyclic AMP is monovalent whereas ATP, ADP, AMP, and P_i are all multivalent anions at neutral pH (White and Zenser, 1971; Ramachandran, 1971). These methods are somewhat more convenient and give better recovery of cyclic AMP than the method of Krishna *et al.* (1968).

One important pitfall in such methods arises from the fact that adenylate cyclase preparations invariably contain sufficient ATPase activity to rapidly deplete the substrate concentration, thereby shortening the period of time over which adenylate cyclase activity can be observed and precluding even the simplest kinetic analysis. This problem has been approached by the addition of an ATP-regenerating system to the adenylate cyclase assay medium (Birnbaumer *et al.*, 1969). A recent finding of great promise is that 5′-adenylylimidodiphosphate (AMP-PNP) is a substrate for adenyl cyclase but not for ATPase (Rodbell *et al.*, 1971e). A second pitfall arises from the presence in some adenylate cyclase preparations of cyclic 3′5′-phosphodiesterase, which hydrolyzes cyclic AMP to 5′AMP (Sutherland and Rall, 1957; Butcher and Sutherland, 1962). The effect of this enzyme is usually negated by adding either a large excess of unlabeled cyclic AMP or by adding a cyclic phosphodiesterase inhibitor such as theophylline to the assay medium. Fortunately, the purified rat liver plasma membranes contain such low activity of cyclic phosphodiesterase that these precautions are not required (Tomasi et al., 1969).

Adenylate cyclase is substrate specific for ATP since, with the exception of AMP-PNP (see above), no other nucleotide has been shown to be either a substrate or a competitive inhibitor. The enzyme requires a divalent cation, Mg^{2+} or Mn^{2+} and is inhibited by Ca^{2+} (Pohl et al., 1971a; Birnbaumer et al., 1969; Sutherland et al., 1962). The active form of the substrate is probably ATP-Mg. There is some evidence that free ATP is an inhibitor of the enzyme (Birnbaumer et al., 1969; Drummond and Duncan, 1970; Drummond et al., 1971; Severson et al., 1972). The K_m for ATP is of the order of 10^{-4} M (Drummond et al., 1971; Severson et al., 1972), but the presence of ATPase activity has prevented adequate kinetic study of the enzyme. The Mg^{2+} dependence of the enzyme is sigmoidal (Pohl et al., 1971; Birnbaumer et al., 1969; Severson et al., 1972) and optimal at an ATP : Mg^{2+} ratio greater than 1, suggesting that the enzyme may have an allosteric regulatory site for a divalent cation. Birnbaumer et al. have presented evidence that hormones activate adenylate cyclase in the fat cell ghost by increasing the apparent affinity of this site for Mg^{2+} (Birnbaumer et al., 1969). However, in heart and skeletal muscle, hormonal stimulation of the enzyme is associated with changes in V_{max} only with respect to Mg^{2+} (Drummond and Duncan, 1970; Drummond et al., 1971; Severson et al., 1972).

The specific activity of adenylate cyclase in purified liver plasma membranes is about twenty-fold greater than that of the total liver homogenate (Pohl et al., 1971), a purification which correlates well with 5'-nucleotidase, a plasma membrane marker. All attempts at removing the liver adenylate cyclase from the membrane structure have been unsuccessful (Pohl et al., 1971) including the lubrol method devised for heart by Levey (1970). The activity of the enzyme can be destroyed by treatment of the membranes with detergents, phospholipase A, and trypsin (Pohl et al., 1972; Birnbaumer et al., 1971). All of these observations indicate that adenylate cyclase is an integral part of the liver plasma membrane.

Stimulation of the enzyme occurs in the presence of 10^{-10} M glucagon, is half-maximal at 4×10^{-9} M, and is maximal at 10^{-7} M (Pohl et al., 1971a). This dose response relationship is very similar to that of the effects of glucagon on cyclic AMP levels in a perfused rat liver system (Exton et al., 1971). Epinephrine at 10^{-5} M produces less than 10% of the adenylate cyclase stimulation produced by glucagon at 10^{-6} M, but cyclic AMP levels obtained with epinephrine in the rat liver perfusion system are less than 10% of those obtained with glucagon.

The time course of adenylate cyclase in the presence or absence of glucagon is linear and extrapolates to the origin indicating that the process of activation of adenylate cyclase is very rapid. In addition, it has recently been shown that the activation of the enzyme by glucagon is rapidly reversible; a competitive inhibitor of glucagon decreases the activity of the glucagon stimulated adenylate cyclase to the basal rate in less than 1 min (Birnbaumer *et al.*, 1972). Thus, the adenylate cyclase system responds very rapidly to changes in the composition of the medium. This property fits well with the rapid onset of action in the perfused liver and the rapid decay of glucagon effects when the hormone is removed from the perfusate (Exton *et al.*, 1971; Sutherland and Robinson, 1969).

Adenylate cyclase is also stimulated by F^- (Sutherland *et al.*, 1962). The mechanism of this stimulation is unknown; however differential effects of Mn^{2+} and PP_i on the glucagon and fluoride stimulated activities make it clear that these agents stimulate the same enzyme by different mechanisms (Birnbaumer *et al.*, 1969; Wolff and Jones, 1970). Although the physiologic importance of the fluoride effect is questionable, it has proved to be a very useful way of testing the integrity of the catalytic site of adenylate cyclase. In the liver membrane preparation, the unstimulated adenylate cyclase activity is so low that in the absence of a stimulator the difference from zero activity approaches the error of the assay method (Pohl *et al.*, 1971a).

Using the fluoride stimulated activity as a marker for the catalytic site, it is possible to alter the membranes with either digitonin or phospholipase A in such a way that catalytic activity of the system remains intact but is no longer sensitive to glucagon (Birnbaumer *et al.*, 1969). Furthermore, glucagon sensitivity can be partially restored by incubating the membranes in suspensions of phospholipids (Pohl *et al.*, 1971c; Levey, 1971; Réthy *et al.*, 1971). When the membranes are digested with much lower concentrations of phospholipase A, the apparent affinity of the system for glucagon is decreased ten-fold with little or no effect on the maximal glucagon stimulation (Pohl, unpublished). Under certain conditions, glucagon can protect the system against the effects of phospholipase A (Pohl, unpublished). All of these observations indicate that membrane lipids are involved in the process of stimulation of adenylate cyclase by glucagon.

VIII. Insulin Effector Systems

Once bound to its receptor, insulin results in the modification of a variety of plasma membrane functions including glucose transport, membrane potential, ion flux and perhaps adenyl cyclase activity (for review, see Steiner and Freinkel, 1972). Furthermore, the hormone appears to do so in a concerted fashion, in so far as the alterations in these functions occur with a relatively similar time course. The mechanism whereby this is accomplished is unknown, but several current hypotheses as to the nature of the effector system(s) may be examined in the light of available evidence.

The possibility that there exist multiple effector systems for insulin should be considered. Thus, the insulin receptor site may be an allosteric site associated with each glucose transport assembly, adenylate cyclase molecule, etc. To be consistent with the insulin binding data, these heterogeneously situated receptor sites would all be required to exhibit the same binding affinity for insulin (Cuatrecasas, 1971a). While this would appear unlikely, such a hypothesis cannot be discarded in the absence of more direct data.

A more widely discussed mechanism suggests that the insulin receptor is coupled to one of the *known* insulin-sensitive membrane functions, and the alteration in this function is responsible for all of the other observed actions of the hormone. Shortly after the discovery that insulin enhanced hexose transport (Levine et al., 1950), it was suggested that this function fulfilled such a role (Levine and Goldstein, 1955; Levine, 1957). It soon became apparent that, for example, insulin-induced antilipolysis in adipose cells (Jungas and Ball, 1963), as well as a large number of insulin effects in a variety of tissues, were demonstrable in the absence of glucose (for review, see Levine, 1966). More recently it has been suggested that adenyl cyclase is the primary effector of insulin actions, since under certain conditions, insulin can be shown to reduce the intracellular concentration of cyclic AMP. The data bearing on the effects of insulin on cyclic AMP will be evaluated below. However, it is reasonably clear that while an insulin-induced reduction in cyclic AMP may account for certain effects of the hormone, such as antilipolysis and the activation of glycogen synthesis in adipose tissue, it cannot account for all the observed actions of the hormone. This is best appreciated through examination of the interrelationships between cyclic AMP and insulin stimulated glucose transport in adipocytes and muscle. If insulin stimulated glucose transport in adipocytes by reducing intracellular cyclic AMP, it should decrease the concentration of the nucleotide below the basal level. In fact, in the absence of lipolytic hormones insulin augments glucose transport without any demonstrable effect on cyclic AMP (Sneyd et al., 1968; Butcher et al., 1966; Manganiello, 1967). While insulin can be shown to cause a 50% drop in the cyclic AMP levels induced by lipolytic hormones (Butcher et al., 1968), the resultant cyclic AMP levels are still three-to-ten-fold greater than under basal conditions. Despite the presence of such markedly elevated intracellular cyclic AMP levels, insulin-stimulated glucose uptake remains unimpaired (Sneyd et al., 1968). Theophylline (which inhibits cyclic AMP phosphodiesterese) and dibutyryl cyclic AMP both inhibit basal glucose utilization by the adipose cell: however the increment in glucose utilization due to insulin is not diminished by maximal concentrations of either lipolytic hormones or methylxanthines (Blecher, 1967a; Rodbell, 1967b). Finally, cyclic AMP, added to plasma membrane vesicles isolated from adipocytes, does not alter glucose transport by these particles nor the insulin induced increment in glucose transport (Carter *et al.*, 1972a). Thus, the conclusion that insulin modifies glucose transport via cyclic AMP does not appear tenable. Furthermore in skeletal muscle, insulin enhances glycogen synthetase activity without a demonstrable effect on intracellular cyclic AMP levels (Goldberg *et al.*, 1967; Craig *et al.*, 1969). At present, no single known effect of insulin can be demonstrated to underlie all of the other actions of the hormones.

The controversial question of the effects of insulin on cyclic AMP metabolism merits further examination. Firstly, does insulin alter intracellular concentrations of cyclic AMP? In liver and adipose tissue the bulk of evidence indicates that insulin lowers cyclic AMP under certain conditions; in skeletal (and probably cardiac) muscle, no significant change in cyclic AMP levels occurs on exposure to insulin. Rats injected with anti-insulin serum show a progressive rise in the hepatic content of cyclic AMP (Jefferson *et al.*, 1968). Using isolated perfused livers from normal rats and animals pretreated with anti-insulin serum, addition of insulin to the perfusate *in vitro* lowers hepatic content of cyclic AMP in both circumstances, and inhibits the glucagon induced increment in cyclic AMP as well (Jefferson *et al.*, 1968). Insulin can inhibit the increase in cyclic AMP content in whole rat epididymal fat pads (Butcher *et al.*, 1966) or isolated fat cells (Butcher *et al.*, 1968) caused by submaximal concentrations of lipolytic hormones; in the absence of lipolytic hormones, insulin does not lower basal levels of the nucleotide (Butcher *et al.*, 1966, 1968; Manganiello *et al.*, 1967). Diaphragmatic muscle from rats treated *in vivo* with insulin show a rise in glycogen synthetase activity in the glucose-6-phosphate independent form (Goldberg *et al.*, 1967). Paradoxically, this is asociated with a small rise in cyclic AMP content of the muscle (Goldberg *et al.*, 1967). Isolated diaphragms treated with insulin *in vitro* show no change in cyclic AMP content (despite the usual increase in glycogen synthetase in the (I) Form), nor did insulin affect the epinephrine-induced increase in cyclic AMP content and phosphorylase activity (Craig *et al.*, 1969). In the isolated perfused rat heart, the level of cyclic AMP is not altered by insulin (Robinson *et al.*, 1968). Next, in those tissues where insulin does appear to modify cyclic AMP concentrations, how is this accomplished? The initial observation that insulin increased cyclic AMP phosphodiesterase activity (Senft *et al.*, 1968) has been followed by several reports which fail to show any change in this activity (Blecher *et al.*, 1968; Mueller-Oerlinghausen *et al.*, 1968; Menahan *et al.*, 1968; Hepp *et al.*, 1969). The reports concerning the effects of insulin on adenyl cyclase are conflicting. Jungas found that pretreatment of fat pads with insulin led to a 30% decrease in homogenate adenyl cyclase activity (Jungas, 1966). Several subsequent studies examining this activity in broken cell preparations of adipose tissue found no (Cryer *et al.*, 1969; Vaughn and Murad, 1969) or inconsistent (Rodbell *et al.*, 1968) effects with insulin. A recent report (Illiano and Cuatrecasas, 1972) indicated that in membrane preparations from liver and adipocytes, insulin at concentrations of 5–50 microunits/ml inhibited the epinephrine and glucagon stimulation of adenyl cyclase (but not the basal activity); at higher concentrations of insulin no inhibition was found, and perhaps a slight stimulation was observed at very high insulin levels. Most prior studies of the interaction between insulin and adenyl cyclase have not explored the effects of insulin at concentrations less than 100 microunits/ml, and further data on this point is required.

Finally, certain observations suggest that insulin may antagonize the effects of cyclic AMP independently of inducing changes in total cyclic AMP concentration. This hypothesis is based on the effects of insulin added to various tissues in the presence of exogenous cyclic AMP or dibutyryl cyclic AMP. Since exogenous cyclic AMP production is bypassed, and insulin does not appear to enhance diesterase activity, antagonistic effects presumably reflect interference by insulin with a cyclic AMP activated effector pathway. A series of conflicting reports have emerged. Several investigators have reported that lipolysis in fat cells induced by exogenous AMP (Hepp *et al.*, 1969) or dibutyryl cyclic AMP (Goodman, 1969) was unaffected by insulin (1000 microunits/ml). Conversely, Solomon *et al.* (1971), found that while insulin at 1000 microunits/ml did not inhibit the lipolytic effect of exogenous cyclic AMP or dibutyryl cyclic AMP (and in fact appeared to have enhanced it), physiologic concentration of insulin (2–20 microunits/ml) significantly inhibited nucleotide induced lipolysis. In skeletal muscle, antagonistic effects of dibutyryl cyclic AMP and insulin on a number of functions have also been reported (Chambaut *et al.*, 1969). The interpretation of such contradictory data is inappropriate at present; however the possibilities that insulin alters the subcellular distribution of cyclic nucleotides or accelerates the production of an inhibitor of cyclic AMP (Manganiello *et al.*, 1967; Ho and Sutherland, 1971) should be mentioned.

The last hypothesis concerning insulin action which will be considered suggests that the multiple changes in plasma membrane function (and subsequently, intracellular metabolism) are simultaneously instituted by an alteration in the structure of the plasma membrane, induced by the binding of insulin. This concept has been suggested in various forms by numerous authors (Rodbell *et al.*, 1968; Peters, 1956; Hechter, 1955; Krahl, 1961). The supportive evidence is largely indirect and comes from the observed functional effects induced by relatively non-specific perturbations of the surface of hormonally responsive cells. Incubation of fat cells with phospholipase C under sublytic conditions results in increased stereospecific glucose transport, antilipolysis and enhanced incorporation of C^{14} histidine into protein, thus mimicking some of the effects of insulin (Blecher, 1965; Rodbell, 1966). Various preparations of phospholipase A also promote glucose transport and antilipolysis in fat cells (Blecher, 1967b, 1969). A number of surface active agents, including lysophosphatides (Blecher, 1967b) and the polyene antifungal antibiotic filipin (Kuo, 1968) (which binds to both membrane sterols and phospholipids), can also exhibit insulinomimetic effects at sublytic concentrations. Limited proteolysis of fat cells, with a variety of proteases of differing specificity, can result in metabolic changes which simulate the response to insulin (Kuo et al., 1966a,b, 1967a). While it is likely that the functional alterations induced by these agents arise from changes in plasma membrane structure, the nature of these structural changes is unknown, and their relationship to insulin action is inferential.

Purified plasma membrane vesicles from adipose tissue have been examined for alterations in structure which might accompany the functional activation by insulin. Purified plasma membrane particles isolated from adipocytes contain a functionally intact glucose transport system (Carter et al., 1972a). If intact fat cells are exposed to insulin prior to homogenization, the subsequently isolated plasma membrane vesicles exhibit enhanced glucose transport, relative to a parallel preparation of plasma membrane derived from the same pool of cells, but which were not exposed to the hormone (Avruch et al., 1972). The structural characteristics of such "insulin activated" and control adipocyte plasma membrane vesicles were compared using a number of spectroscopic techniques, including infrared spectroscopy, fluorescence probe analysis and native membrane protein fluorescence. Despite the presence of significantly enhanced glucose transport in the "insulin activated" membranes, no structural differences between these and the unstimulated membranes were detected by the spectroscopic methods employed (Avruch et al., 1972). These results indicate that the action of insulin does not involve large-scale changes in the structure of the plasma membrane. The possibility remains however that structural changes may occur in relatively circumscribed areas of the plasma membrane (i.e. changes which are not widely propagated) such as in the vicinity of the insulin receptor. In order for such structural changes to fulfil the role of an effector, it would be required that the membrane functions modified by insulin reside in close proximity to the receptor. The resolution of this problem awaits the subfractionation of the plasma membrane of an insulin sensitive cell.

IX. The Transducer System

It is evident that some alteration must occur in the hormone, the receptor, or both consequent to their interaction. These structural changes constitute a signal which is received by the effector, resulting in a change in the activity of the effector. Although the mechanisms operative in these transduction phenomena are almost entirely unknown, the possibilities include only the formation or breakage of covalent bonds and conformational changes. Abundant biochemical models are available for both kinds of mechanisms, for example phosphorylations of regulatable enzymes and membrane transport systems (Holzer and Duntze, 1971; Avruch and Fairbanks, 1972) and conformational changes in allosteric enzymes (Koshland and Neet, 1968). Application of the methods used for these analogous problems to the study of hormone action may provide insight regarding the nature of transduction. In the following section, the limited evidence regarding transducer systems will be reviewed.

X. The Glucagon Transducer

The need to postulate and attempt to observe a transducer system is clear cut for glucagon because of the evidence, reviewed above, that the receptor and effector systems for this hormone are functionally, and probably molecularly, distinct. There appears to be a covalent change in the glucagon molecule consequent to interaction with plasma membranes resulting in inactivation of the hormone (Pohl *et al.*, 1972), and it is possible that the free energy of hydrolysis of a peptide bond is the signal which increases the activity of adenylate cyclase. However, the relationship of the inactivation of glucagon to the glucagon initiator has not been established, and there is evidence that glucagon can act without being inactivated (Johnson *et al.*, 1972). Spectrophotometric evidence for a conformational change in glucagon upon binding of the peptide to detergents and phospholipids has been presented (Bornet and Edelhoch, 1971). However, this system is not necessarily a valid model for events occurring at the receptor, both because the state of the lipids is different in the membrane, and because of the high concentrations of glucagon employed in the spectrophotometric studies.

An interesting and experimentally approachable possibility for the glucagon transducer is a change in the state of membrane lipids (Pohl *et al.*, 1971c; Perkins and Moore, 1971). As noted above, there is good evidence that membrane lipids are intimately involved in the glucagon sensitive adenylate cyclase system. Furthermore, this system can be viewed as a means for transmitting information across a lipid barrier, and it is possible that these lipids serve a functional as well as a structural role. However, experimental support for this hypothesis is lacking.

The finding that des-His-glucagon is a competitive inhibitor of glucagon (Rodbell *et al.*, 1971c) may, if substantiated, provide clues regarding the nature of the transducer for glucagon. Since des-His-glucagon binds to the receptor but does not act, the terminal histidine can be considered to participate in an active site. In chymotrypsin and certain other endopeptidases, an active site histidine interacts with a nearby serine rendering the hydroxyl group uniquely reactive (for review see Dickerson and Geis, 1969). This reactivity can be demonstrated by phosphorylation with diisopropylphosphofluoridate. If the histidine of glucagon is involved in a similar mechanism, it may be possible to identify glucagon specific, uniquely reactive amino acid residues in the plasma membrane. In hemoglobin, two globin histidines are involved in the binding of the protein to the heme iron (for review see Dickerson and Geis, 1969). The glucagon histidine may also form a complex with a membrane bound cation. Thus, although the nature of the glucagon transducer is unknown, the problem is not unapproachable.

XI. Insulin Transducer

As discussed above, it has been suggested that an alteration in membrane structure is the mechanism by which insulin coordinately modifies a number of membrane functions, and thus this structural change would represent the insulin effector system itself. Despite this semantic difficulty, some evidence does exist that the insulin receptor is linked to glucose transport (and presumably other insulin sensitive membrane functions) through distinct and separable membrane components. Under certain conditions, the coupling between glucose transport and insulin binding can be abolished whereas both functions remain individually intact. As noted earlier, crude membrane preparations from adipocytes retain the full cellular complement of insulin binding sites (Crofford et al., 1970; Cuatrecasas, 1971c). In addition, purified plasma membrane preparations can remove insulin from the medium, presumably through binding (Crofford and Okayama, 1970); as with binding, this removal is abolished by trypsin. As mentioned previously, isolated adipocyte plasma membrane vesicles retain an intact glucose transport system (Carter et al., 1972a). However, when insulin is added directly to the purified plasma membrane, no stimulation of glucose transport ensues (Avruch et al., 1972). Furthermore anti-insulin serum added to "insulin activated" plasma membranes, prepared as described above, fails to reverse the enhanced glucose transport, whereas such anti-sera rapidly abolish this effect in the intact cell (Avruch et al., 1972). These findings suggest that the coupling between insulin binding and glucose transport has been disrupted during cell homogenization. More compelling and direct evidence on the nature of the membrane species which transduce insulin binding into glucose transport has been presented by Cuatrecasas (1971a). Incubation of intact adipocytes with neuraminidase, at low concentrations, results in stimulation of glucose transport without alteration in insulin binding. However, with somewhat higher neuraminadase concentrations, the insulin stimulatable component of glucose transport is irreversibly abolished without alteration in basal stereospecific glucose uptake, cellular leakiness, or insulin binding. This effect was not modified by the presence of insulin in the medium, or by the addition of a variety of protease inhibitors or substrates other than those susceptible to sialidase. These results strongly indicate that sialic acid residues of cell membrane glycoproteins play a critical role in the processes which transmit the binding of insulin to glucose transport. It is anticipated that more specific information on these cell surface glycoproteins will become rapidly available.

The role of membrane SH groups in insulin action requires some comment. Although a number of reports (Bewsher *et al.*, 1966; Cadenas *et al.*, 1961; Whitney *et al.*, 1963; Fong *et al.*, 1962; Edelman *et al.*, 1963; Mirsky and Perisutti, 1962) suggested that membrane sulfhydryl groups were critical to insulin binding, these studies are subject to serious criticism, and recent direct evidence (Cuatrecasas, 1971c) conclusively eliminates this possibility. Nevertheless, SH reactive compounds (both thiols and SH inhibitors) modify insulin action in a variety of ways. These effects have been extensively evaluated using the isolated fat cell system, and although the quantitative aspects vary with the specific reagent, the results are qualitatively similar. In general, three phases of altered metabolic response are seen with increasing sulfhydryl reagent concentrations. At low concentrations, these agents stimulate glucose utilization (Lavis and Williams, 1970; Kuo *et al.*, 1967a; Carter and Martin, 1969; Minemura and Crofford, 1969; Minemura *et al.*, 1967) an effect not seen in homogenates of these treated cells (NEM, p-chloromercuribenzene sulfonate, iodoacetimide, L-cysteine, reduced glutathione, 2-mercaptoethanol and dithiothreitol); these agents inhibit the lipolytic response to submaximal levels of ACTH and epinephrine but not to maximal levels of lipolytic hormones (Lavis and Williams, 1970; Minemura and Crofford, 1969) a pattern similar to that observed for insulin (p-chloromercuribenzene sulfonate, L-cysteine, reduced glutathione 2-mercaptoethanol and dithiothreitol); and they diminish intracellular cyclic AMP levels (P-chloromercuribenzene sulfonate) (Minemura and Crofford, 1969). The maximal increase in glucose uptake produced by p-chloromercuribenzene sulfonic acid (PCMBS) is 50–90% of the maximal response to insulin, but incubation of such PCMBS treated cells with insulin can enhance glucose uptake to levels beyond that which can be achieved with insulin alone (Minemura and Crofford, 1969). In contrast, fat cells exposed to concentrations of NEM (Carter and Martin, 1969) or cysteine (Lavis and Williams, 1970) which stimulate basal glucose uptake, on subsequent addition of maximal levels of insulin, exhibit glucose uptake identical to that of insulin treated cells which have not been exposed to sulfhydryl-active reagents. As the concentrations of these agents is increased, the insulin-induced increment in glucose uptake is inhibited, with no change in basal uptake (NEM, maleimide) (Crofford et al., 1968; Carter and Martin, 1969). Under these conditions, glucose transport and intracellular utilization per se appear to remain intact, since glucose oxidation can be stimulated by increasing the extracellular glucose concentration (Crofford et al., 1968; Carter and Martin, 1969). Finally, at high levels of sulfhydryl reagents, basal glucose utilization is depressed in intact cells as well as homogenates (Carter and Martin, 1969; Lavis and Williams, 1970). These complex actions may be partially sorted out. Clearly, the effects of the highest concentrations of these reagents reflect inhibition of intracellular metabolism as well as glucose transport and require no further consideration. The stimulation of glucose transport observed at the lowest concentrations is certainly due to an effect at plasma membrane. It is unlikely that the locus of this action is directly on the glucose transport system. The effect of NEM on glucose transport in purified membrane particles from isolated fat cells (Carter et al., 1972a), and red blood cells (Carter et al., 1972b; Dawson and Widdas, 1963) is only inhibitory; no stimulation of transport is seen at any concentration of this reagent. Is the stimulatory effect due to modification of a sulfhydryl group normally involved in the sequence of insulin action? This appears unlikely for PCMBS, where the combined effects with insulin exceed those found with insulin alone (Minemura et al., 1969); however this possibility remains tenable for NEM. Furthermore, the inhibition of insulin induced increment in glucose transport (but not basal glucose transport) by intermediate levels of NEM reinforces this hypothesis, since Cuatrecasas has shown (Cuatrecasas, 1971c) that NEM at hundred-fold greater concentrations does not alter the binding of insulin to its receptor. At present no specific information is available concerning the distribution or function of protein sulfhydryl groups in the membrane of an insulin sensitive cell. In fact, the only quantitative study of tissue sulfhydryl groups in relation to insulin appeared in 1958; Ungar and Kadis (1959) reported that insulin treatment of rat hemi-diaphragms diminished the total free sulfhydryl content (measured by amperometric titration of extracts of frozen homogenates) by 22·1%. Certainly, much important information concerning insulin action lurks in this area.

XII. Conclusion

The search for the primary site of peptide hormone action has led to the plasma membrane of the target cell. The plasma membrane structures which intervene between the "first message", i.e. the hormone, and the "second message", whatever its nature, are collectively called the "initiator" system for peptide hormone action. The functional elements of the initiator systems for glucagon and insulin can be described in general terms. First, the hormone

K

is specifically bound to a plasma membrane site, the receptor system. A signal is thereby generated, which is propagated by components we call the transducer system, to one or more other plasma membrane constituents. The function of these latter components, which we call the effector system(s), is thereby altered, and this (these) alterations mediate all the subsequently observed effects of the hormone. The best defined of the functional components of this plasma membrane bound initiator system is the receptor function. Indeed, attempts at the isolation and structural characterization of this component are currently underway. In the case of glucagon, the effector system, i.e. adenylate cyclase, is also clearly defined, and its characterization is approachable by currently available techniques. The areas of greatest uncertainty pertain to the nature of the effector system for insulin, and the manner in which the receptor and effector functions are linked. The solution of these questions will be of great significance, not only to endocrinologists, but to all interested in the regulatory functions of biological membranes, and the relationship of these functions to membrane structure.

XIII. Acknowledgement

The authors wish to thank Ms. Mary Beth Jovanovich for her editorial services. S.L.P. acknowledges the support of the American Diabetes Association.

References

Ahlquist, R. P. (1948). *Am. J. Physiol.* **153,** 586.
Ahlquist, R. P. (1959). *Pharmacol. Rev.* **11,** 441.
Ahlquist, R. P. (1968). *A. Rev. Pharmac.* **8,** 259.
Arquilla, E. R., Ooms, H. and Mercola, K. (1968). *J. clin. Invest.* **47,** 474.
Avruch, J. Unpublished observations.
Avruch, J. and Fairbanks, G. (1972). *Proc. natn. Acad. Sci. U.S.A.* **69,** 1216.
Avruch, J. and Wallach, D. F. H. (1971). *Biochim. biophys. Acta.* **233,** 334.
Avruch, J., Carter, J. R., Jr., and Martin D. B. (1972). *Biochim. biophys. Acta.* **288,** 27.
Bär, H.-P. and Hechter, O. (1969). *Proc. natn. Acad. Sci. U.S.A.* **63,** 350.
Benedetti, E. L. and Emmelot, P. (1968). *In* "Ultrastructure in Biological Systems" (A. J. Dalton and F. Hagvenav, eds.), Vol. 4, p. 33. Academic Press, New York.
Berson, S. A. and Yalow, R. S. (1957). *Ann. N.Y. Acad. Sci.* **70,** 56.
Bewsher, P. D., Hillman, C. C. and Ashmore, J. (1966a). *Molec. Pharmac.* **2,** 227.
Birnbaumer, L. Personal communication.
Birnbaumer, L. and Rodbell, M. (1969). *J. biol. Chem.* **244,** 3477.
Birnbaumer, L., Pohl, S. L. and Rodbell, M. (1969). *J. biol. Chem.* **244,** 3468.
Birnbaumer, L., Pohl, S. L., Krans, H. M. J. and Rodbell, M. (1970). *In* "Advances in Biochemical Psychopharmacology" (E. Costa and P. Greengard, eds.), Vol. 3, p. 185. Raven Press, New York.
Birnbaumer, L., Pohl, S. L. and Rodbell, M. (1971). *J. biol. Chem.* **246,** 1857.

Birnbaumer, L., Pohl, S. L., Rodbell, M. and Sundby, F. (1972). *J. biol. Chem.* **247,** 2038.

Blanchard, M. H. and King, M. V. (1966). *Biochem. biophys. Res. Commun.* **25,** 298.

Blecher, M. (1965). *Biochem. biophys. Res. Commun.* **21,** 202.

Blecher, M. (1967a). *Biochem. biophys. Res. Commun.* **27,** 560.

Blecher, M. (1967b). *Biochim. biophys. Acta.* **137,** 557.

Blecher, M. (1969). *Biochim. biophys. Acta.* **187,** 380.

Blecher, M., Merlino, N. and Ro'Ane, J. T. (1968). *J. biol. Chem.* **243,** 3973.

Bornet, H. and Edelhoch, H. (1971). *J. biol. Chem.* **246,** 1785.

Bromer, W. W., Sinn, L. G. and Behrens, O. K. (1957). *J. Am. chem. Soc.* **79,** 2807.

Bromer, W. W., Boucher, M. E. and Koffenberger, J. E., Jr. (1971). *J. biol. Chem.* **246,** 2822.

Brunfeldt, K. (1965). *Science Tools* **12,** 17.

Butcher, R. W. and Sutherland, E. W. (1962). *J. biol. Chem.* **237,** 1244.

Butcher, R. W., Sneyd, J. G. T., Park, C. R. and Sutherland, E. W., Jr. (1966). *J. biol. Chem.* **241,** 1651.

Butcher, R. W., Baird, C. E. and Sutherland, E. W., Jr. (1968). *J. biol. Chem.* **243,** 1705.

Cadenas, E., Kaji, H., Parks, C. R. and Rasmussen, H. (1961). *J. biol. Chem.* **236,** PC 63.

Carter, J. R., Jr. and Martin, D. B. (1969). *Biochim. biophys. Acta* **177,** 521.

Carter, J. R., Jr., Avruch, J. and Martin, D. B. (1972a). *J. biol. chem.* **247,** 2682.

Carter, J. R., Jr., Avruch, J. and Martin, D. B. (1972b). Submitted for publication.

Chambaut, A. M., Eboue-Bonis, D., Hanoune, J. and Clauser, H. (1969). *Biochem. biophys. Res. Commun.* **34,** 283.

Craig, J. W., Rall, T. W. and Larner, J. (1969). *Biochim. biophys. Acta* **177,** 213.

Crofford, O. B. (1968). *J. biol. Chem.* **243,** 362.

Crofford, D. B. and Okayama, T. (1970). *Diabetes* **19,** Suppl. 1, 369.

Crofford, O. B., Trumbo, D. S. and Minemura, T. (1968). *J. clin. Invest.* **47,** 23a.

Crofford, O. B., Minemura, T. and Kono, T. (1970). *In* "Advances in Enzyme Regulation" (G. Weber, ed.), Vol. 8, p. 219, Academic Press, New York.

Cryer, P. E., Jarett, L. and Kipnis, D. M. (1969). *Biochim. biophys. Acta* **177,** 586.

Cuatrecasas, P. (1969). *Proc. natn. Acad. Sci. U.S.A.* **63,** 450.

Cuatrecasas, P. (1971a). *Proc. natn. Acad. Sci. U.S.A.* **68,** 1264.

Cuatrecasas, P. (1971b). *J. biol. Chem.* **246,** 6522.

Cuatrecasas, P. (1971c). *J. biol. Chem.* **246,** 6532.

Cuatrecasas, P. (1971d). *J. biol. Chem.* **246,** 7265.

Cuatrecasas, P. (1972). *Proc. natn. Acad. Sci. U.S.A.* **69,** 318.

Cuatrecasas, P. and Anfinsen, C. (1971). *A. Rev. Biochem.* **40,** 259.

Cuatrecasas, P. and Illiano, G. (1971). *J. biol. Chem.* **246,** 4938.

Cuatrecasas, P., Desbuquois, B. and Krug, F. (1971). *Biochem. biophys. Res. Commun.* **44,** 333.

Davoren, P. R. and Sutherland, E. W. (1963). *J. biol. Chem.* **238,** 3016.

Dawson, A. C. and Widdas, W. F. (1963). *J. Physiol.* **168,** 644.

DeZoeten, L. W. and Van Strik, R. (1961). *Recl. Trav. chim. Pays-Bas Belg.* **80,** 927.

Dickerson, R. E. and Geis, I. (1969). "The Structure and Action of Proteins." Harper and Row, New York.

Drummond, G. I. and Duncan, L. (1970). *J. biol. Chem.* **245,** 976.

Drummond, G. I., Severson, D. L. and Duncan, L. (1971). *J. biol. Chem.* **246,** 4166.

Edelhoch, H. and Lippold, R. E. (1969). *J. biol. Chem.* **244,** 3876.

Edelman, P. M., Rosenthal, S. L. and Schwartz, I. L. (1963). *Nature, Lond.* **197,** 878.

Epand, R. M. (1972). *J. biol. Chem.* **247,** 2132.

Exton, J. H., Robison, G. A., Sutherland, E. W. and Park, C. R. (1971). *J. biol. Chem.* **246,** 6166.

Fain, J. N. and Loken, S. C. (1969). *J. biol. Chem.* **244,** 3500.

Felts, P. W., Ferguson, M. E. C., Hagey, K. A., Stitt, E. S. and Mitchell, W. M. (1970). *Diabetologia* **6,** 44.

Ferrebee, J. W., Johnson, B. B., Mithoefher, J. C. and Gardella, J. W. (1951). *Endocrinology* **48,** 277.

Fong, C. T., Silver, L., Popenoe, E. A. and Debons, A. F. (1962). *Biochim. biophys. Acta* **56,** 190.

Freychet, P., Roth, J. and Neville, D. M. (1971a). *Proc. natn. Acad. Sci. U.S.A.* **68,** 1833.

Freychet, P., Roth, J. and Neville, D. M. (1971b). *Biochem. biophys. Res. Commun.* **43,** 400.

Freychet, P., Kahn, R., Roth, J. and Neville, D. M. (1971c). *In* "Proc. of the Second Symp. on Protein and Polypeptide Hormones." *Excerpta Medica,* Amsterdam. (In press.)

Furchgott, R. F. (1959). *Pharmac. Rev.* **11,** 429.

Garratt, C. J. (1964). *Nature, Lond.* **201,** 1324.

Garratt, C. J., Cameron, J. S. and Menzinger, G. (1966). *Biochim. biophys. Acta* **115,** 179.

Glieman, J. (1968). *Acta physiol. Scand.* **72,** 481.

Gliemann, J. (1969). *Danish Med. Bull.* **16,** Suppl. 4, 1.

Goldberg, N. D., Villar-Palasi, C., Saska, H. and Larner, J. (1967). *Biochim. biophys. Acta* **148,** 665.

Goldfine, I. D., Roth, J. and Birnbaumer, L. (1972). *J. biol. Chem.* **247,** 1211.

Goodman, H. M. (1969). *Proc. Soc. exp. Biol. Med.* **130,** 97.

Gratzer, W. B. and Beaven, G. H. (1969). *J. biol. Chem.* **244,** 6675.

Gratzer, W. B., Bailey, E. and Beaven, G. H. (1967). *Biochem. biophys. Res. Commun.* **28,** 914.

Gratzer, W. B., Beaven, G. H., Rattle, H. W. E. and Bradbury, E. M. (1968). *Eur. J. Biochem.* **3,** 276.

Hechter, O. (1955). Concerning possible mechanisms of hormone action. *In* "Vitamins and Hormones". Vol. 13, p. 293.

Hepp, K. D., Menahan, L. A., Wieland, O. and Williams, R. H. (1969). *Biochim. biophys. Acta* **184,** 554.

Ho, R. J. and Sutherland, E. W. (1971). *J. biol. Chem.* **246,** 6822.

Holzer, H. and Duntze, W. (1971). *A. Rev. Biochem.* **40,** 1971.

Hunter, W. M. and Greenwood, F. C. (1962). *Nature, Lond.* **194,** 495.

Illiano, G. and Cuatrecasas, P. (1972). *Science, N.Y.* **175,** 906.

Izzo, J. L., Bale, W. F., Izzo, M. J. and Roncone, A. (1964a). *J. biol. Chem.* **239,** 3743.

Izzo, J. L., Roncone, A., Izzo, M. J. and Bale, W. F. (1964b). *J. biol. Chem.* **239,** 3749.

Jarett, J., Reuter, M., McKeel, D. W. and Smith, R. M. (1971). *Endocrinology,* **89,** 1186.

Jefferson, L. S., Exton, J. H., Butcher, R. W., Sutherland, E. W. and Park, C. R. (1968). *J. biol. Chem.* **243,** 1031.

Johnson, L. B., Blecher, M. and Giogio, N. A. (1972). *Biochem. biophys. Res. Commun.* **46,** 1035.

Jost, J. P. and Rickenberg, H. V. (1971). *A. Rev. Biochem.* **40,** 741.

Jungas, R. L. (1966). *Proc. natn. Acad. Sci., U.S.A.* **56,** 757.

Jungas, R. L. and Ball, E. G. (1963). *Biochemistry* **2,** 383.

King, M. V. (1965). *J. molec. Biol.* **11,** 549.

Kono, T. (1968). *Biochemistry* **7,** 1106.

Kono, T. (1969a). *Biochim. biophys. Acta* **178,** 397.

Kono, T. (1969b). *J. biol. Chem.* **244,** 1772.

Koshland, D. E., Jr. and Neet, K. E. (1968). *A. Rev. Biochem.* **37,** 359.

Krahl, M. E. (1961). "The Action of Insulin on Cells". Academic Press, New York.

Krishna, G., Weiss, B. and Brodie, B. B. (1968). *J. Pharmac. exp. Ther.* **163,** 379.

Krug, F., Desbuquois, B. and Cuatrecasas, P. (1971). *Nature New Biology* **234,** 268.

Kuo, J. F. (1968). *Archs Biochem. Biophys.* **127,** 406.

Kuo, J. F., Holmund, C. E. and Dill, I. K. (1966a). *Life Sci.* **5,** 2257.

Kuo, J. F., Holmund, C. E., Dill, I. K. and Bohonos, N. (1966b). *Arch. biochem. Biophys.* **117,** 269.

Kuo, J. F., Dill, I. K. and Holmund, C. E. (1967a). *J. biol. Chem.* **242,** 3659.

Kuo, J. F., Dill, I. K. and Holmund, C. E. (1967b). *Biochim. biophys. Acta* **148,** 683.

Lande, S., Gorman, R. and Bitensky, M. (1972). *Endocrinology* **90,** 597.

Lavis, V. R. and Williams, R. H. (1970). *J. biol. Chem.* **245,** 23.

Levey, G. S. (1970). *Biochem. biophys. Res. Commun.* **38,** 86.

Levey, G. S. (1971). *Biochem. biophys. Res. Commun.* **43,** 108.

Levey, G. S. and Epstein, S. E. (1969). *Circ. Res.* **24,** 151.

Levine, R. (1957). *Surv. Biol. Progr.* **3,** 185.

Levine, R. (1966). *Am. J. Med.* **40,** 691.

Levine, R. and Goldstein, M. S. (1955). *In* "Recent Progress in Hormone Research", **11,** 343.

Levine, R., Goldstein, M. S., Huddlestun, B. and Klein, S. P. (1950). *Am. J. Physiol.* **163,** 70.

Malaisse, W. J., Malaisse-Lagae, F. and Mahew, D. (1967). *J. clin. Invest.* **46,** 1724.

Malaisse, W. and Franckson, T. R. M. (1965). *Arch. int. Pharmacodyn. Thér.* **155,** 484.

Mandl, I. (1961). *Adv. Enzymol.* **23,** 163.

Manganiello, V. C., Murad, F. and Vaughn, M. (1967). *J. biol. Chem.* **246,** 2195.

McDonald, J. K., Callanan, P. X., Zeitman, B. B. and Ellis, S. (1969a). *J. biol. Chem.* **244,** 6199.

McDonald, J. C., Zeitman, B. B., Reilly, T. J. and Ellis, S. J. (1969b). *J. biol. Chem.* **244,** 2693.

McKeel, D. W. and Jarett, L. (1970). *J. Cell. Biol.* **44,** 417.

Menahan, L. A., Hepp, K. D. and Wieland, O. (1969). *Eur. J. Biochem.* **8,** 435.

Minemura, T. and Crofford, O. B. (1969). *J. biol. Chem.* **244,** 5181.

Minemura, T., Crofford, O. B. and Neuman, E. U. (1968). *Fed. Proc.* **27,** 566.

Mirsky, E. A. and Perisutti, G. (1962). *Biochim. biophys. Acta* **62,** 490.

Moran, N. C. (1966). *Pharmac. Rev.* **18,** 503.

Mueller-Oerlinghausen, B., Schwabe, U., Hasselblatt, A. and Schmidt, F. H. (1968). *Life Sci.* **7,** 593.

Murad, F. and Vaughan, M. (1969). *Biochem. Pharmac.* **18,** 1129.

Mutt, V. and Jorpes, J. E. (1967). *Recent Prog. Horm. Res.* **24,** 539.

Neville, D. M., Jr. (1960). *J. biophys. biochem. Cytol.* **8,** 413.

Neville, D. M., Jr. (1968). *Biochim. biophys. Acta.* **154,** 540.

Newerly, K. and Berson, S. A. (1957). *Proc. Soc. exp. Biol. Med.* **94,** 751.
Patel, D. J. (1970). *Macromolecules* **3,** 448.
Perkins, J. P. and Moore, M. M. (1971). *J. biol. Chem.* **246,** 62.
Peters, R. A. (1956). *Nature, Lond.* **177,** 426.
Pohl, S. L. Unpublished observations.
Pohl, S. L., Birnbaumer, L. and Rodbell, M. (1969). *Science, N.Y.* **164,** 566.
Pohl, S. L., Birnbaumer, L. and Rodbell, M. (1971a). *J. biol. Chem.* **246,** 1849.
Pohl, S. L., Birnbaumer, L. and Rodbell, M. (1971b). *In* "Annual Reports in Medicinal Chemistry, 1970", (C. K. Cain, ed.), p. 233. Academic Press, New York.
Pohl, S. L., Krans, H. M. J., Kozyreff, V., Birnbaumer, L. and Rodbell, M. (1971c). *J. biol. Chem.* **246,** 4447.
Pohl, S. L., Krans, H. M. J., Birnbaumer, L. and Rodbell, M. (1972). *J. biol. Chem.* **247,** 2295.
Pohl, S. L. and Rodbell, M. Unpublished observations.
Ramachandran, J. (1971). *Analyt. Biochem.* **43,** 227.
Reik, L., Petzold, G. L., Higgins, J. A., Greengard, P. and Barnett, R. J. (1970). *Science, N.Y.* **168,** 384.
Réthy, A., Tomasi, V. and Trevisani, A. (1971). *Archs. Biochem. Biophys.* **147,** 36.
Rieser, P. (1967). "Insulin, Membranes and Metabolism", Williams and Wilkins, Baltimore.
Robison, G. A., Butcher, R. W. and Sutherland, E. W. (1967). *Ann. N.Y. Acad. Sci.* **139,** 703.
Robison, G. A., Butcher, R. W. and Sutherland, E. W. (1968). *A. Rev. Biochem.* **37,** 149.
Rodbell, M. Personal communication.
Rodbell, M. (1964). *J. biol. Chem.* **239,** 375.
Rodbell, M. (1966). *J. biol. Chem.* **241,** 130.
Rodbell, M. (1967a). *J. biol. Chem.* **242,** 5744.
Rodbell, M. (1967b). *J. biol. Chem.* **242,** 5751.
Rodbell, M., Jones, A. B., DeCingolani, G. E. C. and Birnbaumer, L. (1968). *Recent Prog. Horm. Res.* **24,** 215.
Rodbell, M., Birnbaumer, L. and Pohl, S. L. (1969). *In* "The Role of Adenyl Cyclase and Cyclic 3′,5′-AMP in Biological Systems", (T. W. Rall, M. Rodbell and P. Condliffe, eds.), p. 59. U.S. Government Printing Office.
Rodbell, M., Birnbaumer, L. and Pohl, S. L. (1970a). *J. biol. Chem.* **245,** 718.
Rodbell, M., Birnbaumer, L., Pohl, S. L. and Krans, H. M. J. (1970b). *Acta Diabetologica Latina* **7,** Suppl 1, 9.
Rodbell, M., Krans, H. M. J., Pohl, S. L. and Birnbaumer, L. (1971a). *J. biol Chem.* **246,** 1861.
Rodbell, M. Krans, H. M. J., Pohl, S. L. and Birnbaumer, L. (1971b). *J. biol. Chem.* **246,** 1872
Rodbell, M., Birnbaumer, L., Pohl, S. L. and Sundby, F. (1971c). *Proc. natn. Acad. Sci. U.S.A.* **68,** 909.
Rodbell, M., Birnbaumer, L., Pohl, S. L. and Krans, H. M. J. (1971d). *In* "Structure-Activity Relationships of Protein and Polypeptide Hormones", (M. Margoulies and F. C. Greenwood, eds.), p. 199. Excerpta Medica, Amsterdam.
Rodbell, M., Birnbaumer, L., Pohl, S. L. and Krans, H. M. J. (1971e). *J. biol. Chem.* **246,** 1877.
Rosa, U., Massaglia, A., Pennisi, F., Cozzani, I. and Rossi, C. A. (1967). *Biochem. J.* **103,** 407.

Rudman, D., Garcia, L. A., DiGirocamo, M. and Shank, P. W. (1966). *Endocrinology* **78**, 169.

Rudman, D., Garcia, L. A., Del Rio, A. and Akgus, S. (1968). *Biochemistry* **7**, 1864.

Schiffer, M. and Edmundson, A. B. (1970). *Biophys. J.* **10**, 293.

Senft, G., Schultz, G., Munske, K. and Hoffman, M. (1968). *Diabetologia* **4**, 322.

Severson, D. L., Drummond, G. I. and Sulakhe, P. V. (1972). *J. biol. Chem.* **247**, 2949.

Sneyd, J. G. T., Corbin, J. D. and Park, C. R. (1968). *In* "Pharmacology of Hormonal Polypeptides and Proteins", p. 367 (N. Back, L. Martini and R. Paoletti, eds.). Plenum Press, New York.

Solomon, S. S., Brush, J. S. and Kitabchi, A. E. (1971). *Biochim. biophys. Acta.* **218**, 167.

Spiegel, A. M. and Bitensky, M. W. (1969). *Endrocrinology* **85**, 638.

Srere, P. A. and Brooks, G. C. (1969). *Archs Biochem. Biophys.* **129**, 708.

Stadie, W. C., Haugaard, N., Marsh, J. B. and Hirs, A. G. (1949). *Am. J. Med. Sci.* **218**, 265.

Stadie, W. C., Haugaard, N. and Vaughn, M. (1952). *J. biol. Chem.* **199**, 729.

Stadie, W. C., Haugaard, N. and Vaughn, M. (1953). *J. biol. Chem.* **200**, 745.

Stagg, B. H., Temperley, J. M., Rochman, H. and Morley, J. S. (1970). *Nature, Lond.* **228**, 58.

Steck, T. L. and Wallach, D. F. H. (1970). *In* "Methods in Cancer Research", Vol. 5, Chap. IX, p. 121, (H. Busch, ed.). Academic Press, New York.

Stein, O. and Gross, J. *Endocrinology* **65**, 707.

Steiner, D. F. and Freinkel, N. (Eds.). (1972). "Handbook of Physiology". *Amer. Physiol. Soc.*, Washington, D.C.

Sundby, F. (1970). *In* "Abstracts, VII Congress of the International Diabetes Federation", p. 80. (R. R. Rodrigues, F. J. G. Ebling, I. Henderson aqd R. Assan, eds.). ICA No. 209, Excerpta Medica, Amsterdam.

Sutherland, E. W. and Rall, T. W. (1957). *J. biol. Chem.* **232**, 1077.

Sutherland, E. W. and Rall, T. W. (1960). *Pharmac. Rev.* **12**, 265.

Sutherland, E. W. and Robison, G. A. (1969). *Diabetes* **18**, 797.

Sutherland, E. W., Rall, T. W. and Menon, T. (1962). *J. biol. Chem.* **237**, 1220.

Swann, J. C. and Hammes, G. G. (1969). *Biochemistry* **8**, 1.

Thomsen, J., Kristiansen, K. and Brunfeldt, K. (1972). *FEBS Letters* **21**, 315.

Tomasi, V., Koretz, S., Ray, T. K., Dunnick, J. and Marinetti, G. V. (1970). *Biochim. biophys. Acta* **211**, 31.

Ungar, G. and Kadis, S. (1959). *Nature, Lond.* **183**, 49.

Vallence-Owen, J. and Hurlock, B. (1954). *Lancet* **1**, 68.

Vaughn, M. and Murad, F. (1969). *Biochemistry* **8**, 3092.

White, A. A. and Zenser, T. V. (1971). *Analyt. Biochem.* **41**, 372.

Whitney, J. E., Cutler, O. E. and Wright, F. E. (1963). *Metabolism* **12**, 352.

Wohltmann, H. J. and Narahara, H. T. (1966). *J. biol. Chem.* **241**, 4931.

Wolff, J. and Jones, A. B. (1970). *Proc. natn. Acad. Sci. U.S.A.* **65**, 454.

Worthington, W. C., Jr., Jones, D. C. and Buse, M. G. (1964). *Endocrinology* **74**, 914.

Yalow, R. S. and Berson, S. A. (1966). *Trans. N.Y. Acad. Sci.* **28**, 1033.

Chapter 5

Phospholipid Dynamics in Plasma Membranes

E. FERBER

Max Planck-Institut für Immunbiologie,
Freiburg/Brg., West Germany

I. Introduction

Investigations concerning the structure and function of isolated cytoplasmic membranes have recently gained in interest. This is because membranes are responsible for the integrity of the cell contact between cells, as well as for

Abbreviations: GP = glycerol-3-phosphate; GPC = glycerol-3-phosphorylcholine; GPE = glycerol-3-phosphorylethanalamine.

221

the transport of metabolites. Changes in the composition of the essential structural components of membranes, such as the phospholipids, should, therefore, affect the function of these membranes. Relatively little work is known concerning the phospholipid metabolism of isolated plasma membranes. One exception is the membranes of the mammalian erythrocytes.

Recently there have been reviews concerning the general metabolism of phospholipids (Hill and Lands, 1970; Lennarz, 1970) and also about the membrane structure (Korn, 1969) as well as the technique for the isolation of plasma membranes (Steck and Wallach, 1970). In this chapter the phospholipid metabolism of those membranes will be critically compared.

II. Erythrocytes

A. INTRODUCTION

With purified mammalian erythrocytes, it is possible to obtain cytoplasmic membranes by the relatively simple method of hypertonic shock, because these cells do not possess any other type of membrane system. For this reason, these membranes have been used as a model for biochemical investigations. Comparative investigations of whole cells and isolated membranes show, at least qualitatively, that plasma membranes of erythrocytes contain the entire lipids of the cell, and also all of the biochemical reactions of the phospholipid metabolism (see van Deenen and de Gier, 1964). It is, however, difficult to interpret the earlier results concerning the phospholipid metabolism of these cells, since they were often contaminated with metabolically active reticulocytes and leucocytes. This is certainly true for the postulated biosynthesis of long chain fatty acids from acetate, as also for the *de novo* synthesis of complex phospholipids, such as lecithin. Today, it can be established as certain (van Deenen and de Gier, 1964; Mulder and van Deenen, 1965a; Shohet and Nathan, 1970; Pittman and Martin, 1966) that neither the synthesis of long chain fatty acids in mature erythrocytes nor the *de novo* synthesis of complex lipids like lecithin are of any great significance in these cells. Thus, the reactions discovered in purified whole erythrocytes and their isolated membranes, as reported by several research groups, gain in significance.

These are:

(a) The incorporation of inorganic phosphate into phosphatidic acid.
(b) The incorporation of inorganic phosphate into polyphosphoinositides.
(c) The incorporation of long chain fatty acids into phosphatidylcholine or phosphatidylethanolamine; i.e. the synthesis of lecithin from lysolecithin and of phosphatidylethanolamine from lysophosphatidylethanolamine.
(d) The degradation of phospholipids.

B. DIGLYCERIDE KINASE

Hokin and Hokin (1961, 1963) were able to prove the existence of a biochemical reaction in human erythrocytes in which diglycerides are phosphorylated to phosphatidic acid in the presence of an ATP generating system. The reaction is: 1,2-diglyceride + ATP $\rightleftharpoons$ phosphatidic acid + ADP. This kinase shows a high specificity for diglycerides, because the synthesis of phosphatidic acid from monoglycerides has only 1/40 of the rate of turnover as the diglycerides (see Table I). The synthesis of phosphatidic acid from monoglycerides is probably accomplished with lysophosphatidic acid serving as an intermediate because this compound is accumulated when acylation ceases, e.g. if CoA is omitted. Thus the acylation of monoglycerides as the primary step in the reaction, and the following phosphorylation of the

TABLE I

Relative rates of phosphatidic acid synthesis in ghosts

Pathway	Substrates	Phosphatidic acid synthesis (nmoles × mg^{-1} × h^{-1})
Phosphorylation of diglyceride	[γ^{32}P] ATP, diglyceride	33·6
Acylation of α-glycerophosphate	α-glycero [^{32}P]-phosphate, CoA, oleate, palmitate, ATP	0·013
Phosphorylation of monoglyceride followed by acylation of lysophosphatidic acid	[γ^{32}P] ATP, monoglyceride, CoA, palmitate	0·84

(from Hokin and Hokin, 1963)

diglycerides, is the less probable pathway for the synthesis of phosphatidic acid. Table I also shows comparative measurements for the synthesis of phosphatidic acid from glycerol-3-phosphate, which shows only 1/2500 of the rate of phosphorylation as that observed for the diglycerides. Since whole erythrocytes (Raderecht et al., 1962; van Deenen et al., 1961; Mulder and van Deenen, 1965a) are able to incorporate radioactive phosphate exclusively into phosphatidic acid, it is very probable that the phosphate is incoporated by the diglyceride kinase pathway. A turnover of diglyceride and phosphatidic acid is also possible in erythrocyte membranes, since Hokin et al. (1963) have demonstrated the existence of phosphatidic acid phosphatase (1-α-phosphatidate phosphohydrolase EC 3.13.4) in such membranes. However, the proposed hypothesis by these authors concerning the meaning of this phosphatidic acid–diglyceride cycle for membrane transport phenomena will not be discussed here.

C. SYNTHESIS OF POLYPHOSPHOINOSITIDES

The inositol phosphatides also have the capability of acting as an acceptor for phosphate, in that they can be phosphorylated with ATP to di- and triphosphoinositides. Hokin and Hokin (1964) were able to show, that in the presence of P^{32}-ATP produced by an ATP-generating system radioactively labeled polyphosphoinositides are synthesized. In degradation experiments with phosphomonoesterase it could be further shown that 90% of the labeled phosphate was incorporated into the terminal phosphate groups of di- and triphosphoinositides. However, since this system also contains inorganic ^{32}phosphate, the possibility of a passive exchange reaction cannot be ignored. The synthesis of polyphosphoinositides was almost completely independent of the addition of exogenous inositolphosphatides functioning as acceptor. The addition of inositolphosphatide increased the incorporation rates of phosphate into diphosphoinositide by a maximum of only 30%, and inhibited the synthesis of triphosphoinositides. Experiments designed to synthesize polyphosphoinositides through the increased synthesis of endogenous inositolphosphatides by the addition of CTP, phosphatidic acid, inositol and ATP were also negative. Further kinetic studies brought no proof for the possible relationship of this reaction with the ATPase system.

D. INCORPORATION OF LONG CHAIN FATTY ACIDS AND ACYLATION OF LYSOPHOSPHATIDES

1. *Acyl-transfer from Acyl-CoA*

Mature human erythrocytes are not able to synthesize long chain fatty acids from acetate, because they do not contain an acetyl-CoA-carboxylase, which catalyzes the synthesis of malonyl-CoA (Pittman and Martin, 1966). On the other hand, they exhibit enzyme activities for the activation and incorporation of long chain fatty acids (acid : CoA-ligase and acyl-CoA : lysolecithin acyltransferase). The first reports concerning the incorporation of long chain fatty acids in ghosts were published by Oliveira and Vaughan (1962), and then later described by them in detail (Oliveira and Vaughan, 1964).

Oliveira and Vaughan were already able to show that neither glycerol-3-phosphate nor CDP-choline were utilized. They proved further, that the employed fatty acids (linoleic acid and also palmitic acid) were incorporated into position 2 of the lecithin molecule. However, they had no direct evidence for the acylation of lysolecithin, because the incorporation did not increase after the addition of lysolecithin.

There are several possibilities for this type of behavior:

1. The authors employed palmitic acid for this experiment, which has a low affinity to 1-acyl-GPC-acyltransferases (Waku and Lands, 1968a).
2. The limiting step using free fatty acids in erythrocytes is the acid : CoA-ligase reaction, so that endogenous lysolecithin is already present in saturating concentrations (as the authors discussed).

In contrast to other research groups (Munder et al., 1965; Mulder and van Deenen, 1965b), Oliveira and Vaughan were also able to find differences in the linoleic acid incorporation in different species, e.g. sheep erythrocyte ghosts had an essentially lower turnover (1/10) than did the ghost preparations from rats and humans. Since other researchers, in opposition to Oliveira, were not able to find such differences in various species, when an excess of lysolecithin was present, it is probable that these species differences are due to the different concentrations of endogenous lysolecithin. Besides these discrepancies, the results were, however, confirmed and elaborated (Mulder and van Deenen, 1965b; Mulder et al., 1965; Mulder and van Deenen, 1965a; Robertson and Lands, 1964; Waku and Lands, 1968a).

Table II shows qualitatively, that from different lipid precursors exclusively lysolecithin can be converted to lecithin in human erythrocyte membranes (Mulder and van Deenen, 1965a). In addition to that, lysophosphatidyl-ethanolamine can also be utilized (Mulder et al., 1965).

In contrast to other authors (Oliveira, see above) Mulder and van Deenen (1965b) reported that palmitic acid was incorporated into position 1 of leci-thin. This result would be easy to explain if erythrocyte membranes were also able to acylate the lysolecithin isomer 2-acyl-GPC. Waku and Lands (1968a) were, however, able to demonstrate that such a reaction plays practically no role in human erythrocyte membranes.

These authors were able to show in detailed studies, that the specificity of

TABLE II

Comparison of incorporation of phospholipid precursors into lecithin of erythrocyte ghosts

Substrate	Incorporation (%)
1. $Na_2H^{32}PO_4$	<0·001
2. Glycero-3-[^{32}P] phosphate	<0·01
3. [^{32}P] Phosphorylcholine	<0·3
4. Glycero-3-[^{32}P] phosphorylcholine + CDP-choline	<0·3
5. 1-Acyl-2-[1 − ^{14}C] linoleoyl-glycerol + CDP-choline	<1
6. 1-Acyl-2-[1 − ^{14}C] oleoyl-glycerol	<2
7. 1-Acyl-glycero-3-[^{32}P] phosphorylcholine	30–40

(from Mulder and van Deenen, 1965a)

TABLE III

Rate of acyl-CoA acyl-transfer into the 2-position of 1-acyl-GPC in the presence of stromata from human, rat, and cow erythrocytes

Acyl CoA	nmoles × mg protein $^{-1}$ × min^{-1}		
	Human	Rat	Cow
12 : 0	3·0	3·4	2·8
14 : 0	2·7	3·0	2·7
16 : 0	2·0	1·6	2·2
16 : 1	7·5	4·4	3·9
18 : 0	0·7	0·7	0·7
18 : 1c	5·3	2·5	5·0
18 : 1t	3·7	6·0	—
18 : 2cc	14·0	14·0	8·5
18 : 2ct	3·9	6·6	—
18 : 2tc	11·4	13·3	—
18 : 2tt	2·4	3·4	—
18 : 3	12·3	12·3	8·5
20 : 4	8·1	15·5	6·4

Fatty acyl moieties are designated by chain length : no. of double bonds. c = cis, t = trans (double bond nearest the carboxyl group is designated first).
(From Waku and Lands, 1968a.)

lysolecithin acyltransferase has a similar preference for unsaturated acyl-CoA's, as has been observed in other cells (see Table III).

The acylation of alkenyl derivates appear to be also possible in erythrocyte membranes, although the turnover rates were of low order. The activity for the linoleoyl-CoA : 1-alkenyl-GPC-acyltransferase reaction in human erythrocyte membranes, as prepared according to Dodge et al. (1963), is 0·35 nmoles × mg^{-1} × min^{-1} which represents only 2% of the entire linoleoyl-CoA : 1-acyl-GPC-acyltransferase activity (Waku and Lands, 1968b).

It has already been mentioned that free fatty acids can be utilized by erythrocyte membranes, in which it was shown that they possess the fatty acid activating enzyme acid : CoA-ligase.

Using the hydroxamate method and palmitic acid McLeod and Bressler (1967) found very low activities (0·0031 nmoles × mg^{-1} × min^{-1}) in hemoglobin free stroma while Ferber, using the lysolecithin acyltransferase reaction, was able to measure higher activities in stroma prepared by the method of Dodge et al. (1963) (see Table IV). The differences in the enzyme activities in the various preparations of erythrocyte membranes will be discussed later. McLeod and Bressler (1967) were also able to show, that besides lysophosphatides carnitine could also be acylated. Acyl-carnitines are not to be considered as intermediary products in the synthesis of lecithin since the addition of carnitine does not stimulate the production of lecithin, but

TABLE IV

Comparison of enzyme activities of erythrocyte membranes

	Hemolysate	Ghosts, containing hemoglobin	Ghosts, free of hemoglobin according to Dodge et al. (1963)	
	nmoles	nmoles	nmoles	nmoles
	10^{10}cells × min	10^{10}cells × min	10^{10}cells × min	mg protein × min
Acid: CoA-ligase	9.8[a]	4.3[a]	0.4[b]	0.2[b] 0.0031[c]
Oleoyl-CoA : 1-acyl-GPC acyltransferase	34[d]		6.1[b]	3.05[b]
Palmitoyl-CoA : 1-acyl-GPC acyltransferase				5.3[e] 2.0[e]
Lysophospholipase	15.7[a]		2.6[b]	1.3[b]

[a] Ferber et al. (1968b); [b] Ferber, unpublished results; [c] McLeod and Bressler (1967); [d] Ferber et al. (1968c); [e] Waku and Lands (1968a).

actually inhibits, which is probably a result of the competition of lysolecithin and carnitine for acyl-CoA. These authors discuss the possibility of the function of acyl-carnitine acting as a reservoir for acyl groups.

2. *Acyl-transfer from Acyl-GPC*

Marinetti et al. (1958) described a dismutative acyl transfer in preparations with an active lysophospholipase, in which the 1-acyl group of a molecule of lysolecithin is cleaved and transferred to the 2-hydroxyl group of a second molecule of lysolecithin.

The reaction is:

$$2 \text{ Acyl-GPC} \rightarrow \text{diacyl} - \text{GPC} + \text{GPC}$$

Mulder et al. (1965) found only a slight production of lecithin in the absence of acyl-CoA or CoA and ATP, however, the reaction was linear for a short period of time.

Results which suggest that this pathway is not of biological relevance are:

1. Acyltransfer occurs only with non-physiologically high substrate concentrations, which seems to indicate that this reaction is catalyzed by a hydrolase (lysophospholipase).

2. With double labeled lysolecithin (in the fatty acid and choline moieties) Shohet and Nathan (1970) were able to show that lysolecithin is converted as an entire molecule into lecithin.

3. Since naturally occuring 1-acyl-GPC almost exclusively contains saturated fatty acids (palmitic and stearic acid) this reaction should produce saturated lecithins, but such lecithins are only found in low amounts.

E. DEACYLATION OF PHOSPHOLIPIDS

1. *Lysophospholipase*

Mulder et al. (1965) were able to show qualitatively, that hemolyzed rabbit erythrocytes have the capability of hydrolytically cleaving lysolecithin and also lysocephalin to a high degree to free fatty acids and GPC (GPE, respectively). The hydrolytical degradation of the lysophosphatides was higher when there was no acylation producing lecithin (because of the absence of CoA and ATP), which probably is a result of the competition of both enzymes for the same substrate. Later studies confirmed these results (Munder et al., 1965; Fischer et al. 1967) and were able to prove that these reactions are regulated and operate in a similar way in whole erythrocytes (see below).

2. *Phospholipase A*

Several research groups correspondingly reported (Oliveira and Vaughan, 1964; Mulder and van Deenen, 1965a; Robertson and Lands, 1964; Munder *et al.*, 1965; Ferber *et al.*, 1968a) that when using lecithin as a substate mature erythrocytes exhibit no phospholipase A activity. Delbauffe *et al.* (1968a, b) were, however, able to show that lysates from rat erythrocytes degrade phosphatidyl-glycerol. The pH optimum of this phospholipase A was between 6·5 and 7·5. The activities were relatively high with about 30 nmoles/ml × min (20 nmoles × mg protein^{-1} × min^{-1}). At present little can be said concerning the significance of these results with respect to the phospholipid metabolism in erythrocytes.

F. COMPARISON OF ENZYME ACTIVITIES IN DIFFERENT MEMBRANE PREPARATIONS

It should be remembered that in the described studies, as in the measurement of enzyme activities, entirely different methods have been used to prepare the erythrocyte membranes. As well as the production of hemolyzates by freezing and thawing most ghost preparations have been obtained by osmotic lysis, being more or less free of hemoglobin, as well as having different degrees of fragmentation. A few comparative results of enzyme activities are presented in Table IV. Such a comparison of specific activities in preparations which contain hemoglobin are only possible if the activities are calculated in respect to the number of cells present (per 10^{10} cells). In hemoglobin free stroma, as prepared by the method of Dodge *et al.* (1963), the activities are also comparatively given per mg protein. All of the presented activities in hemoglobin free membranes are drastically lowered, especially the acid : CoA-ligase. However, it does not appear that this loss of activity is caused by release of bound enzymes, because the supernatants (with the exception of low lysophospholipase activity) do not demonstrate enzyme activity (Ferber *et al.*, 1968b). For this reason, it is probable that this inactivation of membrane bound enzymes is caused by the dissolution of relatively tightly bound hemoglobin and other membrane proteins. Since free hemoglobin does not bind lipid substrates, it is therefore unable to interfere in such enzyme tests, thus making hemolysates more suitable for these types of studies than purified membrane preparations.

G. CONTROL OF THE LECITHIN SYNTHESIS IN ERYTHROCYTE MEMBRANES

With erythrocyte membranes, it is possible to study the regulation of the few metabolic pathways of the phospholipid metabolism and their biological significance for the entire erythrocyte. However, it is important that the erythrocytes be studied in their natural surroundings, the blood plasma. A few of these aspects will be discussed in the following.

1. *Intracellular ATP Concentration*

An important factor determining the direction of the metabolic pathway in which lysolecithin is metabolized appears to be the intracellular ATP concentration. At high ATP concentrations the lysolecithin acylation reactions dominate and at low concentrations that of the lysophospholipase. This behavior was not only observed in membrane preparations, but also in intact erythrocytes, containing different concentrations of ATP (Mulder *et al.* 1965; Munder *et al.*, 1965). Further proof for these reaction mechanisms are based on the following facts:

1. The ATP dependent fatty acid activating enzyme (acid : CoA-ligase) limits the acylating reactions because of its relatively low activity (see above).
2. In experiments on ghosts reconstituted with ATP it was possible to demonstrate (Ferber *et al.*, 1968a; Fischer *et al.*, 1965; Ferber *et al.*, 1970) that when the physiological ATP concentration was about 1 μmol/ml, the maximum turnover rate of free fatty acids was reached. These experiments show further that only ATP and CoA, which is inside the ghosts serve as substrates of the acid : CoA-ligase, thus allowing the conclusion that this enzyme is located at the inner surface of the erythrocyte membrane.

2. *Plasma Lysolecithin*

Since erythrocytes cannot produce lysolecithin from lecithin, the only source of this substrate is the plasma. Compared with the erythrocyte membrane, the plasma contains relatively high concentrations of lysolecithin. Plasma lysolecithin, and also lecithin, are able to exchange with the erythrocyte lipids. For example, Mulder and van Deenen (1965a) discovered the same level of labeling in lecithin and lysolecithin after the injection of ^{32}P-orthophosphate. About 20% of these labeled plasma lipids could be exchanged *in vitro* in a time span of 5 hours, with lysolecithin being exchanged at a rate twice as high as that of lecithin. Polonovski and Paysant (1963) also came to similar results when they injected ^{32}P-orthophosphate *in vivo* and found the label, after 6 hours, being exclusively located in the lysolecithin fraction of erythrocytes. Similar exchange rates were observed in *in vitro* experiments.

3. *Plasma Fatty Acids*

Although fatty acids in erythrocytes can be liberated from lysolecithin by the lysophospholipase reaction, it appears that the blood plasma is the main source for free fatty acids. This can be displayed in feeding experiments (van Deenen *et al.*, 1962; van Deenen and de Gier, 1964; Mulder *et al.*, 1963; Walker and Kummerow, 1964), but also from the calculations of Waku and

Lands (1968a) concerning the distribution of fatty acids in position 2 of lecithin in erythrocytes. It is interesting to note that these authors were able to calculate the measured lecithin composition through the product of the specific activity of the acyl-CoA : lysolecithin acyltransferase in erythrocyte membranes and the percentage distribution of the corresponding fatty acids in the plasma. There was, however, a noticeable exception in rat erythrocytes, in which the calculated fatty acid distribution strongly deviated from that of the measured values, while there was a good correspondence in human and bovine red cells. This indicates that not only the acyltransferase activities in rat erythrocytes and the fatty acid composition of the plasma, but still other factors must be responsible for the regulation of these processes.

H. EXCHANGE OF CHOLESTEROL

In addition to the exchange of phospholipids between erythrocytes and plasma (see II, G2), similar reactions were also discovered for the membrane cholesterol. *In vitro* experiments with labeled cholesterol showed an exchange occuring between erythrocytes and plasma, to such an extent that after 1 hour 50% of the cholesterol was exchanged in both directions and that after 4–5 hours an equilibrium had been established (Hagerman and Gould, 1951; Porte and Havel, 1961). According to Hagerman and Gould (1951), only free cholesterol is exchanged and not the esterified cholesterol which is at least present in the plasma. This proves too, that only the cholesterol is exchanged and not the whole lipoprotein moiety. It was further shown that the exchange of cholesterol is not based on diffusion, since the concentrations in the erythrocytes and plasma are different and remained unchanged even after longer incubation *in vitro* (Hagerman and Gould, 1951). Murphy (1962) discovered that this exchange was dependent on the temperature, but could not, however, be influenced by the glucose metabolism of the erythrocytes.

Although the exchange occurs in both directions, there are, nevertheless, interesting differences in the velocity of exchange from erythrocytes to plasma and from plasma to erythrocytes. Thus Murphy (1962) was able to show that the loss of cholesterol from erythrocytes, when they are incubated in plasma with a low cholesterol content, is not due to a faster release of cholesterol but to its retarded acceptance from the plasma. As Basford *et al.* (1964) discovered, this process allows the plasma to gain its original cholesterol concentrations while the rate of release of cholesterol from erythrocytes remains constant. On the other hand, the cholesterol pool in erythrocytes with low cholesterol content cannot be compensated by high concentrations in the plasma (Basford *et al.*, 1964). Thus the plasma cholesterol is held constant, even at the expense of the erythrocyte cholesterol, the loss of which, according to these *in vitro* experiments, cannot be compensated, once a deficit has occured.

I. PATHOLOGICAL ERYTHROCYTES

At this point it is not possible to discuss the numerous experiments that have been undertaken in order to relate certain forms of pathological erythrocytes to changes in the phospholipid metabolism or changes of their membrane phospholipids. Only those studies will be mentioned which are concerned with the lysolecithin metabolism, because this is the most important pathway in mature erythrocytes.

Kates *et al.* (1961) reported that whole blood of patients with congenital spherocytic anemia had a higher concentration of monoacyl-GPE. These results, however, could not be confirmed by other researchers (De Gier *et al.*, 1961; Phillips and Roome, 1962). Based on these findings, Robertson and Lands (1964) determined the acylation rates of lysolecithin and lysocephalin in hemolysates in normal and spherocytic erythrocytes. The rates of acylation did not, however, show any significant differences. Since in these experiments free oleic acid was used to measure the rate of incorporation with and without addition of lyso compounds as acceptors the entire acylation process was followed, thus dismissing the explanation that the cause of these negative results is due to changes in the acid : CoA-ligase. These findings further show that spherocytes do not exhibit an altered content of "endogenous" lysolecithin or monoacyl-GPE.

J. *IN VIVO* AGING OF ERYTHROCYTES

Changes of the membrane have been repeatedly made responsible for the limited life-span and for the active elimination of these cells from the blood circulation (Danon 1968, 1970). With respect to the phospholipid metabolism, it appears especially interesting that changes in the fatty acid distribution during aging *in vivo* were observed. Thus van Gastel *et al.* (1965) and Walker and Yurkowski (1967) were able to show that the arachidonic acid content decreases significantly. This finding would be easily explained as a result of the decrease of the lysolecithin acylation activity, because the incorporation of arachidonic acid into phospholipids is mainly brought about by an acylation of lysophosphatides (Hill and Lands, 1968; Kanoh, 1969; van Golde *et al.*, 1968). Ferber (1971) and Ferber *et al.* (1968b, c) were able to show in hemolysates, in washed stroma, and also in whole erythrocytes, that there is a considerable decrease in the activity of enzymes responsible for the acylation of lysolecithin during aging. Thus enzyme activities of the acylating system—with free oleic acid, saturating amounts of substrates and excess of

ATP and CoA—were measured in hemolysates of young erythrocytes of $23 \cdot 5$ nmoles/10^{10}cells × min and in a population of old cells of $7 \cdot 8$ nmoles/10^{10} cells × min, which corresponds to a 67 % decrease in enzyme activity. This great decrease in activity (Ferber *et al.*, 1968c) appeared only when young and aging cells were isolated by a double applied density gradient separation procedure. A single separation brought about an activity decrease of only 43 % (Ferber *et al.*, 1968b). The decrease in acyl-CoA : lysolecithin acyltransferase activity was, on the other hand, lower (31 %). This appears to give further proof to the argument that the limiting step of the acylating system is the acid : CoA-ligase reaction. A decrease in the lysophospholipase was also recorded. It is remarkable that these decreases in activity appear only in *in vivo* aged erythrocytes, and not, however, in stored blood *in vitro*. On the other hand, Shastri and Rubinstein (1969) account for the decreased incorporation rates of long chain fatty acids into lecithin as not being due to a decreased activity of the acid : CoA-ligase or lysolecithin acyltransferase, but rather due to differences in the transport velocity of fatty acids through the membrane. This energy dependent transport is supposed to be, according to Shohet *et al.* (1968) the actual limiting step in the utilization of fatty acids. The decreased transport rate in aged erythrocytes according to these authors is caused by a decrease in the intracellular ATP content, which, however, has not been confirmed by all other researchers (Shojania *et al.*, 1968; Ferber *et al.*, 1968b).

Despite these differences, it can be concluded that the decreased acylation of lysolecithin in aged erythrocytes is a cause for their elimination from the circulatory system, since the function of the cell membrane is greatly altered.

III. Liver

A. INTRODUCTION

Plasma membranes prepared by the method of Neville jr. (1960, 1968) and Emmelot and Bos (1962) are representative of only a part of the entire cell membrane of the liver cell, i.e. "bile fronts". They consist of the bile canaliculi (1000–20,000 Å diameter), because these canaliculi are stabilized by intercellular bridges (desmosomes, junctions), while the other areas of the cell surface disintegrate into small vescicles (500–1000 Å diameter). Although, these preparations of bile canaliculi represent plasma membranes of the liver cell and are easy to identify, because of their morphological structure they can definitely not be expected to be typical for the entire metabolism of the plasma membrane of the liver cell (distribution of enzymes, transport, etc.). Furthermore, comparative studies between individual subcellular particles of the liver cells are difficult, because the parts of the membrane

TABLE V

5'-Nucleotidase and glucose-6-phosphatase in liver membranes (nmoles $\times$ mg protein$^{-1} \times$ min^{-1})

	5'-Nucleotidase			Glucose-6-phosphatase			
Plasma-membrane	Micro-somes	Homo-genate	Plasma-membrane	Micro-somes	Homo-genate	Reference
594	91	63	35	350	—	Victoria et al. (1971)
1366	—	39	—	—	—	Newkirk and Waite (1971)
1300	—	—	9·67	225	—	Kaulen et al. (1970)
—	—	—			—	Stahl and Trams (1968)
96	26	9	20	96	26	Eibl et al. (1969)
1020	139	68	43	214	70	Stein et al. (1968)
643	—	30	7·67	—	21	Torquebiau-Colard et al. (1970)

which disintegrate into small vesicles contaminate the microsome fraction. Another difficulty, is that, by this method of isolation, only liver tissue and not isolated and purified liver cells can be employed, otherwise there would be a destruction of the stabilizing bridges which are essential for the stability of the bile canaliculi. Thus membranes are isolated from a tissue which contains to a considerable degree cells of the reticuloendothelial system (*ca* 10%), which possess highly specific functions (phagocytosis, ameboid movement, etc.). Some of the following studies presented were done on plasma membranes, which were not prepared according to the Neville–Emmelot method (large fragments) but by subfractionation of microsomes (Eibl *et al.*, 1969). Consequently, a comparison of these findings is difficult. For the characterization of the plasma membrane fractions, Table V shows only the values of the 5'-nucleotidase and glucose-6-phosphatase activities, as determined by these authors. Excluding the above mentioned exceptions, the data correspond relatively well.

The main location in the cell for the synthesis of the phospholipids containing nitrogen is the endoplasmic reticulum. The question as to whether mitochondria also possess the ability to synthesize phospholipids is debatable. It appears certain that liver cell plasma membranes cannot complete a *de novo* synthesis of phospholipids from glycerol-3-phosphate, because they are not able to transfer choline or ethanolamine to diglycerides (CDP-choline : 1,2-diglyceride choline phosphotransferase EC 2.7.8.2) (Mudd *et al.*, 1969). But, for this reason, those enzymes of the phospholipid metabolism which have been proven to be present in plasma membranes gain importance.

B. SYNTHESIS OF PHOSPHATIDIC ACID AND OF PHOSPHATIDYL GLYCEROL

(a) sn-glycerol-3-phosphate + 2 acyl CoA → phosphatidic acid + CoASH
(b) phosphatidic acid + CTP → CDP-diglyceride + PPi
(c) CDP-diglyceride + sn-glycerol-3-phosphate → phosphatidyl glycerol-phosphate + CMP
(d) phosphatidyl glycerol-phosphate → phosphatidyl glycerol + Pi.

Stein *et al.* (1968) were able to prove that plasma membranes from liver incorporated glycerol-3-phosphate into lipids which they identified as phosphatidic acid. Since the authors did not report any specific activity, little can be stated concerning the meaning of these findings. However, the activities do appear low, because high enzyme concentrations (1·12 mg protein) were used. It is possible to partially explain the low acylation rates of the glycerol-3-phosphate, in that only the acylation of free oleic acid (with addition of CoA and ATP) was measured. The production of oleoyl-CoA (acid : CoA-ligase) in plasma membranes appears to occur only to a limited extent (see below) (Stahl and Trams, 1968). Similarly, the rate of synthesis of phosphatidyl glycerol from glycerol-3-phosphate and CDP-diglyceride, as discovered by Victoria *et al.* (1971), appears to be of low order. The synthesis of phosphatidyl-glycerol as a primary step in the synthesis of cardiolipin would be of interest, since it has been shown, that cardiolipin occurs in plasma membranes of liver (Ray *et al.*, 1969).

C. INCORPORATION OF LONG CHAIN FATTY ACIDS

Free long chain fatty acids are apparently utilized by plasma membranes to a degree, which is essentially less than that of its acyl-CoA compounds. Thus the incorporation of free palmitic acid respresented only 1 % of the rate of incorporation of palmitoyl-CoA (Stahl and Trams, 1968). This type of behavior has also been found in plasma membranes of other types of cells, such as lymphocytes (Resch *et al.*, 1971), and erythrocytes (Ferber *et al.*, 1968b, c). The acid-CoA ligase (AMP) (EC 6.2.1.3) thus does not appear to be a component of the plasma membrane fraction. Pande and Mead (1968), however, came to exactly opposite results: out of crude nuclear fractions from which purified plasma membranes were obtained, these authors were able to discover relatively high specific activities of acid : CoA-ligase, although, it must be emphasized that these "plasma membranes" were not charactertized by marker enzymes or through any other kind of criteria.

D. ACYLATION OF MONOACYL PHOSPHATIDES (LYSOPHOSPHATIDES)

In 1960 Lands discovered an acyltransferase in microsomal membranes of the liver, which was able to transform lysophosphatides into diacylphosphatides (Lands, 1960). Because of the very high rates for the synthesis of lecithins and phosphatidyl ethanolamines, which can be obtained by these enzymes, and because specific transferases for isomeric lysophosphatides (1-acyl and 2-acyl compounds) were found, the possible functional significance of these enzymes was discussed with respect to the following problems:

(a) The acyltransferases, together with phospholipases are responsible for the redistribution of long chain fatty acids of the phospholipids, thus leading to an asymmetric distribution of saturated and unsaturated fatty acids (Lands and Merkl, 1963; Lands and Hart, 1965; van den Bosch *et al.*, 1967; Stoffel *et al.*, 1967; Sarzala *et al.*, 1970; van den Bosch *et al.*, 1968; Ferber, 1971).

(b) The possible role of acyltransferases for the conversion of cytotoxic lysophosphatides (Mulder and van Deenen, 1965a; Mulder *et al.*, 1965; Waku and Lands, 1968a; Munder *et al.*, 1965; Ferber, 1971; Ferber *et al.*, 1968b).

The first studies concerning the acylation of lysophosphatides by liver

TABLE VI

Acyl-CoA : 1-acyl-GPC acyltransferases in liver membranes

Lysolecithin (acceptor)	Fatty acid (donor)	Specific activity (nmoles × mg protein^{-1} × min^{-1})			Reference
		Plasma membranes	Endoplasmic reticulum	Microsomes	
1-acyl-sn-glycerol-3-phosphorylcholine endogenous	linoleoyl-CoA	12	49	37	Eibl et al. (1969)
	oleoyl-CoA	0·4–0·7	n.d.	n.d.	Stahl and Trams (1968)
1-stearoyl-sn-glycerol-3-phosphorylcholine	arachidonoyl-CoA	8·2	n.d.	65·4	Kaulen et al. (1970)

n.d. = not determined.

plasma membranes were reported by Stein *et al.* (1968). The authors were able to follow the incorporation of natural labeled lysolecithin into lecithin in the presence of free fatty acids (+CoA, ATP) and found equal rates of incorporation for microsomal as well as for plasma membranes. However, no specific activities were stated and the use of free fatty acids in studies with plasma membranes, which have low acid : CoA-ligase activities, does not conform to optimal test requirements. Furthermore, it could not be excluded that the plasma membranes were contaminated to a high degree (20%) with microsomes.

In Table VI several acyl-CoA : lysolecithin acyl transferase activities are presented. Microsomal acyltransferases are known to transfer preferentially highly unsaturated fatty acids to lysophosphatides (Lands and Merkl, 1963; van den Bosch *et al.*, 1967). This phenomenon is particularly present when the enzyme activities of the microsomes are compared. Even when these findings cannot exactly be compared, since the test conditions are not identical, it is apparent that the highest acyltransferase activities are to be found in the endoplasmic reticulum, or in microsomes of the liver cell, but plasma membranes also show distinct activities. The very low enzyme activities found by Stahl and Trams (1968) are the result of the lack of exogenous lysolecithin functioning as an acceptor. In addition, the authors used very high concentrations of oleoyl-CoA (1 μmole/ml) without reaching a saturation of the enzyme. A possible explanation is that high concentrations inhibit, or that large micelles are formed which are unable to act as an active substrate (Zahler and Cleland, 1969). The evaluation of the plasma membrane activities are discussed (Kaulen *et al.*, 1970) in reference to their contamination through microsomal membranes.

On the other hand, it should be emphasized at this point, that because of the formation of small vesicles, as described above, one would expect the microsomal fraction to contain a large portion of plasma membranes. Consequently, at this stage a concluding statement concerning enzyme activities localized in plasma membranes of liver is difficult.

E. DEACYLATION OF DIACYLPHOSPHATIDES

The determination of phospholipase A, especially of the membrane bound enzyme is difficult because of the size and charge (Dawson, 1966; Dawson, 1963a, b) of the macromolecular micelles which phospholipids form in aqueous solution, and because this enzyme demonstrates relatively low activities in animal cells. Membrane bound phospholipase A has been identified in several subcellular structures such as lysosomes (Smith and Winkler, 1968; Mellors and Tappel, 1967; Blaschko *et al.*, 1967a; Stoffel and Trabert, 1969), chromaffin granula (Blaschko *et al.*, 1967b), mito-

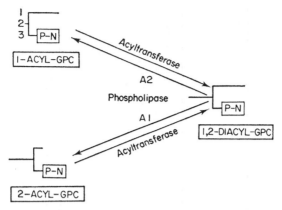

FIG. 1. Metabolism of isomeric lysolecithins. GPC = glycero-3-phosphorycholine; $\boxed{\text{P}-\text{N}}$ = phosphorylcholine

chondria and microsomes (Waite and van Deenen, 1967; Waite et al., 1969), which differ greatly with respect to optimal-pH, substrate specificity (phosphatidyl choline or phosphatidyl ethanolamine) and positional specificity (phospholipase A_1 and A_2). Only recently has it been possible to detect phospholipase A activity in the plasma membranes of the liver (see Table VII). This is, as such, of great significance because the necessary substrates for the reacylation process, the lysophosphatides, are formed, thus allowing a complete de- and reacylation cycle in plasma membranes of the liver to function. The first report concerning phospholipase A activity in liver plasma membranes is from Torquebiau-Colard et al. (1970). The authors used ^{32}P labeled phosphatidyl-glycerol or phosphatidyl-ethanolamine as substrate, and were able to detect the formation of the corresponding lyso compounds. The specific activities, as compared with later studies, were very low probably due to the adverse pH value of 7·5. The positional specificity of the enzyme was, because of the labeling of the phosphate group, not determined. The findings of Newkirk and Waite (1971) and Victoria et al. (1971) all agree that phosphatidyl-ethanolamine is the substrate which can be best hydrolyzed, that the pH-optimum is between 8 and 9 and that the enzyme is activated by Ca^{2+}. Each research group reported opposite results concerning the positional specificity. While Victoria et al. (1971) found a prevailing hydrolysis of the fatty acid at the 2 position of the glycerol moiety (phospholipase A_2), the results of Newkirk and Waite (1971) indicate that the fatty acid in position 1 is cleaved. A requirement for the determination of phospholipase A_1 activity, is that by the use of $[2-^{14}C\text{-acyl}]$-diacylphosphatide the formed lysophosphatide is not degraded further by lysophospholipases, otherwise phospholipase A_2 activity is simulated. However, Victoria et al. found no significant lyso-

TABLE VII

Phospholipase A in plasma membranes of liver

Substrate	Specific activity (nmoles $\times$ mg protein^{-1} $\times$ min^{-1})	K_m	pH optimum	Positional specificity	Remarks	Reference
^{32}P-phosphatidylglycerol	0·27	n.d.	(7·5)	n.d.		Torquebiau-Colard et al. (1970)
Phosphatidylethanolamine = 1-saturated-2-[1 – ^{14}C]-linoleoyl-phosphatidylethanolamine	V_{max}=6·6	5×10^{-4} M	8·0	Predominant phospholipase A_2	Ca^{2+} activates	Victoria et al. (1971)
Phosphatidylglycerol = 1-palmitoyl-2-[1 – ^{14}C]-oleoyl-3-sn-phosphatidyl-1'-sn-glycerol	V_{max}=2·3	$3·6 \times 10^{-5}$ M	8·0	Predominant phospholipase A_2		Victoria et al. (1971)
Phosphatidylethanolamine = [2-^{14}C-linoleoyl]-phosphatidylethanolamine	16·8	n.d.	9	Predominant phospholipase A_1	Ca^{2+} activates	Newkirk and Waite (1971)

n.d. = not determined.

phospholipase activity in plasma membrane fractions. They also employed [1-^{3}H-acyl]-diacyl-phosphatides and received analogue results.

Newkirk and Waite (1971) did not report a determination of lysophospholipase activity, however, they did show formation of labeled lysophosphatidyl-ethanolamine, so that there are obviously no significant lysophospholipase activities in their plasma membrane preparations. Thus, it is not possible to explain the discrepancy between these groups with respect to the determination of the positional specificity. Further studies should show whether or not there was a cross contamination with other subcellular particles. Victoria et al. (1971) reported additional properties of their phospholipase A_2, in that phosphatidyl-choline is practically not utilized, whereas phosphatidyl-glycerol functions as substrate. It is surprising, that phosphatidyl-glycerol had a higher affinity to the enzyme than phosphatidyl-ethanolamine, but only 1/3 of the maximal turnover rates (see Table VII). Deoxycholate and N-ethylmaleinimide did not effect the enzyme activities.

IV. Lymphocytes

A. INTRODUCTION

Among the eucaryotic cells, lymphocytes are especially suitable for studies on the phospholipid dynamics of the plasma membranes. Because these cells have little cytoplasma and few intracellular membranes, they have an advantageous ratio of plasma membrane to the entire cell. In addition, there are now methods available, which allow the isolation of plasma membranes with sufficient yield and purity (Allan and Crumpton, 1970; Ferber et al., 1972). Finally, such studies on the plasma membranes of lymphocytes are interesting, in that these cells can be stimulated out of a resting metabolic activity when an antigen or even an unspecific mitogen binds to its outer membrane. Although there are a few studies concerning the phospholipid metabolism of whole lymphocytes (Kay, 1968; Fisher and Mueller, 1968; Fisher and Mueller, 1969; Lucas et al., 1971; Resch et al., 1971), there are little data about the enzymes of the phospholipid metabolism of the plasma membrane at our disposal.

B. DIGLYCERIDE KINASE AND METABOLISM OF INOSITOL PHOSPHATIDES

Direct measurements of the synthesis of phosphatidic acid were carried out by Fisher and Mueller (1971) with plasma membranes of lymphocytes. These authors were able to prove (Fisher and Mueller, 1968, 1971), that shortly after the stimulation of lymphocytes by phytohemagglutinin there is an increased incorporation of phosphate into phosphatidic acid and inositol phosphatides thus awaking interest in the subcellular localization of these metabolic pathways. At first, Fisher and Mueller studied the incorporation of ^{32}P-phosphate into phosphatidic acid and inositol phosphatide in whole lymphocytes, with a following fractionation into subcellular particles. Moreover, it could be shown that phytohemagglutinin caused the largest increase (six times) in phosphatidic acid of plasma membranes, while in other cell fractions only a three-fold increase was observed. On the other hand, the incorporation of ^{32}P into inositol phosphatides showed no such preference for plasma membranes, with a two to three-fold increase in all fractions after administering phytohemagglutinin. Of the enzymes taking part in these reactions, only diglyceride kinase was measured, and not the CTP : phosphatidic acid cytidyl transferase and the CDP-diglyceride : inositol phosphatidyl transferase.

The activity of the diglyceride kinase in the plasma membranes was surprisingly low (0.5 μmoles $\times$ mg^{-1} $\times$ min^{-1}); there was, however, an increase after the addition of phytohemagglutinin. On the other hand, it is not possible to explain the high incorporation of phosphate in whole cells by the increased rates of the diglyceride kinase. One must further take into consideration, that the addition of phytohemagglutinin in these experiments was carried out directly on isolated membranes, which also caused only slight changes in the activation of other membrane bound enzymes (Ferber and Resch, 1972). These authors discuss the hypothesis that the increased synthesis of phosphatidic acid and inositol phosphatides occurs in plasma membranes of stimulated lymphocytes, but the proof of this hypothesis is based on meager evidence.

C. LYSOLECITHIN-ACYLTRANSFERASE

Stimulated lymphocytes show an increased incorporation of long chain fatty acids into phospholipids (Resch et al., 1971; Resch and Ferber, 1972; Ferber, 1971). In the earlier phases of stimulation by phytohemagglutinin the increase is not based on a stimulation of the de novo synthesis. In addition to this, these fatty acids were mainly incorporated into lecithin, and it was therefore assumed that the incorporation was based on an acylation of endogenous lysolecithin. Measurements of the lysolecithin acyltransferases in subcellular particles of such lymphocytes are also in agreement with these findings (Ferber, 1971; Ferber and Resch, 1972). While there was no increase in the mitochrondrial acyltransferase activity after stimulation, there was a double enhancement of activity in the microsomal fraction 30 minutes after the addition of phytohemagglutinin (Fig. 2). Plasma membranes and endoplasmic reticulum were isolated from these fractions (Ferber et al., 1972a) by a procedure, which was a modification of the method of Wallach and Kamat (1964, 1966).

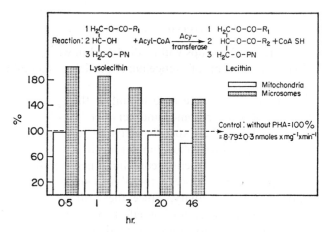

FIG. 2. Stimulation of rabbit lymphocytes with phytohemagglutinin (PHA). Specific activity of the lysolecithin-acyltransferase in mitochondria and microsomes. Cultivation, 0·5–46 hours.

The acyl-CoA : lysolecithin-acyltransferase shows a conspicuous distribution in both fractions (Fig. 3). In contrast to other cells, lymphocytes exhibited higher specific activities of this enzyme in the plasma membrane than in the endoplasmic reticulum. Furthermore, in stimulated lymphocytes only the plasma membrane bound enzyme was activated.

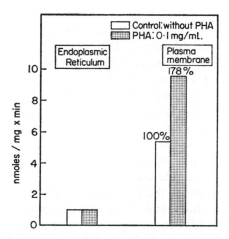

FIG. 3. Stimulation of rabbit lymphocytes with phytohemagglutinin (PHA). Specific activity of the lysolecithin-acyltransferase in endoplasmic reticulum and in plasma membranes. Cultivation, 1 hour.

V. Nervous Tissue

The work of two research groups will be discussed in which the enzymes of the phospholipid metabolism of subcellular particles, including plasma membranes, are measured.

Kai *et al.* (1966a, b, 1968) were able to follow the synthesis of di- and triphosphatidyl-inositol in microsomes and nerve endings from rat brain. In doing so, they found the highest specific activities for the synthesis of the diphosphoinositol phosphatides (phosphatidyl-inositol kinase) in the microsome fraction, followed by the fraction containing nerve endings. Generally, there was a good paralleliom between the 5′-nucleotidase activity and the phosphatidyl-inositol kinase activity. Since the addition of phosphatidyl-inositol had no significant accelerating effect on the turnover rate, the turnover must be primarily due to a phosphorylation of endogenous substrate. Hokin and Hokin (1964) were able to observe the same type of behavior in erythrocyte membranes (see page 224). On the other hand, a typical saturation behavior was found for ATP (substrate saturation at 5 mM ATP). The absolute activities were at 0.05 nmol $\times$ mg^{-1} $\times$ min^{-1} comparatively low.

The subcellular distribution shows, with considerable probability, that the phosphatidyl-inositol kinase is located primarily in the plasma membrane (Kai *et al.*, 1966a). However, the synthesis of triphosphoinositol phosphatide is catalyzed by enzymes (diphosphoinositol phosphatide kinase) which are found in the soluble supernatant fraction (Kai *et al.*, 1968). The authors attribute this to the relatively high polarity of the utilized substrate (diphosphoinositol phosphatide) and the product (triphosphoinositol phosphatide).

Recently, there appeared a study, in which certain enzymes of the phospholipid metabolism of the synaptic membranes of the rat brain were examined (Cotman and Matthews, 1971; Cotman *et al.*, 1971). Through density centrifugation it was possible to isolate two membrane fractions, in which one (fraction (1) contained relatively small vesicles with high 5′-nucleotidase activity, and an other (fraction (2) with relatively large fragments, which were identified morphologically as synaptic membranes. Though this latter fraction possessed a low specific activity of 5′-nucleotidase, it exhibited the highest activities of Na$^+$, K$^+$-ATPase.

It was possible to identify phosphatidic acid phosphatase in these fractions. Unfortunately, the diglyceride kinase, however, was not determined. It was also possible to determine in fraction 1 (also in the microsomal fraction) the formation of inositol phosphatide from activated phosphatidic acid and free inositol (CDP-diglyceride : inositol phosphatidyl transferase). On the other hand, the enzymes for the synthesis of CDP-diglyceride (CTP : phosphatidic acid cytidyl transferase) and for the acylation of carnitine (acetyl-CoA : carnitine acyltransferase) is specifically located in the mitochondria. Thus the work of these two research groups indicate that at least part of the pathway for the synthesis of the inositol phosphatide is located in the plasma membrane.

VI. Bacteria

The phospholipid metabolism of the bacteria shows some special features as compared to that of the eucaryotic cell. Since there has already appeared a detailed review on this subject (Cronan and Vagelos, 1972) only essential differences will be discussed.

All enzymes of the phospholipid metabolism appear to be more or less bound to the cell envelope. Because of this, and because of the possibility of obtaining mutants and inducing enzyme activities, these cells have increasingly gained in importance for investigations concerning the role of membrane lipids and their metabolism. However, most studies have been made with crude particle preparations and only recently have reports appeared concerning the phospholipid metabolism of isolated fractions of the bacterial envelope.

The differences in the phospholipid synthesis are based on the following facts:

(1) The synthesis of the phospholipids in bacteria (*E. coli*) does not occur with free diglyceride as an intermediate but with "activated phosphatidic acid" (CDP-diglyceride).

(2) The main phospholipid phosphatidyl-ethanolamine is not synthesized by a transfer of ethanolamine, but by a transfer of serine (to CDP-diglyceride) and a following decarboxylation step.

(3) Besides acyl-CoA, acylated acyl carrier protein (ACP) also serves as a donor for long chain fatty acids.

In the studies considered here, the cell envelope from *E. coli* and *Salmonella typhimurium* was separated into an outer and inner cytoplasmic membrane and analyzed respectively. In one case the outer lipopolysaccharide and protein containing membrane was freed from the peptidoglycan layer. Table VIII shows a good accordance of the data of White *et al.* (1971) and Bell *et al.* (1971), in that both groups find the synthesizing enzymes exclusively located in the cytoplasmic membrane. The specific activities of White *et al.* (1971) are overall lower than those measured by Bell *et al.* (1971), which is due to the different methods used in preparation of the membranes. Both groups find very low activities of the CDP-diglyceride : 1-serine phosphatidyl transferase in the outer membrane as also in the cytoplasmic membrane. Consistent with this, Kanfer and Kennedy (1964) found this enzyme preferentially in the soluble supernatant. It is, however, probable that this enzyme is initially bound to the cytoplasmic membrane, although essentially less

L

TABLE VIII

Distribution of enzymes of phospholipid metabolism in envelope fractions of *E. coli* (nmoles $\times$ mg protein^{-1} $\times$ min^{-1})

	Cytoplasmic membrane		Outer membrane	
	(Bell et al., 1971)	(White et al., 1971)	(Bell et al., 1971)	(White et al., 1971)
Glycerol-3-phosphate acyltransferase	54·2	3·2	2·5	0·75
1-Acylglycerol-3-phosphate acyltransferase	100·0	—	4·8	—
CTP : phosphatidic acid cytidyltransferase	6·7	2·65	0·73	0·11
CDP-diglyceride: glycerol-3-phosphate phosphatidyltransferase	27·2	14·65	2·17	0·42
CDP-diglyceride: 1-serine phosphatidyl-transferase	0·22	0·117	0·026	0·097
Phosphatidylserine decarboxylase	20·2	4·52	0·97	0·335
Phospholipase A$_1$	0·28	—	8·88	—

tight. Cronan and Vagelos (1972) showed that after careful lysis, this enzyme could be found mainly in the cytoplasmic membrane fraction.

Greater discrepancies are found in the determinations of the enzyme activities for the acylation of glycerol-3-phosphate. Although it is not certain from the data how the determination of glycerol-3-phosphate acyltransferase was carried out, it is possible that these differences are due to the employment of different fatty acids. While White *et al.* obviously used exclusively palmitoyl-CoA it is known from investigations of van den Bosch and Vagelos (1970) and Ray *et al.* (1970), that only in the first step of the acylation of glycerol-3-phosphate (synthesis of 1-acyl-glycerol-3-phosphate) is there a utilization of saturated fatty acids, while the acylation of 1-acyl-glycerol-3-phosphate is preferentially carried out with unsaturated acyl-CoAs, such as palmitoleoyl-CoA or cis-vaccenyl-CoA (van den Bosch and Vagelos, 1970). Therefore, in order to insure optimal test conditions for the synthesis of phosphatidic acid, pairs of saturated and unsaturated acyl-CoA's should be present. White *et al.* (1971) further report, that there are also qualitative differences in the acylation of glycerol-3-phosphate in cytoplasmic and outer membranes. The acylation product consisted, by incubation with cytoplasmic membranes, of up to 32% out of lysophosphatidic acid (1-acyl-glycerol-3-phosphate). On the other hand, the incubations with preparations from the outer wall revealed an acylation product of 77% lysophosphatidic acid the remaining product being almost entirely phosphatidic acid (1,2-diacyl-glycerol-3-phosphate). The most probable explanation for this phenomenon is the activity of the phospholipase A_1, as found by Bell *et al.* (1971) (see Table VIII) which, in contrast to the phospholipid synthesizing enzymes, was mainly identified as part of the outer membrane (wall).

For the determination of phospholipase A activity and especially its positional specificity, Bell *et al.* (1971) used the method of Scandella and Kornberg (1971). This method is based on the use of lecithin or phosphatidylethanolamine, which possess two different types of fatty acids (e.g. 1-oleoyl-2-stearoyl-GPC) and a consecutive determination of the fatty acid present in the lyso-compound. This kind of determination of the positional specificity of phospholipase A is not impaired by the possibility of lysophospholipase being present. There might only be a decrease in the yield of the lyso-compound. Lysophospholipase was identified at least in *Salmonella typhimurium* (Bell *et al.*, 1971) where it also exhibited its highest specific activities in the outer membrane. The failure of lysophospholipase in particulate preparations of *E. coli* is perhaps explained by the findings of Proulx and van Deenen (1966) in which lysophospholipase activity (after sonication) was found in the supernatant fraction. Besides these enzymes, which are known to be localized in the cell envelope, diglyceride kinase (Pieringer and Kunnes, 1965) and phosphatidic acid phosphatase (van den Bosch and Vagelos, 1970) were found as particle-bound enzymes of lipid metabolism. The physiological significance is not, however, known. Since diglyceride is not an intermediate in the synthesis of phospholipids in bacteria, these enzymes could play a role only in the synthesis of triglycerides on the one hand or be part of the phosphatidic acid $\rightleftharpoons$ diglyceride cycle on the other (van den Bosch and Vagelos, 1970).

Isolation techniques of the different membranes of the bacterial envelope are now well developed so these cells have become particularly useful for investigations on the dynamics of phospholipids in cellular membranes. As the enzymes of the entire phospholipid synthesis are localized in the cytoplasmic membrane of the bacterial envelope, it is possible to study on a single defined membrane, the relevance of phospholipids for the physiology of the cell. The exclusive location of the phospholipase A activity in the outer membrane increases the interest in the phospholipid degradation processes, which are observed during the lysis of the cells after phage T_4 infection and after administering colicin K (see Cronan and Vagelos, 1972).

VII. Conclusion

Plasma membranes of the various animal cell types represent highly specialized structures. This is because of the high degree of differentiation of the cells from which the membranes are isolated (erythrocytes, lymphocytes, nervous tissue). The techniques of isolation add to the problem of membrane study. Possible contamination of the employed preparations with other membrane components of the cell lead to further difficulties for comparative investigations. Certainly, the principle question is whether plasma membranes *in situ* exist as separate entities or are a product of numerous transitions, so that a pure preparation is impossible. Despite these limitations, there are some features of the phospholipid metabolism which are obviously characteristic for plasma membranes. For example, in contrast to bacterial membranes, plasma membranes of animal cells do not possess an entire apparatus for the *de novo* synthesis of phospholipids, e.g. they are not capable of synthesizing complex lipids from glycerol-3-phosphate. On the other hand, there is now good evidence that in all plasma membranes acylation reactions of preformed phospholipids (lysophosphatides) and deacylation reactions of phospholipids (phospholipases) are operating. These reactions appear especially interesting because of their possible effects on the physico-chemical state of lipid membranes, which can be influenced by a change in the fatty acid pattern of the phospholipids (see Chapter 2). It is remarkable that there are also similar aspects of the phosphorylation reactions in plasma membranes, such as those leading to the phosphorylation of diglycerides to phosphatidic acid and of inositol phosphatides to di- and triphosphoinositides. Further investigations should reveal the physiological significance of these processes.

References

Allan, D. and Crumpton, M. J. (1970). *Biochem. J.* **120**, 133.

Basford, J. M., Glover, J. and Green, C. (1964). *Biochim. biophys. Acta* **84**, 764.

Bell, R. M., Mavis, R. D., Osborn, M. J. and Vagelos, P. R. (1971). *Biochim. biophys. Acta* **249**, 628.

Blaschko, H., Smith, A. D., Winkler, H., van den Bosch, H. and van Deenen, L. L. M. (1967a). *Biochem. J.* **103**, 30c.

Blaschko, H., Firemark, H., Smith, A. D. and Winkler, H. (1967b). *Biochem. J.* **104**, 545.

Cotman, C. W. and Matthews, D. A. (1971). *Biochim. biophys. Acta* **249**, 380.

Cotman, C. W., McCaman, R. E. and Dewhurst, S. S. (1971). *Biochim. biophys. Acta* **249**, 395.

Cronan, J. E. and Vagelos, P. R. (1972). *Biochim. biophys. Acta* **265**, 25.

Danon, D. (1968). "Bibl. Haemat." (Basel) **29**, 178. *Proc. 11. Congr. Int. Soc. Blood Transf.* 1966.

Danon, D. (1970). *In* "Permeability and Function of Biological Membranes" (L. Bolis, A. Katchalsky, R. D. Keynes, W. R. Loewenstein, B. A. Pethica, eds.), p. 57. North Holland Publishing Co., Amsterdam.

Dawson, R. M. C. (1963a). *Biochim. biophys. Acta* **70**, 697.

Dawson, R. M. C. (1963b). *Biochem. J.* **88**, 414.

Dawson, R. M. C. (1966). *In* "Lipoide". 16th Coll. Gesellschaft Physiologische Chemie Mosbach 1965, p. 29. Springer, Berlin, Heidelberg, New York.

De Gier, J., van Deenen, L. L. M., Geerdink, R. A., Punt, K. and Verloop, M. C. (1961). *Biochim. biophys. Acta* **50**, 383.

Delbauffe, D., Paysant, M. and Polonovski, J. (1968a). *Bull. Soc. Chim. biol.* **50**, 1431.

Delbauffe, D., Paysant, M. and Polonovski, J. (1968b). *Bull. Soc. Chim. biol.* **50**, 1439.

Dodge, J. T., Mitchell, C. and Hanahan, D. J. (1963). *Archs Biochem. Biophys.* **100**, 119.

Eibl, H., Hill, E. E. and Lands, W. E. M. (1969). *Europ. J. Biochem.* **9**, 250.

Emmelot, P. and Bos, C. J. (1962). *Biochim. biophys. Acta* **58**, 374.

Ferber, E. (1971). *In* "The Dynamic Structure of Cell Membranes" (D. F. H. Wallach and H. Fischer, eds.), p. 129. 22nd Coll. der Gesellschaft für Biologische Chemie Mosbach, Baden 1971; Springer, Berlin, Heidelberg, New York.

Ferber, E. and Resch, K. (1972). *Biochim. biophys. Acta.* In press.

Ferber, E., Munder, P. G., Kohlschütter, A. and Fischer, H. (1968a). *Folia haemat.* (*Lpz.*) **90**, 224.

Ferber, E. Munder, P. G., Kohlschütter, A. and Fischer, H. (1968b). *Europ. J. Biochem.* **5**, 395.

Ferber, E., Krüger, J., Munder, P. G., Kohlschütter, A. and Fischer, H. (1968c). *In* "Metabolism and Membrane Permeability of Erythrocytes and Thrombocteys", p. 393. 1st Internat. Symp. Wien 1968. Thieme, Stuttgart.

Ferber, E., Kohlschütter, A., Munder, P. G. and Fischer, H. (1970). *In* "Modern Problems of Blood Perservation" (W. Spielmann, S. Seidl, eds.), p. 14. G. Fischer-Verlag, Stuttgart.

Ferber, E. Resch, K., Wallach, D. F. H. and Imm. W. (1972a). *Biochim. biophys. Acta* **266**, 494.

Fischer, H., Ferber, E., Haupt, I., Kohlschütter, A., Modolell, M., Munder, P. G. and Sonak, R. (1967). *In* "Protides of the Biological Fluids", Vol. 15, p. 175. Proceedings of the 15th Coll. Bruges.

Fisher, D. B. and Mueller, G. C. (1968). *Proc. natn. Acad. Sci. U.S.A.* **60**, 1396.

Fisher, D. B. and Mueller, G. C. (1969). *Biochim. biophys. Acta* **176**, 316.

Fisher, D. B. and Mueller, G. C. (1971). *Biochim. biophys. Acta* **248**, 434.

Hagerman, J. S. and Gould, R. G. (1951). *Proc. Soc. Exp. Biol. Med.* **78**, 329.

Hill, E. E. and Lands, W. E. M. (1968). *Biochim. biophys. Acta* **152**, 645.

Hill, E. E. and Lands, W. E. M. (1970). *In* "Lipid Metabolism" (S. J. Wakil, ed.), p. 185. Academic Press, New York, London.

Hokin, L. E. and Hokin, M. R. (1961). *Nature, Lond.* **189**, 836.

Hokin, L. E. and Hokin, M. R. (1963). *Biochim. biophys. Acta* **67**, 470.

Hokin, L. E. and Hokin, M. R. (1964). *Biochim. biophys. Acta* **84**, 563 (1964).

Hokin, L. E., Hokin, R. M. and Mathison, D. (1963). *Biochim. biophys. Acta* **67**, 485.

Kai, M., White, G. L. and Hawthorne, J. N. (1966a). *Biochem. J.* **101**, 328.

Kai, M., Salway, J. G., Michell, R. H. and Hawthorne, J. N. (1966b). *Biochem. biophys. Res. Comm.* **22**, 370.

Kai, M., Salway, J. G. and Hawthorne, J. N. (1968). *Biochem. J.* **106**, 791.

Kanfer, J. and Kennedy, E. P. (1964). *J. biol. Chem.* **239**, 1720.

Kanoh, H. (1969). *Biochim. biophys. Acta* **176**, 756.

Kates, M., Allison, A. C. and James, A. T. (1961). *Biochim. biophys. Acta* **48**, 571.

Kaulen, H. D., Henning, R. and Stoffel, W. (1970). *Hoppe-Seyler's Z. physiol. Chem.* **351**, 1555.

Kay, J. E. (1968). *Nature, Lond.* **219**, 172.

Korn, E. D. (1969). *Ann. Rev. Biochem.* **38**, 263.

Lands, W. E. M. (1960). *J. Biol. Chem.* **235**, 2233.

Lands, W. E. M. and Hart, P. (1965). *J. Biol. Chem.* **240**, 1905.

Lands, W. E. M. and Merkl, I. (1963). *J. Biol. Chem.* **238**, 898.

Lennarz, W. J. (1970). *A. Rev. Biochem.* **39**, 359.

Lucas, D. O., Shohet, S. B. and Merler, E. (1971). *J. Immunol.* **106**, 768.

Marinetti, G. V., Erbland, J., Witter, J. F., Petrix, J. and Stotz, E. (1958). *Biochim. biophys. Acta* **30**, 223.

McLeod, M. E. and Bressler, R. (1967). *Biochim. biophys. Acta* **144**, 391.

Mellors, A. and Tappel, A. L. (1967). *J. Lipid Res.* **8**, 479.

Mudd, J. B., Golde, L. M. G. and van Deenen, L. L. M. (1969). *Biochim. biophys. Acta (Amst.)* **176**, 547.

Mulder, E., de Gier, J. and van Deenen, L. L. M. (1963). *Biochim. biophys. Acta* **70**, 94.

Mulder, E. and van Deenen, L. L. M. (1965a). *Biochim. biophys. Acta* **106**, 348.

Mulder, E. and van Deenen, L. L. M. (1965b). *Biochim. biophys. Acta* **106**, 106.

Mulder, E., van den Berg, J. W. O. and van Deenen, L. L. M. (1965). *Biochim. biophys. Acta* **106**, 118.

Munder, P. G., Ferber, E. and Fischer, H. (1965). *Z. Naturf.* **20b**, 1048.

Murphy, J. R. (1962). *J. Lab. clin. Med.* **60**, 571.

Neville, D. M., Jr. (1960). *J. biophys. biochem. Cytol.* **8**, 413.

Neville, D. M., Jr. (1968). *Biochim. biophys. Acta* **154**, 540.

Newkirk, J. D. and Waite, M. (1971). *Biochim. biophys. Acta* **225**, 224.

Oliveira, M. M. and Vaughan, M. (1962). *Fedn. Proc.* **21**, 296.

Oliveira, M. M. and Vaughan, M. (1964). *J. Lipid Res.* **5,** 156.

Pande, S. V. and Mead, J. F. (1968). *J. biol. Chem.* **243,** 352.

Phillips, G. B. and Roome, N. S. (1962). *Proc. Soc. exp. Biol. Med.* **109,** 360.

Pieringer, R. A. and Kunnes, R. S. (1965). *J. biol. Chem.* **240,** 2833.

Pittman, J. G. and Martin, D. B. (1966). *J. clin. Invest.* **45,** 165.

Polonovski, J. and Paysant, M. (1963). *Bull. Soc. Chim. biol.* **45,** 339.

Porte, D. and Havel, R. J. (1961). *J. Lipid Res.* **2,** 357.

Proulx, P. R. and van Deenen, L. L. M. (1966). *Biochim. biophys. Acta* **125,** 591.

Raderecht, H. J., Binnewies, S. and Schölzel, E. (1962). *Acta biol. med. germ.* **8,** 199.

Ray, T. K., Skipski, V. P., Barclay, M., Essner, E. and Archibald, F. M. (1969). *J. biol. Chem.* **244,** 5528.

Ray, T. K., Cronan, J. E., Jr., Mavis, R. D. and Vagelos, P. R. (1970). *J. biol. Chem.* **245,** 6442.

Resch, K., Ferber, E., Odenthal, J. and Fischer, H. (1971). *Eur. J. Immunol.* **1,** 162.

Resch, K. and Ferber, E. (1972). *Eur. J. Biochem.* **27,** 153.

Robertson, A. F. and Lands, W. E. M. (1964). *J. Lipid Res.* **5,** 88.

Sarzala, M. G., van Golde, L. M. G., de Kruyff, B. and van Deenen, L. L. M. (1970). *Biochim. biophys. Acta (Amst.)* **202,** 106.

Scandella, C. J. and Kornberg, A. (1971). *Biochemistry* **10,** 4447.

Shastri, K. and Rubinstein, D. (1969). *Can. J. Biochem.* **47,** 967.

Shohet, St. B., Nathan, D. G. and Karnovsky, M. L. (1968). *J. clin. Invest.* **47,** 1096.

Shohet, St. B. and Nathan, D. G. (1970). *Biochim. biophys. Acta* **202,** 202.

Shojania, A. M., Israels, L. G. and Zipursky, A. (1968). *J. lab. Clin. Med.* **71,** 41.

Smith, A. D. and Winkler, H. (1968). *Biochem. J.* **108,** 867.

Stahl, W. L. and Trams, E. G. (1968). *Biochim. biophys. Acta* **163,** 459.

Steck, T. L. and Wallach, D. F. H. (1970). *Methods in Cancer Res.* **5,** 93.

Stein, Y., Widnell, C. and Stein, O. (1968). *J. Cell. Biol.* **39,** 185.

Stoffel, W., de Thomás, M. E. and Schiefer, H. G. (1967). *Z. Physiol. Chem.* **348,** 882.

Stoffel, W. and Trabert, U. (1969). *Z. Physiol. Chem.* **350,** 836.

Torquebiau-Colard, O., Paysant, M., Wald, R. and Polonovski, J. (1970). *Bull. Soc. Chim. biol.* **52,** 1061.

van den Bosch, H., van Golde, L. M. G., Eibl, H. and van Deenen, L. L. M. (1967). *Biochim. biophys. Acta (Amst.)* **144,** 613.

van den Bosch, H., van Golde, L. M. G., Slotboom, A. J. and van Deenen, L. L. M. (1968). *Biochim. biophys. Acta (Amst.)* **152,** 694.

van den Bosch, H. and Vagelos, P. R. (1970). *Biochim. biophys. Acta* **218,** 233.

van Deenen, L. L. M. and de Gier, J. (1964). *In* "The Red Blood Cell" (Ch. Bishpo and D. M. Surgenor, eds.), p. 243. Academic Press, New York, London.

van Deenen, L. L. M., de Gier, J. and Veerkamp, J. H. (1961). Biochem. Probl. Lipids, Proc. Intern. Conf. 5th, Marseille 1960, p. 32.

van Deenen, L. L. M., de Gier, J., Houtsmuller, U. T. M. and Mulder, E. (1962). Biochem. Probl. Lipids, Proc. Intern. Conf. 6th, Birmingham 1962. Elsvier, Amsterdam, p. 413.

van Gastel, C., van den Berg, J., de Gier, J. and van Deenen, L. L. M. (1965). *Br. J. Haemat.* **11,** 193.

van Golde, L. M. G., Pieterson, W. A. and van Deenen, L. L. M. (1968). *Biochim. biophys. Acta* **152,** 84.

Victoria, E. J., van Golde, L. M. G., Hostetler, K. Y., Sherphof, G. L. and van Deenen, L. L. M. (1971). *Biochim. biophys. Acta* **239,** 443.

Waite, M. and van Deenen, L. L. M. (1967). *Biochim. biophys. Acta* **137**, 498.

Waite, M., Sherphof, G. L., Boshouwers, F. M. G. and van Deenen, L. L. M. (1969). *J. Lipid Res.* **10**, 411.

Waku, K. and Lands, W. E. M. (1968a). *J. Lipid Res.* **9**, 12.

Waku, K. and Lands, W. E. M. (1968b). *J. biol. Chem.* **243**, 2654.

Walker, B. L. and Kummerow, F. A. (1964). *Proc. Soc. Exp. Biol. Med.* **115**, 1099.

Walker, B. L. and Yurkowski, M. (1967). *Biochem. J.* **103**, 218.

Wallach, D. F. H. and Kamat, V. B. (1964). *Proc. natn. Acad. Sci. U.S.A.* **52**, 721.

Wallach, D. F. H. and Kamat, V. B. (1966). *Methods Enzymol.* 8, 164.

White, D. A., Albright, F. R., Lennarz, W. J. and Schnaitman, C. A. (1971). *Biochim. biophys. Acta* **249**, 636.

Zahler, W. L. and Cleland, W. W. (1969). *Biochim. biophys. Acta* **176**, 699.

Translated from German by Clay E. Reilly.

Chapter 6

The Role of the Plasma Membrane in Disease Processes

DONALD F. HOELZL WALLACH

*Division of Radiobiology, Tufts–New England Medical Center,
Boston, Massachusetts, U.S.A.*

I. Introduction

As attested by numerous books, reviews, conferences and a huge literature, the field of membrane biology is in a state of ferment. Nevertheless, until the recent recognition that malignant neoplasia involves multiple plasma membrane defects, the medical aspects of membrane biology have received rather little attention. This review* is designed to acquaint membrane specialists with this area, hopefully stimulating thought and experimentation, which would further not only membrane biology, but might also help to solve some of the major health problems on the globe.

II. Cancer

A. INTRODUCTION

The role of the plasma membrane in the fundamental biologic process of

*This article covers selective literature published prior to 1 January 1972.

malignancy, invasiveness and metastasis was recognized long ago (Coman, 1953; Huxley, 1958) and ample evidence now suggests that malignant neo-plasms comprises membrane variants in which the plasma membrane is critically altered. Indeed malignancy is a quasi-clinical definition, sum-marizing a disorder of the social interactions evolving at cell surfaces. However, the aberrant sociology of tumor cells, as well as their often bizarre growth, may be remote from the crucial proximal event which, in turn, may vary with cause. Many of the remote effects may have insignificant conse-quences, but others, such as a possible increased efficiency in the membrane transport of a growth limiting metabolite could give the tumor cells a competitive advantage over normal and other variant cells (Hatenaka *et al.*, 1971; Martin *et al.*, 1971).

Viewing biomembranes as containing relatively ordered lattice systems (Changeux *et al.*, 1967; Changeux and Thiery, 1968; Changeux *et al.*, 1970; Blumenthal *et al.*, 1970), whose behavior could be altered in a cooperative fashion, the membrane alterations in malignancy could result from one or more of the following mechanisms:

(i) Insertion of a *new* subunit by mutation, as a viral gene product of high membrane affinity, or by covalent modification of existing subunits.

(ii) A change in the steady state concentration of a native structure—determining ligand, via mutation, viral infection or viral trans-formation.

(iii) Appearance of a high-affinity isomer of a structure-determining ligand, via mutation, viral infection or viral transformation.

(iv) Exposure to an extrinsic ligand with high membrane affinity.

(v) Alteration of existing membrane proteins and/or lipids by the action of external proteases or intrinsic and/or lipases.

This reasoning is consonant with recent theoretical expositions on co-operativity and dissipative instabilities in membranes (Changeux *et al.*, 1967; Changeux *et al.*, 1968; Changeux *et al.*, 1970; Blumenthal *et al.*, 1970) and has been integrated by the author in these terms into a unifying hypothesis of tumor behavior (Wallach, 1968; Wallach, 1969a; Wallach, 1969b). This hypothesis proposes that oncogenic agents alter the cooperative properties of cellular membranes, modifying numerous membrane functions and, as a consequence, creating the pleimorphic membrane changes seen in neoplasia. Thus the presence of an abnormal structural component could yield morpho-logical changes, produce new membrane antigenicity, change permeabilities, alter binding of metabolites, hormones, etc., and modify the function of otherwise normal, membrane-associated enzymes. One would also anticipate anomalies which, being several steps removed from the proximal event, have no immediately obvious relationship to membranes.

B. CELL CONTACT

Cellular Adhesiveness

The interactions of malignant cells with each other and with normal cells are defective (Coman, 1953; Huxley, 1958; Abercrombie and Ambrose, 1962; Curtis, 1967; Stoker, 1967a) a fact which Coman (Coman, 1953; Coman, 1944; Coman, 1954; Coman, 1960; Coman, 1961; Coman and Anderson, 1955) and others have attributed to decreased "mutual adhesiveness" of tumor cells. However, while malignant cells generally adhere less to each other than do normal cells, this is not an invariant feature of malignancy (Wallach, 1969b).

C. CONTACT INHIBITION OF MOVEMENT

Abercrombie and associates (Abercrombie and Heaysman, 1954; Abercrombie *et al.*, 1957), using time lapse cinematography of cells migrating in culture, noted that contact of these cells immobilized the ruffling and probing of their surfaces at these sites, inhibiting further movement of the cells towards or over each other. The free edges, however, would continue to ruffle and extend processes over free area of the support until a contiguous and immobile monolayer had formed.

Malignant cells commonly behave otherwise and are usually not immobilized by contact. This deviation can occur very early during neoplastic conversion of various cells by oncogenic viruses and other agents (Temin and Rubin, 1958; Vogt and Dulbecco, 1960; Sachs and Medina, 1961; MacPherson and Stoker, 1962; Stoker, 1964) and occasionally tumor cells, which are not inhibited by association with like cells, are blocked by contact with normal cells or tumor cells of different origin (Stoker, 1964; Barski and Belehradek, Jr., 1965; Borek and Sachs, 1966; Stoker *et al.*, 1966; Kohn and Fuchs, 1970). However, contact inhibition of movement is not an invariant characteristic of malignant cells.

D. CONTACT INHIBITION OF GROWTH

Another equally important feature of cell contact is the fact that extensive cellular contact *in vitro* switches off net RNA and protein synthesis and blocks synthesis of new DNA (Stoker, 1967a). This process is thought to be mediated by membrane contact and might clearly represent an important regulatory mechanism. Contact inhibition of growth is typically much less prominent in neoplastic cells than in normal cells.

E. ELECTRICAL COUPLING AND MOLECULAR TRANSFER

Individual normal epithelial and mesenchymal cells commonly communicate electrically through ion flow. These ionic connections depend upon the partition of Ca^{2+} and/or Mg^{2+} across the cell membrane (Rose, 1970; Rose and Loewenstein, 1970; Oliveira-Castro and Loewenstein, 1970). Electrical connections between the hepatocytes in normal or regenerating rat liver are profuse (Penn, 1966; Loewenstein and Penn, 1967) but such coupling is lacking between neoplastic hepatocytes in a number of hepatomas and between malignant thyroid or gastric cells (Kohn and Fuchs, 1970; Jamakosmanovic and Loewenstein, 1968; Kanno and Matsui, 1968). However, ionic communications may be normal between diverse normal and neoplastic fibroblasts in culture (Furshpan and Potter, 1968), and even Novikoff hepatoma (Sheridan, 1970). Indeed, the lack of electrical coupling is thus another common but not invariant aspect of neoplastic growth.

Intercellular communications could serve as an important means for the control of differentiation (Loewenstein, 1969). A model example is the observation that the defective nucleic acid metabolism in a mutant of polyoma-converted BHK21 cells can be corrected by transfer of an undefined substance from wild type cells during cell contact (Stoker, 1967b).

F. CELL FUSION

Cell-fusion is an extreme form of cell contact. It can be brought about *in vitro* by certain myxoviruses (Okada, 1969) or by lysolecithin, and also appears involved in the interaction of sensitized lymphocytes and their targets (Okada, 1962). Virus-induced fusion have been investigated in detail by Okada and associates (Okada, 1962; Okada and Tadokaro, 1962; Okada and Yamada, 1966; Okada and Murayama, 1966) and recently reviewed by Okada (1969) and Poste (1970). A study comparing a number of diverse cell types (Okada and Tadokaro, 1963) suggest that "fusion capacity" parallels malignancy.

G. SURFACE CHARGE

1. *Electrophoresis*

This interaction between individual cells has long been considered to involve the charge density on cell surfaces (Curtis, 1967). This notion arises from the fact that small aqueously dispersed particles subject to random thermal motion, tend to adhere upon contact unless electrostatic repulsions between their surfaces are large enough to make the collisions elastic (Wallach *et al.*, 1966). Accordingly, the altered contact behavior of neoplastic cells has often been ascribed to abnormalities of cell surface charge. However, there has been no convincing correlation of charge density, adhesiveness and malignancy (Table I.).

Nevertheless, there exist several internally controlled experiments showing altered electrophoretic mobilities of cell populations after oncogenic conversion. Thus hamster kidney cells have lesser electrophoretic mobilities than cells from stilbesterol-induced kidney tumors (Ambrose *et al.*, 1956), and the anodic mobility of hepatoma cells is greater than that of normal hepatocytes (Lowick *et al.*, 1961). Moreover, the surface potential of mouse MCIM sarcoma sublines increases as these progress from the solid to the ascites form, a change associated with greater invasiveness (Purdom *et al.*, 1958). Hamster fibroblasts often behave similarly after conversion with polyoma virus (Forrester *et al.*, 1962; Forrester *et al.*, 1964) as do mouse spleen cells in Friend virus disease (Forrester and Salaman, 1967). However, some leukemic mouse cells exhibit anomalously low anodic mobility (Cook and Jacobson, 1968).

There is no justification for the common assumption that malignant cells bear a greater net negative charge than their normal progenitors (Vassar, 1963). However, it appears that neoplastic conversion can change the cell surface potential in many instances (Ambrose *et al.*, 1956; Lowick *et al.*,

TABLE I

Electrophoretic mobilities of normal and Py-converted hamster kidney fibroblasts[a]

Clone	Comment	Electrophoretic mobility (μ/sec/V/cm)
C13	Control-unconverted	$-1\cdot02\pm0\cdot06$
C13	*Neuraminidase-treated*	$-0\cdot64\pm0\cdot05$
P	Py-converted	$-1\cdot26\pm0\cdot06$
Q	Py-converted	$-1\cdot29\pm0\cdot05$
S	Py-converted	$-1\cdot25\pm0\cdot05$
V	Py-converted	$-1\cdot27\pm0\cdot06$
X	Py-converted	$-1\cdot23\pm0\cdot05$
Y	Py-converted	$-1\cdot28\pm0\cdot09$
J	Py-converted	$-0\cdot97\pm0\cdot06$
M	Py-converted	$-1\cdot02\pm0\cdot05$
N	Py-converted	$-1\cdot30\pm0\cdot06$
N	Py-converted *Neuraminidase-treated*	$-0\cdot65\pm0\cdot06$

[a] From Forrester *et al.* (1964).

1961; Purdom et al., 1958; Forrester et al., 1962; Forrester et al., 1964; Forrester and Salaman, 1967; Cook and Jacobson, 1968).

2. Specific Ionogenic Groups

Much of the negative surface potential of mammalian cells is due to sialic acid (N-acetyl-neuraminic acid) (Curtis, 1967), but charged side chains of membrane proteins (Wallach et al., 1966; Cook and Jacobson, 1968) membrane RNA (Weiss and Mayhew, 1966), and polar membrane lipids also contribute to the net cell surface charge.

It has been suggested that increased negative surface charge in malignancy is due to increased surface sialic acid, but there is insufficient evidence for this generalization. Indeed, three lines of virus-converted fibroblasts (including polyoma-converted BHK21 cells) have less total sialic acid than their normal precursors (Ohta et al., 1968). Also, Wu and associates (Wu et al., 1968) report a decrease in the sialic acid (and galactosamine) in the plasma membrane and endoplasmic reticulum of 3T3 cells converted by SV_{40} virus. However, none of these studies can readily be related to surface potential, because the electro-kinetic expression of ionic groups on membranes depends, among more complex variables, upon the effective radius of the charge-bearing site, on other charges and upon the depth of the charge in the membrane, all of which are unknown (Wallach et al., 1966; Wallach and Perez-Esandi, 1964). This point is illustrated by the low anodic mobility of some leukemic mouse cells (Cook and Jacobson, 1968). Here the increased charge due to sialic acid is outweighed by a concomitant rise in surface cationic groups. Recent work on normal Swiss 3T3 (SW3T3) cells, SV_{40} converted variants (3T3SV) and revertant cells (Culp et al., 1971) nevertheless suggest that plasma membrane sialate functions in some manner in regulating cell–cell interaction in this model system (Table II) although sialic acid content is not specified by the

TABLE II

Correlation between contact inhibition, saturation density and sialic acid content in various SW3T3 cells[a]

Cell line[b]	Sialic acid (μg/mg protein)	Saturation density[c] (cells/cm$^2 \times 10^{-4}$)
3T3	$5 \cdot 0 \pm 0 \cdot 6$	$5 \cdot 0$
3T3SV	$3 \cdot 0 \pm 0 \cdot 3$	100
3T3SVR	$4 \cdot 8 \pm 0 \cdot 3$	$8 \cdot 0$

[a] From Culp et al. (1971).
[b] 3T3 = control; 3T3SV = SV_{40}—converted cells; 3T3SVR = reverted to normal growth pattern.
[c] Calculated from cell number per 50 mm diameter plastic dish.

viral genome. Also, it is established that sialic acid removal impairs normal embryonic cell aggregation (Kemp, 1968).

To summarize, we know of only four instances—the conversion of BHK21 cells by polyoma virus (Forrester *et al.*, 1962; Forrester and Stoker, 1964), Friend virus disease (Forrester and Salaman, 1967) and certain leukemias (Cook and Jacobson, 1968), where the abnormal anodic mobility of the con-verted cells is modified by enzymatic (neuraminidase) removal of sialic acid. The possible interrelations between surface charge, sialic acid and/or other ionogenic groups and malignant behavior thus requires more intensive study.

H. LEAKINESS OF THE PLASMA MEMBRANE

The plasma membranes of tumor cells commonly exhibit abnormal "permeability" to certain intracellular enzymes (Hill, 1956; McNair-Scott *et al.*, 1959; Wroblewski, 1958; Wu, 1959; Bissell *et al.*, 1971; Holmberg, 1961; Malmgren *et al.*, 1955; Sylven, 1958; Sylven, 1962; Sylven and Bois, 1960; Sylven and Malmgren, 1958; Sylven and Malmgren, 1955; Sylven *et al.*, 1959). The enzymes released are mostly lysosomal and/or those usually considered free in the cytoplasm. The observed "leakiness' is not due to cell damage. Highly "leaky" cells are normal by the usual criteria of dye exclusion and the phenomenon most probably reflects defects in the plasma and/or lysosomal membranes of malignant cells.

Sylven and associates (1962, 1960, 1955, 1959) have studied this matter, using controlled and elegant micropuncture, combined with ultramicro-analyses. They find a significant release of lysosomal peptidases and hydrolases from intact tumor cells into the interstitial fluid surrounding them, particularly at the invasive zone. They suggest that these enzymes contribute to the destruction of their surroundings by invasive tumors and conjecture that enzymatic modifications of the tumor cell surface might account for their decreased adhesiveness and common loss of contact inhibition.

It is not known whether abnormal plasma membrane "leakiness" is a general property of malignant cells and to what extent it reflects an immune reaction against the tumor, but it is of considerable importance to our under-standing of both the defects common to diverse tumors and the mechanisms of invasiveness.

I. IMMUNOLOGIC CHANGES

1. *New Antigens*

The genetically best defined plasma membrane alterations of neoplastic conversion are immunologic. Plasma membranes of tumors, induced by chemical, physical and viral carcinogens bear transplantation antigens not present in the tissues of origin (Klein, 1968; Boyse and Old, 1969; Burnet, 1970). These new antigens are probably not essential to the cancerous state and oncogenic viruses can effect their appearance without tumor formation (Girardi and Defendi, 1970). However, they play a critical role in the tumor-host relationship, since they can cause the immunological elimination of the malignant cells. The new antigen of chemically induced tumors appear to be tumor specific, but tumors induced by a given virus bear the same new transplantation antigen, regardless of the tissue or species of origin.

2. *Embryonic Antigens*

The neoplastic transformation of cultured hamster cells by various oncogens leads to the appearance of Forssman antigen on the surface of the altered cells (Vogel and Sachs, 1962; Vogel and Sachs, 1964; O'Neill, 1968). This glycolipid antigen is lacking in fibroblasts of neonatal and mature hamsters but is present in embryonic hamster cells. The appearance of embryonic antigens has also been shown to occur in several tumors of the human gastrointestinal tract (Gold and Freedman, 1965a; Gold and Freedman, 1965b; Gold et al., 1968). These glycoproteins associated with the tumor cell surface (Krupey et al., 1968) can be extracted with 0·6 M percholoric acid. A newly developed radioimmunoassay for circulating carcinoembryonic gastrointestinal antigens (Thomson et al., 1969) may greatly assist in the diagnosis of this type of malignancy.

In certain neoplastic cells, but not their normal progenitors, the glycolipid-antigen composition of neoplastic plasma membranes changes (Hakomori and Murakami, 1968) with a drop of hematoside and rise of lactosylceramide as a result of the conversion of BHK21 cells by polyoma virus, suggesting that the carbohydrate moiety of the neoplastic glycolipids is incomplete. This may relate to the observed deletion of blood-group A and B haptens in some human adenocarcinomas (Hakomori et al., 1967) and to abnormal synthesis or carbohydrate chains in tumors, due to a defect in/or inhibition of enzymes participating in carbohydrate metabolism (Hakomori, 1972).

J. ANTIGEN DELETION

Neoplastic conversion often causes deletion of certain organ specific antigens. This has been best studied in hepatocellular carcinomas (Abelev, 1965). These tumors usually lack some but not all of their tissue-specific antigens and one newly induced hepatoma usually differs appreciably from another. However, after many transplant generations, various tumors tend to arrive at a common antigenic pattern. Burtin et al. have also found loss of membrane antigens in human colonic cancer (Burtin et al., 1971).

K. ALTERED AGGLUTINABILITY

In experimental models, neoplastic conversion usually changes the proteins of the transformed cells' plasma membranes, permitting accelerated aggregation by ligands homologous to several distinct chemical groupings namely:

(i)   Certain glycolipids (Hakomori et al., 1968).

(ii)   Certain, as yet undefined, isoantigens (Häyari and Defendi, 1970).

(iii)   Binding sites for wheat-germ phytoagglutinin (Burger, 1969; Burger, 1970).

(iv)   Binding sites for concanavalin A (Con A) (Inbar and Sachs, 1969a; Inbar and Sachs, 1969b; Arndt-Jovin and Berg, 1971; Inbar et al., 1971).

(v)   Binding sites for soybean agglutinin (Sela et al., 1970).

Some of these receptors, notably those for Con A are normally prevalent on many cell types (Edelman and Milette, 1971), but alter the cells' agglutinability at various stages in the cell cycle (Fox et al., 1971) and upon permissive viral infections with polyoma (Benjamin and Burger, 1970) or vaccina (Zarling and Trevethia, 1971). Increased Con A-induced agglutination requires synthesis of new protein but not of cellular or viral DNA.

Recently several laboratories (Inbar et al., 1971; Ozanne and Sambrook, 1971; Cline and Livingstone, 1971; Arndt-Jovin and Berg, 1971) measured Con A binding to normal and malignant cells using tritiated ^{125}I-labelled or [63]Ni-labelled Con A. These studies show that Con A binding is no greater for tumor than normal cells, but that tumor cells are more agglutinable by Con A. It thus appears that the surface membranes of converted cells are topologically altered making them more agglutinable. Further Con A induced agglutination is not an invariant concomitant of other aspects of oncogenic conversion, such as morphologic changes and growth in agar, as well as at high cell density. The same considerations apply to the wheat-germ agglutinin.

The situation can be quite complex: thus Sela et al. (1970) find that the soybean agglutinin (v) agglutinates transformed mouse, rat and human cell lines, but malignant hamster cells became agglutinable only after prolonged pronase treatment. In contrast, normal mouse, rat and human precursor cells became typically agglutinable after short incubation with trypsin or pronase. This finding too can be interpreted in terms of variable reorganization of the cell surface receptors.

Con A bound to transformed cells can cause their death or inhibit their replication—in contrast to the well-known mitogenic action of this substance bound to lymphocytes (Powell and Leon, 1970). However Burger and Noonan (1970) claim that combination of the receptors of transformed cells with "monovalent concanavalin A" restores normal "contact-inhibitory"

TABLE III

Agglutination of normal and neoplastic cells by Con A at low and high temperature after treatment with EDTA or trypsin[a]

Cells	Days after subculture	Treatment with EDTA[b]		Treatment with trypsin[b]	
		4°	24°	4°	24°
Normal	1	−	−	+ + + +	+ + + +
	4	−	−	+ + + +	+ + + +
Polyoma transformed	1	−	−	+ + +	+ + +
	4	−	+ + +	+ + +	+ + +
SV$_{40}$ transformed	1	−	−	+ + + +	+ + + +
	4	−	+ + + +	+ + + +	+ + + +
Rous transformed	1	−	−	+ + + +	+ + + +
	4	−	+ + + +	+ + + +	+ + + +
Dimethyl-nitrosamine transformed	1	−	−	+ + + +	+ + + +
	4	−	+ + + +	+ + + +	+ + + +

[a] From Inbar et al. (1971).
[b] One day or four days after subculture, cells were dissociated with EDTA or trypsin (15 min at 37°C) and tested for agglutination after 30 min of incubation with Con A.

properties to these cells. They relate this observation to an apparent "transient escape from growth control" of normal cultured cells by gentle protease treatment and speculate that if exposure of the surface binding site reflects a change which prevents "contact inhibition", covering these receptors with fragments of concanavalin A might restore "contact inhibition". However, there is little solid support for this contention.

Of utmost importance, however, is the observation of Inbar et al. (1971) which demonstrates specific temperature-sensitive agglutination activity in various neoplastically converted cells in the case of Con A but not all such agglutinins. Tumor cells are agglutinated at 24°C but not at 4°C. Trypsinized normal cells are also agglutinated at 24°. The data point to two "activities"; binding, which is not altered by neoplastic conversion, and agglutinability which is temperature sensitive, activated by trypsin and brought out by neoplastic conversion (Table III).

L. TRANSPORT CHANGES

I have already suggested that the pleiotropic membrane alterations in cancer might include changes in the transport of critical metabolites and thus could confer a selective advantage to the neoplastic cells in a physiologically competitive environment. Several recent studies on model systems are concordant with this view.

1. Sugar Transport

Hatenaka and associates (Hatenaka et al., 1969; Hatenaka et al., 1970; Hatenaka et al., 1971) report mouse embryonic or BALB/3T3 cells, neoplastically converted with murine sarcoma virus, but not SV_{40} or murine leukemia, strikingly increase the uptake of glucose, galactose, mannose and 2-deoxyglucose, but not 3-methyl glucose. This transport modification is associated with a lowered K_M for glucose, galactose, mannose and 2-deoxyglucose (Table IV). These transport changes are to a specific alteration of the plasma membrane sugar transport system and not to increased intracellular hexokinase activity.

TABLE IV
Sugar transport in normal and
murine–sarcoma–virus–converted cells[a]

Sugar	K_M ($M \times 10^4$)	
	Control	Tumor
Glucose	29	3·7
Galactose	25	5·6
Mannose	28	3·4
2-Deoxyglucose	23	5·0
3-Methylglucose	6	11·0

[a] From Hatenaka et al. (1970, 1971).

In an analogous case, Martin et al. (1971) find that neoplastic conversion of chick embryo cells leads to an increased uptake of 2-deoxyglucose under conditions where the two cell types multiply at the same rate. Moreover in cells infected by a temperature-sensitive mutant, 2-deoxyglucose uptake is stimulated at the permissive (36°C), but not at the non-permissive (41·5°) temperature.

This change is distinct from the greater rate of glucose transport in sparsely populated chicken fibroblast cultures compared with highly populated cultures (Sefton and Rubin, 1971).

2. Amino Acid Transport

Foster and Pardee (1969), measuring the accumulation of non-metabolizable amino acids, such as aminoisobutyric acid (AIB) and cycloleucine, as well as metabolites such as arginine, and glutamine, find that confluent

Sw 3T3 cells accumulate the amino acids about 30% less rapidly than non-confluent cells. However, the transport kinetics of confluent and non-confluent polyoma transformed 3T3 cells did not differ significantly. Further, the rate of AIB and cycloleucine accumulation was greater in transformed non-confluent 3T3 cells than in the controls, but this was not true for glutamine and arginine. The authors conclude that amino acid transport in their system depends on cell density but that they had not found any defect clearly related to neoplastic conversion.

III. Erythrocyte Membrane Abnormalities Associated with Abnormal Hemoglobins in Man

A. INTRODUCTION

It is most likely that *in vivo* hemoglobin (Hb) lies in close proximity to the internal surface of the erythrocyte membrane and the physical influences alone of this protein, present in such high concentration, on the membrane interior must be considerable. Moreover hemoglobin can undergo large structural conformational changes, with marked physicochemical consequences when liganded (e.g. ferrous O_2, CO) or ferri-hemoglobin changes to the unliganded ferrous state as during normal physiological oxygenation and deoxygenation. Unfortunately, hemoglobin effects on erythrocyte membranes are little studied, since, in most membrane experiments the protein is removed as completely as possible and since such work starts off with cells, replete with HbO_2, i.e. there are no studies on membranes from deoxygenated cells. Nevertheless, the influences of certain abnormal hemoglobins indicate that the state of this molecule can influence the erythrocyte membrane function.

B. SICKLE CELL HEMOGLOBIN (HB-S)

Membrane distortions during sickling of Hb-S have long been recognized (Ponder, 1951; Bessis *et al.*, 1954) and are attributed to the fact that in deoxygenation, the membrane deformation generally disappears. However, permanent membrane distortion is also seen in sickle cell disease and may reflect irreversible membrane damage during long periods of cell sequestration in cell clumps.

Several studies indicate that the above is more than a gross mechanical phenomenon. Thus preliminary studies by Ponder (1951) and the work of Bessis *et al.* (1954), show that when sickling occurs at low pO_2, the membrane protrusions seen in both normal and diseased cells become rigid, angular rods in the latter case, suggesting a close association of HbS with the membrane. Moreover, in O_2 sickle cells behave normally in response to agglutinating antisera, but not so at low pO_2, although antibody binding does occur.

Sickling also affects ion transport in afflicted cells. Thus homozygous sickle cells (SS cells) show an increase in K^+ permeability in N_2 and a net K^+ loss (Tosteson, 1955). This effect was not related to the plasma, population or cell age, and could be reversed by "liganding" the hemoglobin with O_2 or CO. A specific membrane effect appears involved since concentrated SS solutions have the same Na^+ and K^+ affinities in O_2 as in N_2. Sickling also increases Na^+ permeation (Tosteson *et al.*, 1955). Cs^+ was found to behave as K^+, in so far as active transport was concerned, but could also permeate by simple diffusion. Apparently sickling accelerates the carrier-mediated transport of Na^+, K^+ and, less so, Cs^+, at the same time opening diffusion paths for all of these cations.

C. HEMOLYSIS DUE TO "UNSTABLE" HEMOGLOBINS

Many other hemoglobin variants are associated with hemolytic disease attributable to erythrocyte membrane abnormalities (Table V). These can all be traced to amino acid replacements in the globin moiety of Hb (usually in the β-chain), are always associated with labile hemoglobin, varying degrees of osmotic fragility and accelerated *in vitro* lysis of the erythrocytes. In these variants, of the eleven sites closest to the prosthetic group (Carrell and Lehmann, 1969) seven or eight are defective and more recent evaluations of the extent of the heme binding site suggest the correlation to be 100%. All of these amino acid replacements are ones which weaken the association of

TABLE V

Some hemoglobin variants causing abnormal erythrocyte-membrane fragility[a]

Hemoglobin	Peptide residue involved[b]	Amino acid replacement
Riverdale-Bronx	β–24	Gly → Arg
Genoa	β–28	Leu → Pro
Philly	β–35	Tyr → Pro
Hammersmith	β–42	Phe → Ser
Sydney	β–67	Val → Ala
Santa Ana	β–88	Leu → Pro
Sabine	β–91	Leu → Pro
Gun Hill	β–91–97	5 residues deleted
Köln	β–98	Val → Met
Wien	β–130	Tyr → Asp

[a] A fuller listing is given in Carrell and Lehmann (1969).
[b] The listed cases involve the β-chain but some instances with α-chain abnormalities have been reported.

heme and globin, and the latter tends to unfold when not in its normal tight association with the prosthetic group.

Importantly, virtually all of these diseases involve altered β-chains, and these probably lie close to the erythrocyte membrane. A reasonable explanation for the pathologic process then is that the labile hemoglobins form Hb—S—S-Membrane linkages interfering with membrane properties. That this is very likely so in Hb—Köln has been demonstrated by Jacob and associates (1967).

They were early to show that —SH inhibitors impair the normal cation permeability of erythrocyte membranes, leading to spherocytosis and hemolysis (Jacob and Jandl, 1962) and suggested this to be due to the formation of —S—S— bridges with the —SH at β-93.

These observations are highly pertinent to studies on Hb Köln (Jacob et al., 1967) demonstrating that the replacement of valine at β-98, lowers the affinity of heme to the nearby heme-linked histidine (β-92) and increases the reactivity of the —SH at β-93, which is important in maintaining hemoglobin structure (Carrell and Lehmann, 1969). The authors reasonably argue that the membrane defect in Hb Köln (and related variants) form —S—S— bridges between globin-β 93SH (Heinz bodies, when aggregated) and membrane S—H groups, and, since some of the latter are crucial to membrane stability, membrane destruction.

IV. Metal Toxicity

A. INTRODUCTION

Heavy metals present a serious hazard throughout the world with lead automobile exhaust amounting to thousands of tons per year and mercury an ever-increasing danger, due to its use as a fungicide and catalyst throughout industrialized areas. The binding of many of these substances to biologically reactive groups such as carboxyls, phosphates, imidazoles, phenoxyls, sulfydryls and disulfides has been well studied (Table VI). Clearly the plasma membrane is an important site of heavy metal action, both through its functional susceptibility and the fact that its permeability properties determine the accessibilities of the metals to sensitive sites within the cell. For technical reasons, most basic work on the interaction of plasma membranes with toxic metals has utilized erythrocytes and this area has been well reviewed by Passow (1971).

In the erythrocyte membrane, Pb^{2+} and Hg^{2+} react very rapidly with sensitive membrane sites before reaching into the cell interior. Cu^{2+} and Zn^{2+} permeate slowly and show no exceptional affinity for membrane components. Thallium enters cells via the Na^+, K^+ transport system, and uranyl ion appears

TABLE VI

Binding constants (Log K_1)

for some toxic metal cations and small molecules

Cation	RS⁻	NH_3	Imidazole	CH_3COO^-
Hg^{2+}	20	8·8	—	10·3
Ag^+	15	3·2	—	3·7
Cu^{2+}	—	4·2	4·4	8·2
Pb^{2+}	11	—	—	5·5
Cd^{2+}	8	2·7	2·8	3·9
Ni^{2+}	—	2·8	3·3	5·8
Co^{2+}	9	—	—	4·6

(from Passow 1971)

to impair membrane function by complexing with phosphate and carboxyl groups (Rothstein, 1970) Fe^{3+} and Cr^{3+} *per se* are hardly bound or absorbed but iron is taken up by reticulocytes at the rate of 5×10^4 atom/min from trans-ferrin [where two Fe^{3+} atoms are chelated by three phenoxyls] (Jandl and Simmons, 1957; Jandl *et al.*, 1959; Jandl and Katz, 1963).

B. EFFECTS OF TOXIC METALS ON APPARENT MEMBRANE SURFACE CHARGE

1. *Electrophoresis*

At neutral pH most cells bear a net negative charge and accordingly migrate toward the cathode in an applied DC field. Heavy metals reverse these mobilities in the order $Th > UO_2 > La > Cu$, $Ni > Ca$, Sr, Ba, Mn (Bangham *et al.*, 1958). Although much of the negative surface charge on cells arises from —COO⁻s, especially N-acetylneuraminate alpha-glycosidically linked

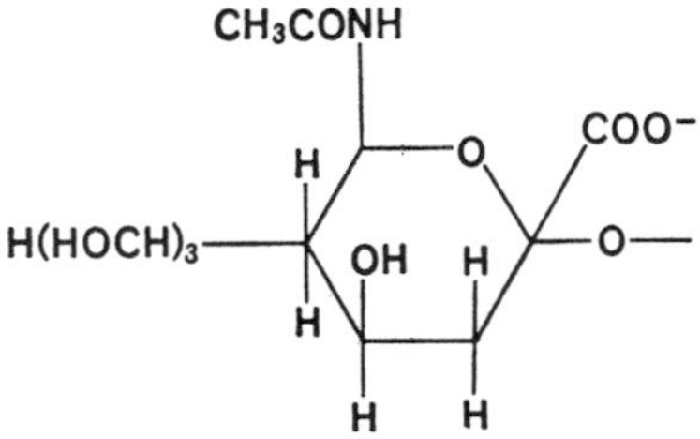

to membrane proteins (Winzler, 1970), determination analysis of absolute surface charge is not readily accessible to analysis by any method, let alone cell electrophoresis (Wallach *et al.*, 1966). The metal effects observed thus do not yield unequivocal data.

2. *Agglutination*

Contact between cells is hindered by their like surface charges (Wallach *et al.*, 1966) and becomes more prevalent at high ionic strengths and upon binding multivalent cations, often causing agglutination. Moreover poly-carboxylic acids block such agglutination, as do plasma proteins. In addition, Cr^{3+}, Fe^{3+}, Th^{3+}, and Al^{3+} bind certain proteins firmly to the erythrocytes of some species, which can then be agglutinated by divalent immunoglobulin, specific for the attached protein. This "sensitizing" effect occurs only when the metal and protein are added to the cell surfaces simultaneously.

C. SPECIFIC METAL EFFECTS

1. Mercury

Inorganic and organic mercurials can profoundly impair active and passive membrane transport by reacting with critical membrane protein SH-groups. In erythrocytes, which have been most fully studied, mercurials effect *both* rapid K^+ efflux and Na^+ influx across the membrane. In other cells amino acid transport may also be afflicted.

The affinity of Hg for —SH is so great that these groups are probably the metal's primary site of action. However, it is possible that two or more other chemical groups are so arranged geometrically to allow formation of Hg-chelates with similar stabilities as Hg-mercaptides. Moreover, the affinity of a given membrane —SH complex may for steric reasons, or because of the proximity of charged amino acid residues, differ from that of a free mercaptide (Webb, 1966).

At physiological pH the ratio $[HgCl_2/Hg_{total}]$ is about 0·33, and under these circumstances Hg reacts with thiols to form either: R—S—Hg or R′—S—Hg—S—R″. Organic mercurials are examples of the latter, the R′ representing the organic residue of the reagent.

Because of the chelation possibilities already noted, one cannot be certain that membrane-bound Hg is associated only with mercaptides. About 4×10^{-15} moles of metallic Hg react per ghost (van Stevenick et al., 1965), about five times the number of sites available to organic mercurials. In intact cells the principal binding sites are saturated in the order: membrane, Hb, gluthathione—SH. From various amino acid analyses of erythrocyte membranes (Rosenberg and Guidotti, 1969) about 1% of the amino acid residues of erythrocyte membranes arise from —SH, enough to account for the binding of Hg^{2+}. The discrepancy could be analytical, due to an underestimate of available —SH through formation of —S—Hg—S— within the membrane, or through participation of other types of ligands.

Organic mercurials inhibit the active transport of Na^+ and K^+ by erythro-cyte membranes, as well as their Na^+, K^+-activated ATPase (Passow, 1971), but inorganic Hg behaves differently. It also inhibits both the ion-specific and cation-insensitive ATPases of these membranes and increases passive per-meability to Na^+, K^+, etc. However, Hg causes greater permeability alterations, which occur more rapidly and pass through a peak with increasing Hg concentration (Weed, 1962). Passow (1971) suggests that the phenomenon might depend on the interaction of the *bifunctional* Hg^{2+} with closely-spaced —SHs. Then, when $[SH] > [Hg^{2+}]$, the complex —S—Hg—Cl would predominate, giving way to —S—Hg—S— at intermediate Hg levels and returning to —S—Hg—Cl at maximum doses.

2. *Lead*

Lead impairs biological functions when present at concentrations of 10^{-4} M or less, where Pb^{2+} salts, other than phosphates (solubility product = 10^{-30}) are quite soluble. In comparison to its high affinity for phosphates, lead binding to —SH is at least 100 times less than that of Hg, Au and Cu (Table VI). One therefore suspects diverse phosphate groups in the toxic actions of lead.

Lead has long been known to impair the functional and physical properties of plasma membranes (Passow, 1971). Erythrocyte membranes are particularly sensitive to the action of this metal and have recently yielded several clues as to its mode of action on membranes (Passow, 1971). At low Pb^{2+} levels (10^{-7} moles Pb^{2+}/g cells) increases K^+ efflux by 10^3. This K^+ loss is very rapid at the start but shifts to a slow rate after a fixed time. The number of reactive sites is much less than the total amount of lead bound per cell (Grigarzick and Passow, 1958). Also, Hb-free membranes require only 10^6 Pb^{2+} atoms/cell for a maximal response, compared with 10^7/intact cell; the former figure could represent the number of membrane binding sites, assuming no pertinent reorganization of membrane ligands during Hb-removal.

Unlike mercury, Pb^{2+} increases only K^+ efflux, but not Na^+ influx, leading to cell shrinkage. Moreover, measurements of the volume distributions in a cell population show that increasing lead concentration causes an increasing proportion of the cells to shrink, the other cells retaining their original volume. Also the shrunken cells are osmotically less fragile and contain less K^+ than normal. Apparently, a given cell either leaks K^+ rapidly or stays intact. The former account for the high efflux rate, while the remaining cells, retaining normal K^+ concentration, retain the normal slow rate of K^+ efflux, yielding the slow phase observed. At high Pb^{2+} levels, all cells in the population lose K^+ at the rapid rate and under this condition there is a massive loss of intracellular ATP, which is not reversed by the simultaneous increase in 2,3-diphosphoglycerate, a known Pb^{2+} complexer (Laris *et al.*, 1962; Lindeman and Passow, 1960).

The specificity of the Pb^{2+} effect is remarkable in that 10^3 fold increment in K^+ or Rb^+ flux occurs without essential change in Na^+ flux. Also, immediately after Pb^{2+} addition $^{42}K^+$ enters the cells against its own concentration gradient. The ratio [$^{42}K^+$ inside/$^{42}K^+$ outside] increases rapidly toward 3 but then drops to 1·0. The energy for the transient $^{42}K^+$ accumulation derives from the simultaneous efflux of non-radioactive K^+ from the cell by a carrier mediated counter transport. Possibly the Pb^{2+} activates normally unused carriers, thus providing a highly specific alteration in K^+ permeability.

3. *Copper*

Cu^{2+} specifically impairs the carrier-mediated transport of glycerol, without, however, affecting the glucose system (Stein, 1967). The marked decrease of glycerol transport with small increases in [H^+] near neutrality suggests involvement of critical histidines. Inhibition very likely requires two copper atoms at the active site (Stein, 1967). Glycerol transport can also be somewhat impaired by Ni and Au (Wilbrandt, 1941), but less specifically than with Cu.

4. *Thallium*

The crystal radius of Tl^+ lies between those of K^+ and Rb^+ and, at concentrations below 1 millimolar, Tl^+ behaves as K^+ in its action on squid giant axons (Mullins and Moore, 1960). At higher levels, it becomes toxic, stimulating K^+ efflux (Mullins and Moore, 1960; Truhaut, 1960).

Tl^+ toxicity derives from its competition for the physiological Na^+, K^+-transport system. Concordantly, Tl^+ accumulation requires metabolic energy. Moreover, Tl^+ uptake occurs in two phases, an initial, rapid one, which can be inhibited by cardiac glycosides (0·1 mM) or by addition of extracellular K^+, and a rather insensitive, slow phase (Gehring and Hammond, 1964).

5. *Uranium*

Uranium toxicity is due primarily to UO_2^{2+}, which does not form at neutral pH, because of formation of complexes with OH^-. For this reason, most of the work on uranium toxicity has been carried out on acid stable yeasts. This topic, as well as the various toxic effects observed in animals, have been recently reviewed by Rothstein (1970). The toxic action of UO_2^{2+} is at cell surfaces; there it forms stable, but reversible complexes with phosphate and carboxyl groups, without penetrating the membrane. The major consequence of this binding is an impairment of sugar transport into the cell. Judging from the extensive studies on yeast, phosphoryl complexing interferes with an active transport mechanism with rather high affinity for glucose, while carboxyl-complexing may interfere with passive or facilitated diffusion. It is likely that additional transport processes are impaired in mammalian systems, e.g. galactose transport by intestinal mucosa (Newey *et al.*, 1966).

In mammals, kidney damage is the major toxic effect of uranium because the normal acidification of urine in the renal tubules causes UO_2^{2+} production, which impairs glucose and amino acid resorption and eventually causes destruction of renal tubules (Passow *et al.*, 1961).

6. *Platinum*

cis-[PtCl$_2$(NH$_3$)$_2$] and several other Pt compounds strongly bind to the plasma membrane and thereby impair the entry of nucleotide precursors, etc., into cells. At the same time the interaction of *cis*-[PtCl$_2$(NH$_3$)$_2$] erythrocyte ghosts (Hoerer and Nicolau, 1971) demonstrate marked quenching of membrane tryptophane fluorescence at 350 μ by the platinum complex, which absorbs between 330–390 μ.

V. Membrane Aspects of Immunology

A. INTRODUCTION

The immune response involves the plasma membrane (a) as carrier of the specific groups of surface topography, distinguishing self from not-self, tissue from tissue, various stages of differentiation and diverse surface domains of a given same cell; (b) in immunocytes, as bearers of the apparatus necessary to determine whether colliding cells or molecules are "self" and, if not, activation of the machinery which initiates the immune responses; (c) as catalytic surfaces for the generation of complement, and (d) as the targets of immunologic cell killing.

B. ANTIGEN RECOGNITION

The clonal selective theories of the immune response contend that given immunocompetent cells are predestined to respond to at most a limited set of antigens, and that this commitment is defined by the presence of their plasma membranes of antigen-binding receptors of a single specificity (Burnet, 1970; Jerne, 1971; Burnet, 1959; Mitchinson, 1969; Siskind and Benacerraf, 1969). They also imply that the antigen-receptors have the characteristics of antibodies.

Considerable new evidence shows that such precommitted lymphocytes do exist and that the antigen-binding receptors are indeed immunoglobulins. Thus:

(a) Numerous investigators have shown that nonimmune animals bear small numbers of lymphocytes which can bind a specific radioactively or fluorescently labelled antigen to their surfaces where it can be measured by radioautography or microfluorescene (Naor and Saltzinau, 1967; Byrt and Ada, 1969; Davie and Paul, 1971; Davie *et al.*, 1971).

(b) Treatment of lymphoid cells from non-immunized mice with radio-iodinated antigens of very high specific activity, depletes the capacity of the cell population to transfer immune responsiveness to isogeneic, irradiated recipients (Davie *et al.*, 1971), presumably by specific irradiative killing the antigen-binding cells.

(c) After percolation of the lymphocytes from non-immunized animals over columns of glass or plastic beads bearing adsorbed specific antigen, the eluted cells population is markedly impaired in its ability to adoptively transfer primary immune responsiveness to that antigen in syngeneic mice, made immunoincompetent by X-irradiation, i.e. receptor-bearing cells have been removed (Wigzell and Anderson, 1971).

(d) The receptor-bearing cells are highly hapten specific (Davie and Paul, 1971) and appear to be mostly immunoglobulins of the gamma$_2$-heavy chain class (Taylor *et al.*, 1971). However, although their incidence is only about 0·05%, it appears that there is more than one type of receptor, those mediating cellular immunity revealing more complex specificities than the hapten-specific receptors which foster antibody production.

When followed by immunofluorescence, using fluorescein-labelled anti-immunoglobulins, the distribution of the presumed antigen receptors has been found sometimes diffuse, sometimes spotty or patchy, occasionally in the form of a ring or even a polar cap (Taylor *et al.*, 1971), have recently clarified this matter: Apparently the receptors are scattered randomly over the surface of resting lympocytes, but the interaction of the anti-Ig with the membrane Ig receptors membrane first induces a polar distribution following which the Ig molecules are interiorized.

This topographic relocation varies with temperature. At 0°, surface fluorescence is entirely ring-like, whereas incubation at 37° fostered cap formation. "Ring cells" stained at 0° change into "cap cells" when warmed, the transition proceeding more rapidly near 37°C; the process also requires metabolic energy.

C. COMPLEMENT ACTION

The complement system involves the plasma membrane in (a) as a catalytic support for the generation of the effector(s) of complement action, and (b) as the prime target of these effectors (Müller-Eberhardt, 1969). Complement lysis of an erythrocyte requires the attachment of only one molecule of immunoglobulin M (or two properly placed molecules of IgG) per cell. As an early consequence of a complement action ion permeability and/or transport, are impaired, ultimately causing cell lysis.

As a first step of complement activation (Fig. 1) antibody (A) becomes

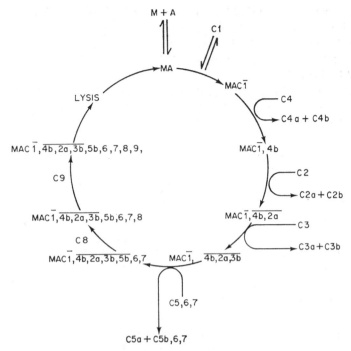

FIG. 1. Membrane role in complement activation.

membrane (M) bound (AM), the C1 component of complement* then
attaches to AM, converting component C1s to an esterase C1. This complex
may transfer to other membrane sites. C1 fosters membrane binding of
components C4 as well as C2, and to C4a and C4b and cleavage of C4 to
C4a and C4b, 5–10% of the latter remaining membrane associated. C2 is
peptidolysed, and thereby activated. The larger fragment, C2a attaches to
membrane-bound C4b, yielding C4b, 2a, which splits C3 into C3a and C3b,
the latter binding firmly to the membrane. C1, 4, 2, 3 regulates the subsequent
steps in an unresolved fashion. C5 is cleaved into at least two components,
C5a and C5b, of which 1% of C5b remains membrane bound and the rest is
inactivated. Thereafter C6 and C7 interact and attach to bound C5, followed
by adsorption of C8. The cell membrane becomes fragile with or without
component 9, which interacts with membrane bound C8. The final reaction
leading to membrane lysis is still uncertain.

   Biomembranes or artificial lipid membranes, negatively stained after
complement action in conjunction with an antibody directed against a
membrane component, reveal micromorphologic imperfections, consisting of

* The numbering of complement components is according to the chronology of their
discovery

an opaque center surrounded by a light ring 85–110 Å in diameter, which are independent of the membrane antigen or the source of antibody. Saponin and filipin produce a similar but not identical defect in various membranes.

Little biochemical effort has been concerned with the action of activated complement on the cell membrane, and most of the published work deals with the lipids of target cell membranes or artificial lipid membranes. This area has not been particularly fruitful, despite early encouraging results.

del Castillo and co-workers (del Castillo et al., 1966; Toro-Goyco et al., 1966) were the first to use artificial lipid membranes to study immune phenomena and observed that the combination of certain pure macromolecular antigens, and their antibodies, could decrease the resistance of thin lipid bilayer. Complement was not required and analogous resistance changes occurred in their systems when certain enzymes in the aqueous phases combined with their substrates. However, Barfort et al. (1968) noted a drastic and rapid decrease in the electrical resistance of sphingomyelin–alpha tocopherol bimolecular lipid membranes in 0·1 M NaCl, when minute amount of antibody and complement were present on one side of the membrane and homologous antigen (e.g. lysozyme or insulin) on the other. Finally, Haxby and associates (1968) showed conclusively that artificial liposomes containing Forssman glyco-lipid antigen become abnormally permeable to glucose, when exposed to specific antibody plus active complement. However, the careful study of Inoue and Kinsky (1970) on artificial liposomes provides a strong argument that the terminal action of complement does not involve cleavage of membrane lipids. In fact these workers were unable to detect any lipid degradation in an artificial liposome system, sensitive to complement. This does not exclude the possibility that the combination of antigen, antibody and active complement can change the organization of lipid arrays, without altering the chemical character of the lipid molecules.

Although proteins are the major components of membranes, they have been little studied as possible targets of complement action. Recently, however, the use of electrophoretic molecular sieving in detergent laden polyacrylamide before and after the action of complement on sensitized erythrocyte membranes reveals significantly peptidolysis in the membrane proteins of high molecular weight, producing new electrophoretic components of low molecular weight (Knueferman et al., 1971). These results are of a preliminary nature, and suggest, but do not prove, that membrane proteins are among the targets of complement action.

D. CELL-MEDIATED CYTOTOXICITY

Cell-mediated cytotoxicity, recently reviewed by Perlmann and Holm (Perlmann and Holm, 1969) is poorly understood as far as the participating

membrane-mediated processes are concerned; both (a) the possible cooperation between immunologically aggressive cells and (b) the action of an
aggressor immunocyte on the plasma membrane of its target are controversial
topics. Concerning the former, Hülser and Peters (1971) have clearly demonstrated that ionic communications form between lymphocytes stimulated by
phytohemagglutinin within minutes after activation, because of the decrease
in electrical potential (measured by microelectrodes) between contacted
stimulated cells. They find no such communication after simple contact or
cell agglutination; stimulation is essential.

Concerning the action of immunocytes upon their targets, Sellin, Wallach
and Fisher (Sellin *et al.*, 1971; Sellin, 1972) find that fluorescein permeable
junctions form between H2^b immunocytes, sensitized against H2^d mastocytes.
In their studies lymphoid cells *or* target cells were preloaded intracellularly
with fluorescein, by exposing them to permeant dye, fluorescein-diproprionate,
which is rapidly esterolyzed by healthy cells into poorly permeant fluorescein.
When junctions form between labelled and unlabelled cells, this is detected
and quantified by the fluorescein transfer between the cells. No junctions form
between autochthonous lymphocytes and mastocytes, but significant fluorescein transfer occurs when H2^d mastocytes are reacted with H2^b lymphocytes
from nonimmune animals, possibly due to precommitted immunocytes.
Significantly, junctions are common when lymphocytes from sensitized
animals were used.

These data indicate that in cell-mediate cytotoxicity, the lymphocyte and
target cell membrane fuse, to form cytoplasmic bridges. These are transitory,
since a given lymphoid cell moves freely from one target to another. Whether
such contacts are essential for cell mediated immunity is not yet known, nor
is the nature of the material or information transferred.

VI. Demyelination

A. INTRODUCTION

Myelin, the most widely studied plasma membrane derivative, is often
afflicted by several devastating disorders, the demyelinating diseases, of which
multiple sclerosis is the most common in the Western world. Possible related
conditions are: multifocal leukoencephalopathy, likely of viral etiology (Zu
Rhein, 1969), Kuru and Scrapie.

Although often reviewed (Pette and Westphal, 1969; Millar, 1971), the
causes and molecular biology of these conditions remain obscure; indeed
we probably face a number of diseases with several etiologies. One possibility
is that they arise from inherent or acquired defects in the production and/or
turnover of myelin components, particularly certain proteins. However, some

demyelinating disorders may be autoimmune diseases such as the experimental allergic encephalomyelitides (EAE). Possibly, certain protein components of the myelin membranes are not ordinarily surveilled by the immunologic system; then immunocyte clones capable of reacting with these substances are not eliminated and when they are presented to the immune system, perhaps during a viral infection, they are labelled "not-self", engendering an immune response. Because of its special morphology, myelin may be a particularly suitable candidate for such a hypothetical mechanism.

B. EARLY BIOCHEMICAL DEFECTS IN DEMYELINATING CONDITIONS

In multiple sclerosis, even morphologically normal myelin regions may be defective in lipid composition with a diminished proportion of long-chain fatty acids (Gerstl et al., 1961), and a shift toward greater saturation in the lecithins (Baker et al., 1963). Myelin cerebrosides also appear to be defective (Gerstl et al., 1970), but other components are present in normal amounts and proportions (Gerstl et al., 1967: Eng et al., 1968). These data are not inconsistent with the hypothesis of Gerstl et al. (1965), that impairment of fatty acids elongation is an early metabolic defect in multiple sclerosis, but one can also postulate that normal lipids may not fit well with an abnormal myelin protein.

The possible role of myelin proteins in myelin disorder has been generally ignored, but Benjamins et al. (1970), injecting puromycin intracerebrally into myelinating young rats to block protein synthesis decreased ^{3}H— leucine incorporation into myelin proteins four times after a few hours, while phosphatide and sulfatide metabolism remained unchanged. Two to three weeks later, amino acid incorporation was still less than normal, although lipid metabolism remained normal. The authors reason that an "acceptor protein" for myelin lipids is laid down very early and that lipid deposition follows later. They also suggest that many myelin disorders could arise from abnormal "acceptor proteins" which could produce lipid abnormalities also.

C. DEMYELINATING DISEASES AS AUTOIMMUNE PROCESSES

Certain basic myelin proteins when injected into isogenic animals produce experimental allergic encephalomyelitis (EAE), which mimic multiple sclerosis, but has not been proven identical to that disorder (Levine, 1971; Eylar, 1971; Martenson et al., 1971; Kibler et al., 1971). Proteins of this type, from various mammalian sources have been well characterized (Chao and Einstein, 1970; Eylar, 1970). Typical molecular weights are 18,200 (bovine) and 19,400 (human); there are no carbohydrate components. Eylar (1971) finds that the proteins derived from various sources are uniformly

M

encephalomyelogenic in guinea-pigs and that Trp and the peptic segment —Trp—Gly—Ala—Glu—Gly—Gln— are essential for biologic activity.

Chao and Einstein (1970) have extended these findings using bovine basic protein and guinea-pigs and rabbits as test animals. The active protein has only one Trp, which they too find essential for biologic activity in the guinea-pigs, but not in the rabbit, i.e. there is some species specificity. However, cyanogen bromide cleavage at the two met residues produces segments with 20, 143 and 3 amino acids respectively, of which only the large one is active.

Relating EAE to multiple sclerosis, Kibler *et al.* (1971) observe that encephalomyelogenic proteins, or active tryptic fragments therefrom, can induce blast activation or rabbit lymphocytes, although they found no differences between the materials from normal and multiple sclerosis patients. However, Caspary and Chambers (1970) note a specific interaction between encephalomyelogenic protein and the antibodies in the sera of some patients with multiple sclerosis, and it is equally important that peripheral lymphoid cells from patients with multiple sclerosis and from rabbits with EAE are toxic for neuroglia (Berg and Kallen, 1963).

VII. Intracellular Parasitism

A. INTRODUCTION

Penetration of host-cell plasma membranes by parasites is crucial to the survival of those with an obligatory intracellular phase; information about the necessary molecular processes would assuredly further the management of the millions of diseases caused by such agents.

B. PASSIVE ENTRY

Viruses usually adsorb to characteristic membrane receptors and penetrate animal cells by phagocytosis, or mechanisms (probably pinocytosis) not requiring cellular energy production (Holland and McLaren, 1959; Cohen, 1963). Some myxoviruses contain enzymes (e.g. neuraminidase) and act on cell surface carbohydrates, which are required for attachment, but appear non-essential for entry. The possible role of viral proteases or lipases in membrane penetration has been little studied although Zhdanov and Bukrinskaya (Zhdanov and Bukrinskaya, 1967) postulate a "cell wall destroying enzyme" for Sendai virus. Passive contact between cells and small particles, between cells and, perhaps, cell fusion, probably depends upon the attractive and repulsive forces between charged particles, as detailed elsewhere (Curtis, 1967; Wallach *et al.*, 1966; Pethica, 1961). Viruses, parasites and cells bear net negative surface charges. Electrostatic repulsions tend to keep them apart while diverse attractions, particularly hydrophobic and London–van der Waals forces, tend to foster contact. Recent studies (Wolpert and Gingell, 1968; Brewer and Bell, 1969; Gingell, 1967; Gingell, 1968) on the relation between membrane surface potential and structure, suggest that changes of surface potential reorient polar groups within the membrane core. Gingell's work (1967), demonstrating diminished surface potential upon cell contact fits this hypothesis, but even more important perhaps is the demonstration by infrared spectroscopy that surface potential changes are likely to reorient apolar membrane domains (May *et al.*, 1970).

Mycobacteria, such as tubercle and leprosy bacilli, which initiate their pathogenic action intercellularly, enter by phagocytosis (Sheppard, 1955; Trager, 1960), as do smooth variants of *Brucella* (Holland and McLaren, 1959). *Leishmania donovani* is commonly phagocytosed by macrophages, which, if intact, foster the morphologic changes preceding proliferation Miller and Twohy, 1969; Lamy *et al.*, 1966; Lamy *et al.*, 1967; Akiyama and Taylor, 1970; Adler, 1964; Akiyama and Haught, 1971).

C. PENETRATION OF PLASMODIA INTO ERYTHROCYTES

Malarial parasites must penetrate erythrocytes by mechanisms other than phagocytosis, which often involve remarkable host specificities; e.g. *Plasmodia* generally proliferate in (or penetrates) only those cells with high intracellular K^+. In their electron-microscopic study of erythrocyte penetration by *P.berghei* and *P.gallinaceum*, J. Ladda *et al.* (1969) find that the merozoites collide randomly with erythrocytes, but entry is initiated only when their conical tip ("conoid") contacts the erythrocyte membrane. The "conoid" indents contacted erythrocytes, deepening the depression as the parasite moves further into the cell with expansion of the erythrocyte membrane forming a cavity surrounding the invader and leaving the two plasma membranes closely apposed. This eventually leads to full endocytosis, the parasite undergoing ameboid differentiation in a membrane-enclosed vacuole. If the "conoid" secretes proteases these must be rather unique, since prior treatment of erythrocytes with trypsin, chymotrypsin and neuraminidase does not affect parasite penetration (Sherman, 1966).

D. ENTRY OF TOXOPLASMA

Toxoplasma merozoites of the specialized "conoid" regions resemble plasmodial merozoites and induce vacuolization of target cell plasma membranes during entry (Jadin and Creemers, 1968). Penetration of *T. gondii* into cultured HeLa cells increases when intact parasites are mixed with extracts of disintegrated organisms (Norrby and Lycke, 1967). Also penetration is enhanced by lysozyme, hyalouronidase, beta-glucuronidase and beta-galactosidase, so that the real nature of the membrane-active factor is not yet specified (Lycke et al., 1965).

E. ADDITIONAL MEMBRANE EFFECTS

1. Transport

Malarial parasitemia involves the erythrocyte membrane in *primary* way, but other possibly secondary membrane effects may also be important. Thus, both parasitized and uninfested erythrocytes in various malarias show abnormal osmotic fragilities (Fogel et al., 1966); moreover, the rate of hemolysis in malaria exceeds that expected from the number of infested cells (Zuckerman, 1964). Finally Dunn (1969) has extended and confirmed the observations of Overman (1948) of reversible changes of cation permeability in erythrocytes of monkeys infected with P. knowlesi. The erythrocyte [Na^+] of diseased monkeys is twice normal, presumably due to observed anomalies in active and passive Na^+ fluxes.

Intracellular [Na^+] is also elevated in muscle and liver; however, intracellular [K^+] is depressed in erythrocytes, unchanged in muscle and elevated in liver (Boehm and Dunn, 1970). This suggests that malarial parasites effect the production and/or release of some unknown membrane-active agent(s), but it is unclear how such might relate to the infestation and destruction of erythrocytes by the parasites.

The amino acid transport and incorporation in the red cells of normal and P. knowlensii-infected Rhesus monkeys differ significantly. McCormick (1970), following the distribution of ^{14}C-labelled leu,. isoleu, cys., met. and his. between normal plasma and cells from diseased animals finds the steady state amino acid concentration generally higher in cells from parasitized animals and a proportionately higher incorporation into protein. Since normal cells were not separated from parasitized ones in these studies, one cannot discern possible effects of the parasite on amino acid transport mechanisms. P. lophurae behaves analogously but with different amino acids (Sherman et al., 1967): proline accumulates most, al., ser., thr. less so, while met., his, and leu. were as in the normal cells.

F. LIPID CONTENT OF PARASITIZED ERYTHROCYTES

Intracellular growth of malarial parasites increases cellular lipid (Lawrence and Cenedella, 1969). Total cholesterol tends to double and phospholipids increase up to fourfold. Parasitized cells contain significantly higher levels of phosphatidyl ethanolamine than normal, but a lesser proportion of choline-lipids. This is not due to reticulocytosis. Moreover, since the parasite lipids

(*P. berghei* in rats) contribute only 5·8% of the total and 2·3% of the phospholipid and the parasites contain 40% free fatty acids, compared with traces in red cells, and since the predominant parasite phospholipids are cholinelipids, which decline in infected cells, the lipid changes must represent a change in the host cell membrane (Rao *et al.*, 1970). Concordantly Gutierrez (1966) and Cenedella (1968) find increased incorporation of labelled free fatty acids and glucose into the phosphatides (especially phosphatidylethanolamine) of erythrocytes infested with malarial parasites. Their observation of increased lyso-phosphatidylethanolamine, suggests phospholipase-A activation and/or decreased membrane acyl-transferase activity, both of which may relate to cell lysis.

G. ENTRY OF LYMPHOCYTES INTO CELLS

Although not necessarily an aspect of parasitism, lymphocytes have long been noted to penetrate into cells and wander about therein, in a process termed "emperipolesis" (Humble *et al.*, 1956; Pulvertaft, 1959). The phenomenon commonly occurs in tissue culture when lymphocytes contact other cells. Both normal (Stanfield *et al.*, 1963) and malignant lymphocytes (Dreyer *et al.*, 1964) can penetrate their own cell type, epithelial, mesenchymal, normal, embryonic, or malignant cells, even from other species. That the lymphoid cells are totally within their "host" cells has been shown electron microscopically for lymphoma cells inside macrophages (Shelton and Dalton, 1959) and for normal lymphocytes passing through endothelial cells (Marchesi and Gowans, 1964) and into Chang hepatoma (Perlmann *et al.*, 1968).

The participants in emperipolesis survive long periods in cell culture, but intracellular lymphocytes sometimes fragment suddenly, which would allow transfer of genetic information to the host cells; however, there is no transmittal of tritiated thymidine to the host cells (Ioachim, 1965).

Some lymphocytes seem to be attracted to certain fibroblasts which they enter repeatedly, leaving an identical neighbor untouched, but cellular immunological differences appear unessential to emperipolesis since this is seen also in preparation of fibroblasts and lymphocytes from the same individual (Moore and Hinka, 1969).

H. CODA

The group of diseases caused by intracellular parasites, representing a most serious threat to world health, involve the plasma membrane in at least three ways: (a) cellular recognition, (b) penetration of cell into cell, and (c) membrane fusion. The mechanisms underlying all of these processes remain to be elucidated.

VIII. Fertilization

A. INTRODUCTION

Although a normal biologic event, uncontrolled fertility is a major world health problem. Crucial to the process is the penetration of the sperm through the plasma membrane of the ovum and an understanding of this step might offer a unique means for regulation of population growth.

B. THE ROLE OF THE "ACROSOME"

During maturation, spermatozoa develop cap-like, membrane-bounded organelle, the acrosome, atop their nuclei. This contains numerous "lysosomal" enzymes (Bedford, 1970). Acrosomal proteases and the hyalouronidase are important to sperm-penetration of the zona pellucida (Stambaugh and Buckley, 1969) and since lysolecithin, a product of phospholipase A, is known to induce fusion between diverse cells (Lucy, 1970), one might suspect its participation in the penetration of the sperm into the egg. Striking membrane alterations in the acrosomal region occur during capacitation, and sperm adhere and fuse with the vitelline membrane only after acrosomal breakdown (Teichmann and Bernstein, 1969).

The acrosome disintegrates during fertilization with multiple fusions between its outer membrane and the sperm plasma membrane, so that the sperm head becomes covered by a cluster of membrane vesicles. These detach before or during penetration while the inner acrosomal membrane fuses with the penetrating surface of the sperm head; in pathological cases, however, the plasma and acrosomal membranes do not combine in a co-ordinated fashion (Bedford, 1969).

C. PENETRATION OF THE OVUM

The egg plasma membrane is surrounded by a "vitelline" envelope, through which the sperm must bore before it can reach the membrane proper. The mechanisms by which sperm penetrate the true plasma membrane of the ovum are obscure. Various sperm lysins have been proposed (Stambaugh and Buckley, 1969) but it is not known whether the components of the plasma membranes of either ovum or sperm change molecularly during penetration. It appears certain that the acrosomal contents and outer acrosomal membrane are shed before penetration, but possibly some essential enzymes remain attached to the inner acrosomal membrane (Teichmann and Bernstein, 1969). Morphologic details of the interactions of gamete membranes in fertilization are given in (Bedford, 1970).

After reaching the vicinity of the egg membrane the sperm stimulates this structure and ceases its tail movements. Only the sperm head contacts the egg surface but, since the entire sperm is bounded by a single membrane, a cooperative reaction propagated after the initial contact appears not unlikely.

Recently, several workers have described the process of sperm penetration, by measuring the change of electrical potential across the egg membrane during fertilization, following up the prediction of Lillie (1911) that the activation of the ovum might simulate the responses of excitable cells. Thus, Morrill *et al.* (1971) report depolarization of the egg membrane at the time of fertilization, as do Steinhardt *et al.* (1971). The latter authors, working with sea urchins and sand dollars, note that the eggs depolarize within three seconds after sperm addition, but that this process takes a long period to go to completion. Other reported changes, e.g. increased respiration (Nakazawa, *et al.*, 1970) and altered light scattering (Paul and Epel, 1971) occur much later.

It appears that the first stage of penetration necessitates multiple small fusions of the post-acrosomal sperm plasma membrane with bulbous protrusions on the egg surface. The plasma membrane of the sperm then disappears and this region of the sperm becomes enveloped by ooplasm. Whether the sperm membrane normally incorporates with that of the egg is unknown; however, it is known that it can do so from hybridization studies. Most likely sperm penetration involves two membrane processes: (a) membrane fusion caudally, and (b) ingestion of the sub-acrosomal region as a vacuole with an inner acrosomal membrane and an outer membrane derived from the ovum plasmalemma.

There is essentially no information as to the changes, if any, in the membrane proteins and/or lipids which follow and accompany sperm penetration. However, it appears that this induces a propagated cooperative membrane modification, such as suggested by Changeux and associates (1967), which leads to the subsequent rejection of other sperm.

A whole armamentarium of molecular probes are now being applied to the biology of spermatozoal membranes. Particularly pertinent is the work of Edelman and Milette (1971). Rodent sperm, gently agitated and treated with trypsin, chymotrypsin, subtilisin or pronase, rapidly separated into sperm heads and [tails + midpieces]. These can then be separated by centrifugation in density gradients for further study of their general membrane properties. Mouse spermatozoa contain about 10^7 Concanavalin A receptors per cell, most of these located on the plasma membrane overlying the acrosome region; this substance could agglutinate sperm with each other as well as with somatic cells bearing the Con A receptor.

Edelman and Milette also applied the fluorescent probe 8-anilino-1-naphthalene sulfonic acid (ANS) to examine the sperm plasma membrane, and that it labels and immobilizes the whole sperm. This suggests that the plasma membrane of the sperm is unusually sensitive, because of its role in the control of sperm mobility.

D. CODA

Fertilization is a unique case of intracellular parasitism, and in common with other such phenomena, involves the plasma membrane in at least three ways, namely (a) cell-cell recognition, (b) penetration of cell into cell, and (c) membrane fusion.

IX. Transport Defects

A. INTRODUCTION

Many membrane transport defects have been recognized and those due to radiation damage and metal poisoning have already been discussed. Here I will summarize several other such conditions of major biomedical importance.

B. GENETIC

Amino Acid Transport

In man four hereditary, autosomal recessive conditions involving impaired amino acid transport have been characterized. In all cases amino acid absorption in the intestine and the kidney are impaired concurrently. They are:

(a) *Cystinuria* (Milne *et al.*, 1961), which can be a unique defect or accompany other abnormalities of amino acid transport (Milne *et al.*, 1961; Crawhall and Watts, 1968). (b) *Abnormal ab- and re-sorption* of *cystine, lysine arginine and ornithine* (Crawhall and Watts, 1968). (c) *Iminoprolinuria*, affecting the transport of proline, hydroxyproline and glycine (Goodman *et al.*, 1967; Scriver, 1968), and (d) *Hartnup's disease* involving impaired transport of nearly all neutral amino acids (Milne *et al.*, 1960; Navab and Asatoor, 1970).

C. DEFECTIVE RENAL WATER RESORPTION (HEREDITARY DIABETES INSIPIDUS)

This arises from an impaired sensitivity to vasopressin (Orloff and Burg, 1966) which normally fosters water resorption in the distal nephron.

D. DEFECTIVE GLUCOSE-TRANSPORT; RENAL GLYCOSURIA

This appears to be an autosomal dominant defect, due to impaired glucose resorption in the proximal renal tubule, mainifested by a diminished T_{max} for glucose permeability of the luminal membranes of the proximal tubular cells and/or impaired carrier properties in these cells (Krane, 1966; Steward and Steward, 1969).

E. DEFECTIVE RENAL H^+-RESORPTION—RENAL ACIDURIA

The hereditary cases are genetically dominant. The defect is due to the inability of the tubular cells to maintain a steep pH gradient across their membranes, due to an abnormal permeability of the luminal membranes to H^+ (Seldin and Wilson, 1966; Gentz et al., 1969).

F. HYPOPHOSPHATEMIA (WITH HEREDITARY VITAMIN-D RESISTANT RICKETS)

This condition is X-linked dominant. The most important defect is an abnormally low T_{max} for phosphorous, leading to inadequate phosphate resorption from the glomerular filtrate and hence hypophosphatemia (Williams et al., 1966).

G. CYSTIC FIBROSIS

This common disease is expressed only in the homozygous condition. It involves impaired secretory function of serous and mucous glands, including the pancreas, eccrine sweat glands, salivary glands and mucous-secreting glands. The defect is also apparent in non-endocrine cells, i.e. erythrocytes (Balfe et al., 1968). Thus a clear impairment of Na^+ transport has been demonstrated in the erythrocytes of children with cystic fibrosis, as well as in the cells of their parents. The patients' cells furthermore exhibit a decrease in ouabain-sensitive ATPase activity. Defective Na^+ transport has also been demonstrated in the sweat glands (Schulz, 1969) of patients with the disease. Accordingly, Marsden (1969) has proposed that the disease with its abnormal epithelial glycoprotein secretion and its Na^+ transport defects can be fully explained in terms of the defective sodium transport.

H. HEREDITARY SPHEROCYTOSIS

This erythrocyte defect is clinically characterized by a hyperpermeability to Na^+, leading to increased Na^+ pumping, energized through increased glycolytic ATP production. However, the primary defect is probably one of membrane structure, very likely involving membrane proteins (Jandl, 1967). In associated with morphological and permeability alterations is a marked propensity for lipid loss (Reed and Swischer, 1966; Jacob, 1966) as well as a reversible depletion of cholesterol. This latter effect very probably accounts for the abnormal cell morphology and the overall contraction of cell surface area.

I. TRANSPORT ANOMALIES DUE TO BACTERIAL TOXINS

Several bacterial toxins severely impair various membrane transport mechanisms. Probably the most important are those involving tissues in extensive contact with the individual's external environment. Among such bacterial toxins the action of cholera toxin is pre-eminent.

J. CHOLERA

This disease is caused by the choleragen, secreted by the microorganism. The choleragen is a protein of about 61,000 molecular weight (Finkelstein and Lospallo, 1969) which is physiologically active at levels of about 0·1 $\mu g/Kg$. Some strains of *E. coli* produce similar substances (Chen *et al.*, 1971), which are possibly responsible for many debilitating and sometimes lethal diarrheas.

Choleragen alters the function of plasma membranes, in the intestinal cells causing excessive fluid secretion, without changing membrane morphology or passive permeability. It induces the secretion of and inhibits normal, non-glucose-mediated Na^+ transport (Field *et al.*, 1969) which, in the intestine, produces large fluid loss. Its action is not limited to intestinal cells; less than 10^{-9} g of the agent causes adipocytes to release stored lipids, an effect that can be blocked by low levels of insulin or prostaglandin E (Vaughn *et al.*, 1970). Finally, it has now been shown (Kimberg *et al.*, 1971) that cholera toxin stimulates adenyl cyclase in intestinal cells, thus giving rise to the increased levels of cAMP. The precise action of choleragen has yet to be defined, but it could (a) act as a hormone analogue, (b) stimulate adenyl cyclase independently, or (c) mimic adenyl cyclase.

K. BOTULISM

Botulism, caused by the potent exotoxin of *Cl. botulinum*, is due to impaired transmission of neuro-impulses between the synaptic membranes of the entire nervous system. The exotoxin is released as a protein complex of about mol wt 900,000 including the active neurotoxin with about mol wt 150,000. This can be further split into a 70,000 mol wt unit, and thereby activated (Lamanne and Sakaguchi, 1971). Until recently it was believed that botulinus toxin acts by blocking the release of acetylcholine at presynaptic membranes. However, Simpson and Marimoto (1969) suggest that its function may be otherwise and show that the toxin is definitely not an acetylcholineesterase inhibitor. Significantly the exotoxin of *Cl. botulinum* induces development of numerous Ach receptors, but on muscle membranes; these disappear after recovery (Simpson and Marimoto, 1969).

L. TETANUS

Tetanus is caused by the exotoxin of a similar organism, *Cl. tetani*, and is also a neuroactive agent, although more limited to the cerebrospinal axis in its reactivity. It is a protein with molecular weight of about 67,000 with a strong binding affinity for nervous system gangliosides. It appears that tetanus toxin acts specifically on inhibitory synapses, possibly blocking the release of glycine, the neurotransmitter in these synapses, from the presynaptic membrane (Johnson *et al.*, 1969; Felinec and Shank, 1971).

X. Acknowledgement

Supported by Grant CA 12178 from the United States Public Health Service, and Award PRA-78 of the American Cancer Society.

References

Abelev, G. I. (1965). *Prog. Exp. Tumour Res. (Basel)* **7**, 104–157.
Abercrombie, M. and Ambrose, E. J. (1962). *Cancer Res.* **22**, 525–548.
Abercrombie, M. and Heaysman, J. E. M. (1954). *Expl. Cell Res.* **6**, 293–306.
Abercrombie, M., Heaysman, J. E. M. and Karthauser, H. M. (1957). *Expl Cell Res.* **13**, 276–291.
Adler, S. (1964). *Adv. in Parasitol.* **2**, 35.
Akiyama, H. J. and Haught, R. D. (1971). *Am. J. trop. Med. Hyg.* **20**, 539.
Akiyama, H. J. and Taylor, J. C. (1970). *Am. J. trop. Med. Hyg.* **19**, 747.
Ambrose, E. J., James, A. M. and Lowick, J. H. B. (1956). *Nature, Lond.* **177**, 576–577.
Arndt-Jovin and Berg, P. (1971). *J. Virol.* **8**, 716.
Balfe, J. W., Cole, C. and Welt, L. G. (1968). *Science, N.Y.* **162**, 689.
Baker, R. W., Thompson, and Zrlja, K. J. (1963). *Lancet* **1**, 26.
Bangham, A. D., Pethica, B. A. and Seaman, G. V. F. (1958). *Biochem. J.* **69**, 12.
Barfort, P., Arquilla, E. R. and Vogelhut, P. O. (1968). *Science, N.Y.* **160**, 1119.
Barski, G. E. and Belehradek, Jr., J. (1965). *Expl. Cell Res.* **37**, 464–480.
Benjamin, R. and Burger, M. M. (1970). *Proc. natn. Acad. Sci. U.S.A.* **67**, 929.
Benjamins, J. A., Kreutz, D. and McKhann, G. M. (1970). *Trans. Am. Neurol. Assn.* **95**, 78.
Bedford, J. M. (1969). *In* "Advances in Biosciences". Vol. 4, p. 36. Pergamon Press.
Bedford, J. M. (1970). *Biol. Reprod. Supp.* **2**, 128.
Berg, O. and Kallen, A. (1963). *Acta path. microbiol. scand.* **52**, 23.
Bessis, M., Bricka, M., Breton-Gorius, J. and Tabuis, J. (1954). *Blood* **9**, 39.
Bissell, M. J., Rubin, H. and Hatie, C. (1971). *Expl. Cell Res.* **68**, 404.
Blumenthal, R., Changeux, J. P. and Lefever, R. (1970). *J. Memb. Biol.* **2**, 351.
Boehm, T. M. and Dunn, M. J. (1970). *Proc. Soc. exp. Biol. Med.* **133**, 370.
Borek, C. and Sachs, L. (1966). *Proc. natn. Acad. Sci. U.S.A.* **56**, 1705–1711.
Boyse, E. A. and Old, L. J. (1969). *Ann. Rev. Genet.* **3**, 269–290.
Brewer, J. E. and Bell, L. G. (1969). *J. Cell Sci.* **4**, 17.
Burger, M. M. (1969). *Proc. natn. Acad. Sci. U.S.A.* **62**, 994.

288   DONALD F. HOELZL WALLACH

Burger, M. M. (1970). *Nature, Lond.* **227**, 170.
Burger, M. M. and Noonan, K. D. (1970). *Nature, Lond.* **228**, 512.
Burnet, M. F. (1959). *In* "The Clonal Selection Theory of Acquired Immunity",
p. 209. Vanderbilt University Press, Nashville, Tenn.
Burnet, M. F. (1970). *Nature, Lond.* **226**, 124.
Burtin, P., von Kleist, S. and Sabine, M. C. (1971). *Cancer Res.* **31**, 1038.
Byrt, P. and Ada, G. L. (1969). *Immunology,* **17**, 503.
Carrell, R. W. and Lehmann, H. (1969). *Seminars in Hematol.* **6**, 116.
Caspary, E. A. and Chambers, M. (1970). *Europ. Neurol.* **3**, 206.
Cenedella, R. J. (1968). *Am. J. trop. Med. Hyg.* **17**, 680.
Changeux, J. P., Thiery, J., Tung, Y. and Kittel, C. (1967). *Proc. natn. Acad. Sci.
U.S.A.* **57**, 334–341.
Changeux, J. P. and Thiery, J. (1968). *In* "Regulatory Functions of Biological
Membranes", BBA Library, No. 11 (J. Jarnefelt, ed.) Elsevier, Amsterdam.
Changeux, J. P., Blumenthal, R., Kasai, M. and Podleski, T. (1970). *In* "Molecular
Properties of Drug Receptors" (R. Porter and M. O'Connor, eds.) p. 197.
Churchill, London.
Chao, L. P. and Einstein, E. R. (1970). *J. biol. Chem.* **245**, 6397.
Chen, L. C., Rhode, J. E. and Sharp, G. W. (1971). *Lancet* **1**, 939.
Cline, M. J. and Livingstone, D. C. (1971). *Nature New Biology.* **232**, 155.
Cohen, A. (1963). *In* "Mechanisms of Viral Infection" (W. Smith, ed.), p. 151.
Academic Press, New York.
Coman, D. R. (1944). *Cancer Res.* **4**, 625–629.
Coman, D. R. (1953). *Cancer Res.* **13**, 397–404.
Coman, D. R. (1954). *Cancer Res.* **14**, 519–521.
Coman, D. R. (1960). *Cancer Res.* **20**, 1202–1204.
Coman, D. R. (1961). *Cancer Res.* **21**, 1436–1438.
Coman, D. R. and Anderson, T. F. A. (1955). *Cancer Res.* **15**, 541–543.
Cook, G. M. W. and Jacobson, W. (1968). *Biochem. J.* **107**, 549–557.
Crawhall, J. C. and Watts, R. W. (1968). *Am. J. Med.* **45**, 736.
Culp, L. A., Grimes, W. J. and Black, P. H. (1971). *J. Cell Biol.* **50**, 682.
Curtis, A. S. G. (1967). *In* "The Cell Surface", pp. 206, 211, 214, 238, 259, 266.
Academic Press, New York.
Davie, J. M. and Paul, W. E. (1971). *J. exp. Med.,* **134**, 000.
Davie, J. M., Rosenthal, A. S. and Paul, W. E. (1971). *J. exp. Med.* **134**, 000.
del Castillo, J., Rodriquez, Romero, C. A. and Sanchez, V. (1966). *Science, N.Y.*
153, 185.
Dreyer, D. A., Schullenberger, C. C. and Mochowski, L. (1964). *Tex. Rep. Biol.
Med.* **22**, 61.
Dunn, M. J. (1969). *J. clin. Invest.* **48**, 674.
Edelman, G. M. and Milette, C. F. (1971). *Proc. natn. Acad. Sci. U.S.A.* **68**, 2436.
Eng, L. F., Chao, F. C., Gerstl, B., Pratt, D. and Tavaststjerna, M. G., *Biochemistry*
7, 4455.
Eylar, E. H. (1971). *In* "Immunologic Disorders of the Nervous System" (L. P.
Rowland, ed.), p. 50. Williams-Wilkins, Baltimore.
Eylar, E. H., Callam, J. and Jackson, J. J. (1970). *Science, N.Y.* **168**, 1220.
Felinec, A. A. and Shank, R. P. (1971). *J. Neurochem.* **18**, 2229.
Field, M., Fromm, D., Wallace, C. K. and Greenough, III, W. B. (1969). *J. clin.
Invest.* **48**, 24.
Finkelstein, R. A. and Lospallo, J. (1969). *J. exp. Med.* **130**, 185.

Fogel, B. J., Shields, C. D. and Von Doenhoff, Jr. (1966). *Am. J. trop. Med.* **15,** 269.
Forrester, J. A. and Salamann, M. H. (1967). *Nature, Lond.* **215,** 279–280.
Forrester, J. A. and Stoker, M. G. P. (1964). *Nature, Lond.* **201,** 945–946.
Forrester, J. A., Ambrose, E. J. and MacPherson, J. A. (1962). *Nature, Lond.* **196,** 1068–1070.
Foster, D. O. and Pardee, A. B. (1969). *J. biol. Chem.* **244,** 2675.
Fox, T. O., Sheppard, J. R. and Burger, M. M. (1971). *Proc. natn. Acad. Sci. U.S.A.* **68,** 244.
Furshpan, E. I. and Potter, D. D. (1968). *In* "Current Topics in Developmental Biology". Academic Press, New York.
Gentz, J., Lindblad, B., Lindstedt, S., Zetterstrom, R. (1969). *J. Lab. clin. Med.* **74,** 185.
Gerstl, B., Tavaststjerna, J. K., Hayman, R. B. and Eng, L. F. (1965). *Ann. N.Y. Acad. Sci.* **122,** 405.
Gerstl, B., Kahnke, M. J., Smith, J. K., Tavaststjerna, M. G. and Hayman, R. B. (1961). *Brain* **84,** 310.
Gerstl, B., Eng, L. F., Hayman, R. B., Tavaststjerna, and Bond, P. (1967). *J Neurochem.* **14,** 661.
Gerstl, B., Eng, L. F., Tavaststjerna, J. K., Smith, J. K. and Druse, S. L. (1970). *J. Neurochem.* **17,** 677.
Gehring, P. J. and Hammond, P. B. (1964). *J. Pharm. exp. Therap.* **145,** 215.
Gingell, D. (1967). *J. theor. Biol.* **17,** 451.
Gingell, D. (1968). *J. theor. Biol.* **19,** 340–344.
Girardi, A. J. and Defendi, V. (1970). *Virology* **42,** 688.
Gold, P. and Freedman, S. O. (1965a). *J. exp. Med.* **121,** 439–462.
Gold, P. and Freedman, S. O. (1965b). *J. exp. Med.* **122,** 467–481.
Gold, P., Gold, M. and Freedman, S. O. (1968). *Cancer Res.* **28,** 1331–1334.
Goodman, S. I., MacIntyre, Jr., C. A. and O'Brien, D. (1967). *J. Pediat.* **71,** 246.
Grigarzick, H. and Passow, H. (1958). *Pflügers Arch. ges. Physiol.* **267,** 73.
Gutierrez, J. (1966). *Am. J. trop. med. Hyg.* **15,** 818.
Hakomori, S. (1972). *In* "The Dynamic Structure of Cell Membranes". (D. F. H. Wallach and H. Fischer, eds.) Springer-Verlag, New York.
Hakomori, S. and Murakami, W. (1968). *Proc. natn. Acad. Sic. U.S.A.* **59,** 254–261.
Hakomori, S., Koscielak, J., Block, K. J. and Jeanloz, R. W. (1967). *J. Immunol.* **98,** 31.
Hakomori, S., Teather, C. and Andrews, H. (1968). *Biochem. biophys. Res. Commun.* **33,** 563.
Hatenaka, M., Huebner, R. J. and Gilden, R. V. (1969). *J. natn. Cancer Inst.* **43,** 1091.
Hatenaka, M. Augl, C. and Gilden, R. V. (1970). *J. biol. Sci.* **245,** 714.
Hatenaka, M., Todaro, G. J. and Gilden, R. V. (1971). *Int. J. Cancer.*
Haxby, J. A., Kinsky, C. B. and Kinsky, S. C. (1968). *Proc. natn. Acad. Sci. U.S.A.* **61,** 301.
Hayari, P. and Defendi, V. (1970). *Virology* **41,** 22.
Hill, B. R. (1956). *Cancer Res.* **16,** 460–467.
Hoerer, O. L. and Nicolau, C. (1971). *FEBS Lett.* **14,** 262.
Holland, J. J. and McLaren, L. C. (1959). *J. exp. Med.* **109,** 487.
Holmberg, B. (1961). *Cancer Res.* **21,** 1386–1393.
Hülser, D. F. and Peters, J. H. (1971). Abstract in 3. Tagung der Gesellschaft fur Immunbiologie, Marburg.
Humble, J. G., Jayne, W. H. W. and Pulvertaft, R. J. V. (1956). *Br. J. Haemat.* **2,** 283.

Huxley, J. (1958). *In* "Biological Aspects of Cancer". Harcourt, Brace & Co., New York.
Ioachim, H. L. (1965). *Lab. Invest.* **14,** 1784.
Inbar, M. and Sachs, L. (1969a). *Nature, Lond.* **223,** 710.
Inbar, M. and Sachs, L. (1969b). *Proc. natn. Acad. Sci. U.S.A.* **63,** 1418.
Inbar, M., Ben-Bassat, H. and Sachs, L. (1971). *Proc. natn. Acad. Sci. U.S.A.* **68,** 2748–2751.
Inoue, K. and Kinsky, S. C. (1970). *Biochemistry* **9,** 4767.
Jacobs, H. S., Brain, M. C. and Dacie, J. W. (1967). *J. clin. Invest.* **46,** 1073.
Jacob, H. S. and Jandl, J. H. (1962). *J. clin. Invest.* **41,** 779.
Jacob, H. S. (1966). *Am. J. Med.* **41,** 734.
Jadin, J. and Creemers, J. (1968). *Acta Trop.*, p. 267.
Jamakosmanovic, A. and Loewenstein, W. R. (1968). *Nature, Lond.* **218,** 775.
Jandl, J. H. (1967). *In* "Hereditary Disorders of Erythrocyte Metabolism" (E. Beutler, ed.), p. 209. Grune and Stratton, New York.
Jandl, J. H. and Katz, J. H. (1963). *J. clin. Invest.* **42,** 314.
Jandl, J. H. and Simmons, R. L. (1957). *Br. J. Haemat.* **3,** 19.
Jandl, J. H., Inman, J. K., Simmons, R. L. and Allen, D. W. (1959). *J. clin. Invest.* **38,** 161.
Jerne, N. (1971). *Eur. J. Immunol.* **1,** 1.
Johnson, G. A. R., Groat, W. C. and Curtis, D. R. (1969). *J. Neurochem.* **16,** 797.
Kanno, Y. and Matsui, Y. (1968). *Nature, Lond.* **218,** 775–776.
Kemp, R. B. (1968). *Nature, Lond.* **218,** 1255.
Kibler, R. F., Paty, D. W. and Sherr, V. (1971). *In* "Immunologic Disorders of the Nervous System" (L. P. Rowland, ed.), p. 95. Williams-Wilkins, Baltimore.
Kimberg, D. V., Field, M., Johnson, J., Henderson, A. and Gershon, E. (1971). *J. clin. Invest.* **50,** 2128.
Klein, G. (1968). *Cancer Res.* **28,** 625–635.
Knueferman, H., Wallach, D. F. H. and Fischer, H. (1971). *FEBS Lett.* **16,** 167.
Kohn, A. and Fuchs, P. (1970). *Current Topics in Microbiology & Immunology* **52,** 95.
Krane, S. (1966). *In* "The Metabolic Basis of Inherited Disease". (J. B. Stanbury, J. B. Wyngaarden, and D. S. Fredrickson, eds.), 2nd ed. p. 1221. McGraw-Hill, New York.
Krupey, J., Gold, P. and Freedman, S. (1968). *J. exp. Med.* **128,** 387.
Ladda, J., Aiakawa, M. and Sprinz, H. (1969). *J. Parasitol.* **55,** 633.
Lamanna, C. and Sakaguchi, G. (1971). *Bact. Rev.* **35,** 242.
Lamy, L., Wonde, T. and Lamy, H. (1966). *C. r. hebd. Séanc. Acad. Sci., Paris* (D) **263,** 671.
Lamy, L., Wonde, T. and Lamy-Roux, L. (1967). *C. r. hebd. Séanc Acad. Sci., Paris* (D), **264,** 1889.
Laris, P. C., Ewers, A. and Noviger, G. (1962). *J. cell. comp. Physiol.* **59,** 1945.
Lawrence, C. W. and Cenedella, R. J. (1969). *Expl. Parasit.* **26,** 181.
Levine, S. (1971). *In* "Immunologic Disorders of the Nervous System" (L. P. Rowland, ed.), p. 33. Williams-Wilkins, Baltimore.
Lillie, R. S. (1911). *J. Morphol.* **22,** 695.
Lindeman, B. and Passow, H. (1960). *Pflügers Arch. ges. Physiol.* **271,** 369.
Loewenstein, W. R. L. (1969). *Can. Cancer Conf.* **8,** 162.
Loewenstein, W. R. L. and Penn, R. D. (1967). *J. cell Biol.* **33,** 235–242.
Lowick, J. H. B., Purdom, L., James, A. M. and Ambrose, E. J. (1961). *J. R. microsc. Soc.* **80,** 47–57.

Lucy, J. A. (1970). *Nature, Lond.* **227,** 815.

Lycke, E., Lund, E. and Stannegard, O. (1965). *Br. J. exp. Path.* **46,** 189.

MacPherson, I. and Stoker, M. (1962). *Virology* **16,** 147–151.

Malmgren, H., Sylven, B. and Revesz, L. (1955). *Br. J. Cancer* **9,** 473–479.

Marchesi, V. T. and Gowans, J. L. (1964). *Proc. R. Soc. B.* **159,** 283.

Marsden, J. C. (1969). *Nature, Lond.* **223,** 214.

Martenson, R. E., Biebler, G. E. and Kies, M. W. (1971). *In* "Immunologic Disorders of the Nervous System" (L. P. Rowland, ed.), p. 76. Williams-Wilkins, Baltimore.

Martin, G. S., Venuta, S., Weber, M. and Rubin, H. (1971). *Proc. natn. Acad. Sci. U.S.A.* **67,** 2739.

May, L., Kamble, A. B. and Acosta, I. P. (1970). *J. Membrane Biol.* **2,** 192.

McCormick, F. G. (1970). *Expl. Parasit.* **27,** 143.

McNair-Scott, T. B., Sanford, K. K. and Westfall, B. B. (1959). *Proc. Am. Ass. Cancer Res.* **3,** 41.

Millar, J. H. D. (1971). "Multiple Sclerosis". Charles C. Thomas, Springfield, Illinois.

Miller, H. C. and Twohy, D. W. (1969). *J. Protozool.* **14,** 781.

Milne, M. D., Crawford, M. A., Girao, C. B. and Loughridge, L. W. (1960). *Qt. Jl. Med.* **29,** 407.

Milne, M. D., Asatoor, A. M., Edwards, K. D. and Loughridge, L. W. (1961). *Gut* **2,** 323.

Mitchinson, N. A. (1969). *Cold Spring Harb. Symp. quant. Biol.* **32,** 431.

Moore, A. E. and Hinka, J. (1969). *Cytobiol.* **1B,** 73.

Morrill, G. A., Kostellow, A. B. and Murphy, J. B. (1971). *Expl. Cell Res.* **66,** 289.

Müller-Eberhardt, H. (1969). *A. Rev. Biochem.* **38,** 389.

Mullins, L. J. and Moore, R. D. (1960). *J. gen. Physiol.* **43,** 759.

Nakazawa, T., Asami, K., Shoger, R. and Yasumasu, I. (1970). *Expl. Cell Res.* **63,** 143.

Naor, D. and Saltzinau, D. (1967). *Nature, Lond.* **214,** 687.

Navab, F. and Asatoor, A. M. (1970). *Gut* **11,** 373.

Newey, H., Sanford, P. A. and Smyth, D. H. (1966). *J. Physiol.* **186,** 493.

Norrby, R. and Lycke, E. (1967). *J. Bact.* **93,** 53.

Ohta, N., Pardee, A. B., McAuslan, B. R. and Burger, M. M. (1968). *Biochim. biophys. Acta.* **158,** 98–102.

Okada, Y. (1962). *Expl. Cell Res.* **26,** 98–107.

Okada, Y. (1969). *In* "Current Topics of Microbiology and Immunology" (W. Arber, ed.), p. 102. Springer-Verlag, Berlin.

Okada, Y. and Murayama, F. (1966). *J. exp. Cell Res.* **44,** 527–551.

Okada, Y. and Tadokoro, J. (1962). *Expl. Cell Res.* **26,** 108–118.

Okada, Y. and Tadokoro, J. (1963). *Expl. Cell Res.* **32,** 417–430.

Okada, Y. and Yamada, K. (1966). *Virology* **27,** 115–130.

Oliveira-Castro, G. M. and Loewenstein, W. R. (1970). *J. Memb. Biol.* **5,** 51.

O'Neill, C. H. (1968). *J. Cell Sci.* **3,** 405.

Orloff, J. and Burg, M. B. (1966). *In* "The Metabolic Basis of Inherited Disease" (J. B. Stanbury, J. B. Wyngaarden, and D. S. Fredrickson, eds.), 2nd ed., p. 1247. McGraw-Hill.

Overman, R. R. (1948). *Am. J. Physiol.* **152,** 113.

Ozanne, B. and Sambrook, J. (1971). *Nature New Biology* **232,** 156.

Passow, H. (1971). *In* "Effects of Metals on Cells, Subcellular Elements, and Macromolecules" (J. Maniloff, J. R. Coleman, M. W. Miller, eds.), Ch. 16, p. 291. Charles C. Thomas, Springfield, Illinois.

Passow, H., Rothstein, A. and Clarkson, T. W. (1961). *Pharmacol. Rev.* **13**, 185.

Paul, M. and Epel, D. (1971). *Expl. Cell Res.* **65**, 281.

Penn, R. D. (1966). *J. Cell Biol.* **29**, 171–174.

Perlmann, P. and Holm, G. (1969). *Adv. Immun.* **11**, 117.

Perlmann, P., Holm, G. and Biberfeld, P. (1968). *Expl. Cell Res.* **52**, 672–677.

Pethica, B. A. (1961). *Expl. Cell Res., Supp.* **8**, 123.

Pette and Westphal (1969). *In* "The Pathogenesis and Etiology of the Demyelinating Diseases" (S. Darger, ed.). Basel.

Ponder, E. (1951). *C. r. Séanc. Soc. Biol.* **145**, 1665.

Poste, G. (1970). *Adv. in Virus Res.* **16**, 303.

Powell, A. E. and Leon, N. A. (1970). *Expl. Cell Res.* **62**, 315.

Pulvertaft, R. J. V. (1959). *Proc. R. Soc. Med.* **52**, 315.

Purdom, L., Ambrose, E. J. and Klein, G. E. (1958). *Nature, Lond.* **181**, 1586–1587.

Rao, K. N., Subrahmanyam, D. and Pakrash, S. (1970). *Expl. Parasit.* **22**, 22.

Reed, C. F. and Swischer, S. N. (1966). *J. clin. Invest.* **45**, 77.

Rose, B. (1970). *J. Memb. Biol.* **5**, 1.

Rose, B. and Loewenstein, W. R. L. (1970). *J. Memb. Biol.* **5**, 20.

Rosenberg, S. A. and Guidotti, G. (1969). *In* "Red Cell Membrane Structure and Function" (J. H. Jamieson and T. J. Greenwalt, eds.), p. 93. J. B. Lippincott, Philadelphia.

Rothstein, A. (1970). *In* "Effects of Metals on Cells Subcellular Elements and Macromolecules" (J. Maniloff, J. R. Coleman and M. W. Miller, eds.), p. 365. Charles C. Thomas, Springfield, Illinois.

Sachs, L. and Medina, D. (1961). *Nature, Lond.* **189**, 457–458.

Schulz, I. J. (1969). *J. clin. Invest.* **48**, 1470.

Scriver, C. R. (1968). *J. clin. Invest.* **47**, 823.

Sefton, B. M. and Rubin, H. (1971). *Proc. natn. Acad. Sci. U.S.A.* **68**, 3154.

Sela, B. A., Lis, H., Sharon, N. and Sachs, L. (1970). *J. Memb. Biol.* **3**, 267.

Seldin, D. S. and Wilson, J. D. (1966). *In* "The Metabolic Basis of Inherited Disease" (J. B. Stanbury, J. B. Wyngaarden, and D. S. Fredrickson, eds.), p. 1231. McGraw-Hill.

Sellin, D., Wallach, D. F. H. and Fischer, H. (1971). Abstract in 3. Tagung der Gesellschaft fur Immunbiologie, Marburg.

Sellin, D. (1972). *Europ. J. Immunology* (in press).

Shelton, E. and Dalton, A. J. (1959). *J. biophys. biochem. Cytol.* **6**, 513.

Sheppard, C. W. (1955). *Proc. Soc. exp. Biol. Med.* **90**, 3922.

Sheridan, J. D. (1970). *J. Cell Biol.* **47**, 189a.

Sherman, I. (1966). *Parasitology* **52**, 17.

Sherman, I., Vikar, R. A. and Ruble, J. A. (1967). *Comp. Biochem. Physiol.* **23**, 43.

Simpson, L. L. and Marimoto. (1969). *J. Bact.* **97**, 571.

Siskind, G. W. and Benacerraf, B. (1969). *Adv. Immunol.* **10**, 1.

Stambaugh, R. and Buckley, J. (1969). *J. Reprod. Fertl* **19**, 423.

Stanfield, A. B., Bailey, J. P. and Zellforsch, A. (1963). *Mikrosk. Anat.* **60**, 497.

Stein, W. D. (1967). *In* "The Movement of Molecules Across Cell Membranes". Academic Press, New York.

Steinhardt, R. A., Lundin, L. and Mazia, D. (1971). *Proc. natn. Acad. Sci. U.S.A.* **68**, 2426.

Steward, A. D. and Steward, J. (1969). *Am. J. Physiol.* **1191.**
Stoker, M. G. P. (1964). *Virology* **24,** 165–174.
Stoker, M. G. P. (1967a). *In* "Current Topics in Developmental Biology", Vol. 2. Academic Press, New York.
Stoker, M. G. P., (1967b). *J. Cell Sci.* **2,** 293–304.
Stoker, M. G. P., Shearer, M. and O'Neill, C. (1966). *J. Cell Sci.* **1,** 297–310.
Sylven, B. (1958). *Acta Un. int. Canc.* **14,** 61–62.
Sylven, B. (1962). *In* "Biological Interactions in Normal and Neoplastic Growth" (M. J. Brennan and W. L. Simpson, eds.). Little, Brown & Co., Boston, Mass.
Sylven, B. and Bois, I. (1960). *Cancer Res.* **20,** 831–836.
Sylven, B. and Malmgren, H. (1955). *Acta radiol.* Suppl. **154,** 1–124.
Sylven, B., Ottoson, R. and Revesz, L. (1959). *Br. J. Cancer* **13,** 551–565.
Taylor, R. B., Duffus, P. H., Raff, M. C. and de Petris, S. (1971). *Nature New Biology* **233,** 225.
Teichmann, R. J. and Bernstein, M. H. (1969). *Anat. Rec.* **163,** 343.
Temin, H. M. and Rubin, H. (1958). *Virology* **6,** 669–688.
Thomson, D. M., Knipey, J. and Freedman, S. O. (1969). *Proc. natn. Acad. Sci. U.S.A.* **64,** 161.
Toro-goyco, E., Rodriquez, A. and del Castillo, J. (1966). *Biochem. biophys. Res. Commun.* **23,** 344.
Tosteson, D. C. T. (1955). *J. gen. Physiol.* **39,** 55.
Tosteson, D. C., Carlsen, E. and Dunham, E. T. (1955). *J. gen. Physiol.* **39,** 31.
Trager, W. (1960). *In* "The Cell" (J. Brachet and A. Mirshky, eds.), Vol. IV, Ch. 4. Academic Press, New York.
Truthaut, R. (1960). *J. Occup. Med.* **2,** 334.
van Stevenick, J., Weed, R. I. and Rothstein, A. (1965). *J. gen. Physiol.* **48,** 617.
Vassar, P. S. (1963). *Lab. Invest.* **12,** 1072–1077.
Vaughn, M., Pierce, N. and Greenough, III, W. B. (1970). *Nature, Lond.* **226,** 16.
Vogel, M. and Sachs, L. (1962). *J. natn. Cancer Inst.* **29,** 239–252.
Vogel, M. and Sachs, L. (1964). *Expl. Cell Res.* **34,** 448–462.
Vogt, M. and Dulbecco, R. (1960). *Proc. natn. Acad. Sci. U.S.A.* **46,** 365–370.
Wallach, D. F. H. (1968). *Proc. natn. Acad. Sci. U.S.A.* **61,** 868–874.
Wallach, D. F. H. (1969a). *New Engl. J. Med.* **280,** 761–767.
Wallach, D. F. H. (1969b). *J. gen. Physiol.* Suppl. 4, Part 2, p. 35.
Wallach, D. F. H. and Perez-Esandi, M. V. (1964). *Biochim. biophys. Acta* **83,** 363–366.
Wallach, D. F. H., Kamat, V. B. and Gail, M. H. (1966). *J. Cell Biol.* **30,** 601–621.
Webb, J. L. (1966). *In* "Enzyme and Metabolic Inhibitors", p. 734, Vol. 2. Academic Press, New York.
Weed, R. I. (1962). *J. gen. Physiol.* **45,** 395.
Weiss, L. and Mayhew, E. (1966). *J. cell comp. Physiol.* **68,** 345–359.
Wigzell, H. and Anderson, B. (1971). *A. Rev. Microbiol.* **25,** 291.
Wilbrandt, W. (1941). *Pflügers Arch. ges. Physiol.* **244,** 637.
Williams, T. F., Winters, R. W. and Burnet, C. H. (1966). *In* "The Metabolic Basis of Inherited Disease" (J. B. Stanbury, J. B. Wyngaarden, and D. S. Frederickson, eds.), 2nd ed. McGraw-Hill; New York.
Winzler, R. J. (1970). *Int. Rev. Cytol.* **29,** 77.
Wolpert, L. and Gingell, D. (1968). *Symp. Soc. exp. Biol.* **22,** 169–198.
Wroblewski, F. (1958). *Ann. N.Y. Acad. Sci.* **75,** 322–338.
Wu, R. (1959). *Cancer Res.* **75,** 1217–1222.

Wu, H., Meezan, E., Black, P. H. and Robbins, P. W. (1968). *Fed. Proc.* **27** (2), 814.
Zarling, J. M. and Trevethia, S. (1971). *Virology* **45**, 313.
Zhdanov, V. M. and Bukrinskaya, A. G. (1967). *Virologica* **6**, 105.
Zuckerman, A. (1964). *Expl. Parasit.* **15**, 138.
Zu Rhein, H. (1969). *Progr. Med. Virol.* **11**, 185.

Author Index

Numbers in italics refer to the text page on which the complete reference may be found.

A

Abelev, G. I., 261, *287*
Abercrombie, M., 256, *287*
Abodeely, R. A., 150, 152, *179*
Abrahamsson, S., 86, 89, *86*
Abramson, M. B., 35, *86*
Acheson, N. H., 147, 151, 152, *179*
Acosta, I. P., 279, *291*
Ada, G. L., 152, 153, 272, *179*, *288*
Adler, S., 279, *287*
Aggerbeck, L., 47, *88*
Ahlquist, R. P., 185, 186
Akers, C. K., 58, 59, 60, 74, *86*, *88*
Akgus, S., 193, *219*
Akiyama, H. J., 279, *287*
Albright, F. R., 245, 246, 247, *252*
Allan, D., 241, *249*
Allen, D. W., 268, *290*
Allison, A. C., 157, 159, 160, 172, *181*, 232, *250*
Almeida, J. D., 168, *179*
Ambrose, E. J., 256, 258, 259, 260, *287*, *289*, *290*, *292*
Anderegg, J. W., 3, *86*
Anderson, B., 273, *293*
Anderson, T. F. A., 256, *288*
Andrew, E. R., 108, 110, *140*
Andrews, H., *289*
Anfinsen, C., 198, *215*
Aoyagi, T., 168, *182*
Apostolov, K., 156, 169, *179*, *181*
Appleyard, G., 169, *182*
Archibald, F. M., 236, *251*
Armbruster, O., 157, *179*
Arndt-Jovin, D., 262, *287*
Arquilla, E. R., 192, *214*, 275, *287*
Asami, K., 283, *291*
Asatoor, A. M., 284, *291*

Asher, Y., 158, 172, *179*
Ashmore, J., 191, 212, *214*
Atkinson, D., *89*
Atkinson, P. H., 176, 178, *180*
Augl, C., 264, *289*
Avruch, J., 189, 207, 209, 210, 211, 213, *214*, *215*
Azerad, R., 46, *89*

B

Bächi, T., 147, 165, *179*
Bagchi, S. N., 4, 36, *88*
Bailey, A. I., 9, *88*
Bailey, E., 199, *216*
Bailey, J. P., *292*
Baird, C. E., 206, 207, *215*
Baker, R. W., 277, *287*
Bale, W. F., 192, *216*
Balfe, J. W., 285, *287*
Ball, E. G., 206, *217*
Bang, F. B., 147, *182*
Bangham, A. D., 268, *287*
Bar, H. P., *214*
Bär, 188, 202
Barclay, M., 236, *251*
Barfort, P., 275, *287*
Barratt, M. D., 113, *140*
Barrnett, R. J., 195, *218*
Barski, G. E., 256, *287*
Barsukov, L. I., 113, *140*
Basford, J. M., 231, *249*
Bauer, H., 147, 150, 152, *179*, *180*
Bear, R. S., 49, *89*
Beard, D., 147, *179*
Beard, J. W., 147, 158, *179*, *182*
Beaven, G. H., 199, *216*
Becht, H., 148, 151, 152, 165, 167, 168, 169, *179*, *181*

295

N

Elliot, A., 3, *87*
Ellis, S., 197, *217*
Ellison, S. A., 172, *182*
Ellwood, D. C., 169, *182*
Emmelot, P., 202, *214*, 233, *249*
Eng, L. F., 277, *288*, *289*
Engelman, D. M., 4, 14, 36, 48, 49, 75, 76, 77, 78, 79, *89*, 112, 138, *141*, *144*
Engstrom, L. H., 135, *141*
Epand, R. M., 199, *216*
Epel, D., 283, *292*
Epstein, S. E., 202, *217*
Erbland, J., 228, *250*
Esfahani, M., 78, *87*, 136, 138, *141*
Essner, E., 236, *251*
Exton, J. H., 204, 205, 207, *216*
Evans, M. J., 152, *180*
Ewers, A., 270, *290*
Eylar, E. H., 277, *288*

F

Fain, J. N., 203, *216*
Fairbanks, G., 133, *143*, 210, *214*
Falke, D., 147, 172, *183*
Fast, P. G., 19, 20, *89*
Faure, M., 44, 45, 47, *87*, 131, *142*
Feeney, J., 134, *143*
Feigenson, G. W., 105, 106, 138, *140*
Felinec, A. A., 287, *288*
Felts, P. W., 198, *216*
Ferber, E., 225, 226, 227, 228, 229, 230, 232, 233, 236, 241, 242, *249*, *250*, *251*
Ferguson, M. E. C., 198, *216*
Ferrebee, J. W., 191, *216*
Field, M., 286, *288*, *290*
Finean, J. B., 49, 50, 51, 52, 63, 75, 76, 77, *87*, *88*, *89*, 118, *141*, 177, *180*
Finer, E. G., 111, 113, 120, 122, 125, 128, 134, *141*, *142*
Finkelstein, R. A., 286, *289*
Firemark, H., 238, *249*
Fischer, H., 225, 227, 228, 229, 230, 232, 233, 236, 241, *249*, *250*, *251*, 275, 276, *290*, *292*
Fisher, D. B., 241, *250*
Fleischer, S., 4, 27, *89*, 132, *141*, 162, 163, *182*
Flewett, T. H., 169, *179*
Flook, A. G., 109, 111, 113, 120, 122, 125, 134, *141*, *143*

Fluck, D. J., 13, *87*, 112, *141*
Fogel, B. J., 280, *289*
Fong, C. T., 212, *216*
Fontell, K., 35, *87*
Forrester, J. A., 258, 259, 260, *289*
Foster, D. O., 264, *289*
Fournet, G., 3, *87*
Fox, T. O., 262, *289*
Franckson, T. R. M., 191, *217*
Frank, H., 147, 167, 168, *180*, *182*
Franklin, R. M., 82, 83, 84, *87*, 145, 147, 150, 158, 165, 170, *179*, *180*, *183*
Franks, A., 3, *87*
Freedman, R. B., 103, *140*
Freedman, S. O., 261, *289*, *290*, *293*
Freinkel, N., 201, 205, *219*
Freeman, N. K., 152, 153, 157, 159, *180*
Freychet, P., 189, 195, *216*
Friedman, R. M., 157, *180*
Frisch-Niggemeyer, W., 149, *180*
Fromherz, P., 46, *87*
Fromm, D., 286, *288*
Frommhagen, L. H., 152, 153, 157, 159, *180*
Fuchs, P., 256, 257, *290*
Fujiwara, S., 172, *180*
Furchgott, R. F., 185, *216*
Furshpan, E. I., 257, *289*

G

Gafford, L. G., 158, *183*
Gahmberg, C. C., 159, 161, 163, 164, 167, 177, 178, *182*
Gail, M. H., 258, 259, 268, 278, *293*
Gandhi, S. S., 152, *181*
Garcia, L. A., 193, *219*
Garoella, J. W., 191, *216*
Garon, C. F., 152, 156, *180*
Garratt, C. J., 191, 192, *216*
Gehring, P. J., 271, *289*
Geis, I., 211, *215*
Gelderblom, H., 147, *180*
Gentz, J., 285, *289*
Geordink, R. A., 232, *249*
Gerhard, W., 147, 165, *179*
Gershon, E., 286, *290*
Gerstl, B., 277, *288*, *289*
Gilden, R. V., 255, 264, *289*
Gingell, D., 279, *289*, *293*

Subject Index

311